Principles of Warmwater Aquaculture

PRINCIPLES OF
WARMWATER AQUACULTURE

Robert R. Stickney

Associate Professor
Department of Wildlife and Fisheries Sciences
Texas A&M University

A Wiley-Interscience Publication

JOHN WILEY & SONS

New York • Chichester • Brisbane • Toronto

Library of Congress Cataloging in Publication Data:

Stickney, Robert R
  Principles of warmwater aquaculture.

  "A Wiley-Interscience publication."
  Includes bibliographies.
  1. Aquaculture.   2. Channel catfish.
I. Title.
SH135.S75   639'.34   78-25642
ISBN 0-471-03388-X

Printed in the United States of America

10 9 8 7

Preface

Aquaculture, or aquatic agriculture, has been practiced in Asia for at least 3000 years; however as a discipline of science, aquaculture is largely in its infancy. Only a handful of colleges and universities in the United States have courses in aquaculture, let alone programs to train people for the field at either the undergraduate or graduate level. Yet recognition by government that aquaculture is indeed an aspect of agriculture is finally coming. In fiscal 1978 the U.S. Department of Agriculture's farm bill acknowledged for the first time the applicability of aquaculture to the goals of that agency. This does not mean that aquaculture represents the wave of the future in human food production, nor is it a harbinger of countless jobs for trained aquaculturists. There will, however, be expansion of aquaculture in the United States as there has already been in foreign nations, and aquaculture will play a small but significant role in United States' food production.

Information on aquaculture is widely distributed among trade magazines, journals, newsletters, leaflets, pamphlets, and technical reports. The few presently available textbooks either are dated or deal with aquaculture on a global scale. A few books that treat a single species or group of closely related species and some texts on specialized subdisciplines within aquaculture are also available.

This book presents information on aquaculture as it relates to warmwater species of interest to culturists in the United States. Foreign and coldwater species have not been ignored altogether, but these have been incorporated primarily as examples when information on warmwater or native species is unavailable. Foreign and coldwater species are also discussed in the context of their suitability as exotics in warmwater fish growing areas, and they sometimes serve in documentation of generalizations.

The book is not intended as a manual to teach the rearing of any particular species, vertebrate or invertebrate, but the reader will receive a broad overview of warmwater aquaculture. Examples have been drawn from at least three phyla and a large number of species, both marine and freshwater. There is, however, a good deal of emphasis on the culture of channel catfish, for two main reasons. First, channel catfish production is commercially profitable in the southern United States and will continue to be the main warmwater fish cultural crop produced in this country in the foreseeable future. Second, there is more information available on all aspects of channel catfish culture than on any other United States warm-water species. Therefore the channel catfish serves as a thread that winds its way through the text and ties the various topics together.

For the more traditional fishery biologist who is not planning a career as an aquaculturist, this book should still prove useful, since aquacultural management practices often can be incorporated into traditional fishery management on a somewhat less intensive level than that utilized by the fish culturist. Physiologists, pathologists, and others who wish to maintain aquatic animals for experimental purposes will also be able to apply many of the concepts presented here.

This book is dedicated to my wife, Carolan, and to my children, Bobby and Marolan, who have borne the nights and weekends devoted to the raising of fish and the production of this manuscript rather than to them. I express my deep appreciation to my students, both undergraduate and graduate, for their observations, questions, and insight; to Drs. Robert S. Campbell and R. Winston Menzel, who encouraged the development of my interest in the subject and helped shape my career; and to Dr. Robert W. Brick for critical review of one of the drafts of the manuscript.

ROBERT R. STICKNEY

College Station, Texas
February 1979

Contents

7. Disease and Parasitism, 261

Principles of Warmwater Aquaculture

CHAPTER 1

Introduction

AQUACULTURE DEFINED

Aquaculture is the rearing of aquatic organisms under controlled or semi-controlled conditions. Thus aquaculture is underwater agriculture. In this book the emphasis is on organisms, primarily animals, which can be utilized for human food. In the broad sense, aquaculture includes the rearing of tropical fish; the production of minnows and goldfish; the culture of sport fishes for stocking into farm ponds, streams, and reservoirs; the production of animals for augmenting marine stocks; and aquatic plant culture. Minnows, goldfish, and selected sport fishes are discussed in Chapter 9; however the bulk of the material presented concerns the culture of species that are to be sold directly for human consumption. Consideration is given to the control of such nuisance aquatic plants as rooted macrophytes and filamentous algae, and to the promotion of desirable plants such as phytoplankton, but the culture of decorative water plants is not included.

Aquaculture species may be primarily associated with warm water, cold water, or temperatures between those extremes. A great deal of information has been generated over the past several decades on the culture of salmonids, and the concepts behind warmwater aquaculture are based largely on those of trout and salmon production. Because coldwater and midrange animal aquaculture is limited to a few select species, and because salmonid culture is so well documented in various texts, this book concentrates on warmwater aquaculture species. Examples from coldwater and midrange organisms have been utilized to document general principles that apply to a variety of species, independent of temperature preference; to demonstrate where differences occur among species with different temperature tolerances; and to speculate in instances where the only information available comes from species other than those that inhabit warm water.

Recently interest has been aroused in the United States in utilizing the effluent from sewage treatment plants, high density aquaculture facilities, and feed lots for producing various aquatic plants and animals, which can then be eaten by people or recycled back to terrestrial livestock. Though not new in global terms, this concept is somewhat revolutionary to United States thinking. Chapter 2 considers some of the principles behind aquaculture systems of these types.

The definition of aquaculture presented above is not limited to either saline or fresh water, but encompasses both media. The term "mariculture" was coined to distinguish marine aquaculture, but no freshwater equivalent is now in use. If a distinction is required, the best terms to use are "mariculture" and "freshwater aquaculture."

The majority of animals selected for aquaculture around the world come from three phyla: Mollusca, Arthropoda, and Chordata. Table 1.1 presents examples of organisms from each phylum found in the United States. Representatives from other phyla could become important to aquaculture in the future, but most have little value as human food. An exception is the sea urchin (phylum Echinodermata), parts of which are considered delicacies, especially in southern Europe and Japan. Sea urchins would presently find little acceptance in the United States, and it is unlikely that future demands will warrant aquacultural development of these animals in this country. The animals listed in Table 1.1 are currently consumed in large quantities by the American public and either are being cultured or are under consideration by aquaculturists.

Table 1.1 Representatives of the Phyla Mollusca, Arthropoda, and Chordata that Occur in the United States and Are Under Culture or Being Considered by Aquaculturists

Phylum	Common Name	Scientific Name
Mollusca	Quahog clam	*Mercenaria mercenaria*
	American oyster	*Crassostrea virginica*
	Bay scallop	*Aequipecten irradians*
	Blue mussel	*Mytilus edulis*
Arthropoda	Blue crab	*Callinectes sapidus*
	American lobster	*Homarus americanus*
	White shrimp	*Penaeus setiferus*
	Red swamp crayfish	*Procambarus clarkii*
Chordata	Channel catfish	*Ictalurus punctatus*
	Rainbow trout	*Salmo gairdneri*
	Green sea turtle	*Chelonia mydas*

AQUACULTURE AND IMPENDING FOOD CRISES

As maximum production through traditional agriculture is approached around the world, new sources of human food are being sought to prevent the development of famine on a global basis. A logical environment for additional exploitation seemed to be the sea, which has long been believed to be a limitless source of protein for mankind. However Ryther (1969) estimated that the present rate of exploitation cannot be significantly increased because of limits on primary productivity. Idyll (1970) also examined the ability of the sea to provide sufficient food for the welfare of mankind and arrived at a conclusion similar to that of Ryther. Idyll believed that some increase in production could be realized if man began to utilize such exotic creatures as krill (small shrimplike animals in the family Euphausiidae) and phytoplankton directly as food. Some research has gone into the development of such products, but their widespread acceptance is not likely in the near future. Palatability has been one problem, as has consumer education. Contrary to popular opinion, undernourished or even starving people will not eat a given food just because it is available. There are many social and psychological factors involved. This is one of the reasons for the failure of fish protein concentrate (FPC), which was developed in the form of a tasteless, odorless white powder, as a supplement for human diets (Finch, 1969). Despite its innocuous appearance, people in developing nations could not often be convinced of the value of adding FPC to their normal diets, and the benefits that would have accrued from its use were largely unrealized.

Aquaculture has been proposed as a means of supplementing natural marine and freshwater productivity, thus to diminish food shortages. Under the proper circumstances, densities per unit water volume under aquaculture strategies greatly exceed those found in the natural environment. Problems do exist, however. As the density of animals per unit volume of water increases, there is a concomitant need for increased energy input to the culture system. Energy enters aquaculture systems in the forms of electricity and supplemental feed, both of which are very expensive and likely to remain so, perhaps increasing in cost in the foreseeable future. Even if energy costs for large-scale aquaculture were not prohibitive in developing countries, persons who most need the protein produced might be unable to obtain either electricity or supplemental feedstuffs.

Developed nations are not in a position to export their own aquacultural products to developing countries. In the United States the produce of aquaculture must be considered luxury, even gourmet food, which retails in excess of $2/kg in virtually all cases. If export costs were added the cost of the final product would be astronomical relative to the portion of income that the hungry peoples of the world are able to spend on food.

In developing countries the production of luxury foods such as shrimp and channel catfish is geared to export to nations that can afford them. The current best approach for producing food for local consumption in developing countries is through subsistence aquaculture. The technique involves the use of small ponds in which low density culture is conducted with a minimum of management. For example, a single family or small group of neighbors may have a pond they wish to use for fish production. The pond might be drained or poisoned to remove existing fish, after which it would be restocked with a rapidly growing, readily available species that feeds fairly low on the food chain. Management after or in conjunction with stocking might include little more than fertilization with terrestrial animal wastes. Maintenance of the population generally involves natural reproduction in the pond.

Most subsistence aquaculture is conducted in the tropics, where a year-round growing season is often available for warmwater species. Harvest of the crop usually involves the removal of only the quantity of fish required for immediate consumption by the subsistence aquaculturists themselves (although they may be able to augment their incomes by selling small quantities of fish). The general lack of means by which to preserve large quantities of fish rules out large-scale harvesting in most instances. However the use of numerous small ponds will allow complete harvest without burdening the subsistence culturists with unmanageable quantities of fish.

Subsistence scale aquaculture is not generally considered to be practical or desirable in the United States, although there are persons who promote artisan-scale aquaculture for those interested in independence from traditional sources of food (Howard Community College and Foundation for Self-Sufficiency, Inc., 1977). A modification of subsistence level aquaculture is also practiced by the large numbers of farmers and ranchers across the United States who raise channel catfish and other species in small ponds. Most of the fish produced are utilized by the landowner, supplying both food and recreation.

It is not too difficult to visualize the problems associated with finding suitable water for large aquaculture ventures inland, but the vastness of the world ocean seems to indicate that unlimited aquacultural potential exists in the marine environment. However because of the difficulties associated with open ocean aquaculture (construction and maintenance of large pens in deep water is unrealistic economically), mariculture will be restricted to coastal areas except under special circumstances. United States coastal land is becoming less and less available to aquaculturists because of the development that has occurred and is continuing. Competition from industry, land developers, commercial fishermen, and sportsmen, along with attempts to preserve large tracts in their natural states, have dwindled the availability of many parcels of land and the suitability of others.

Because of state and federal laws governing the disruption of coastal marshlands, mariculture must be restricted largely to supratidal and subtidal areas along the coast of the United States. Alternatives to ponds and other land-based mariculture facilities above the high tide line are such techniques as fencing off large areas as rearing pens, hanging strings of mollusks from suitable platforms or rafts, and ocean ranching. The latter technique is possible with anadromous* species, which are produced in land-based hatcheries and released into the ocean. Once they have reached adulthood, they return to the hatchery to spawn and be harvested. This technique is presently being developed for salmon.

BUILDING A NEW SCIENTIFIC DISCIPLINE

The art of rearing aquatic organisms is generally considered to have been developed in China between 3500 and 4000 years ago (Ling, 1974). Written reports on the subject existed in China during the Ying-Shang dynasty (San-Dun, 1975). Yet in parts of the world where fish culture is an ancient art, there was little or no significant development until recent years, when a more scientific approach was adopted.

In the United States aquaculture is practiced largely as a scientific discipline, although certain aspects of art are readily apparent. Science has been incorporated into salmonid rearing for many decades, and more information is currently available on trout and salmon culture than on any other group of fishes in this country. The intensive research effort focused on salmonids was not primarily associated with their potential as aquaculture animals (although a large number of trout farms are presently active in the production of fish for human consumption). Rather, most of the expenditures for research have resulted from attempts to save the anadromous salmonids from extinction as a result of dam construction across home streams, and heavy fishing pressure. Many of the trout reared during the past several decades have been stocked into streams to augment natural stocks and provide recreational fishing. Regardless of why the research was conducted, a vast amount of information applicable to warmwater aquaculture has evolved from studies of the salmonids.

Warmwater aquaculture is relatively new in the United States, although fish hatcheries have existed in this country for more than a century. Channel catfish (*Ictalurus punctatus*) were spawned and released from federal and state hatcheries long before studies at Auburn University demonstrated the economic feasibility of rearing that species for direct human consumption (Swingle, 1957, 1958). The rapid development of channel catfish farming during the 1960s provided the impetus for current interest in warmwater aquaculture in this country.

Channel catfish farmers were fortunate in having a species that is relatively easy to grow under aquaculture conditions. Catfish consume artificial diets immediately after egg sac absorption, withstand fairly wide ranges of

---

* The Glossary at the end of the book provides definitions of terms important to aquaculture.

water quality conditions, grow rapidly, and enjoy a good market, especially in the southern United States. Most other warmwater aquatic organisms do not share all these advantages; however the successes demonstrated by catfish farmers have inspired work on other species.

The development of commercial channel catfish culture and scientific research on that species occurred simultaneously. The primary research goal has been to improve culture strategies for a species that has been reared profitably since the middle of the century. In 1955 there were less than 50 hectares (ha)* of channel catfish in production in the United States. By 1975 production had increased to 18,809 ha (National Academy of Sciences, 1978). Production figures for 1975 revealed a total of 38,000 metric tons by 2000 farmers in 13 states, with about 80% of the fish being reared in Mississippi, Arkansas, and Louisiana (National Academy of Sciences, 1978).

For other aquaculture candidates commercial interest has often developed without an adequate data base from which to plan effective culture strategies. Some aquacultural enterprises have failed because promoters and investors neglected to determine whether the species selected could be grown in captivity before they embarked on commercial ventures. On the whole, aquaculture is less advanced in the United States than in many foreign nations. A recent study predicted that aquaculture will play only a minor role in United States food production in the near future, but that role could be expanded through an increase in research and development funds (National Academy of Sciences, 1978). The report lists limited public support, the need for a comprehensive government policy, and lack of political power on the part of aquaculturists as major constraints to aquacultural development in this country. United States aquacultural production was only 65,000 metric tons in 1975, but the National Academy of Sciences (1978) predicted that 250,000 metric tons could be produced in 1985, and possibly a million metric tons by 2000 if support for this area of agriculture by government and private sources is increased to required levels.

A setback to the development of warmwater aquaculture in the United States came in the late 1960s when a great amount of interest was generated in penaeid shrimp culture. Three species of penaeid shrimps are widely sought in the Gulf of Mexico and along the coast of the southeastern United States: the white shrimp (*Penaeus setiferus*), brown shrimp (*P. aztecus*), and pink shrimp (*P. duorarum*). The high market price and excellent flavor of wild shrimp indicated to many prospective investors that these animals would be naturals for aquaculture. Great sums were inves-

* Appendix 1 lists conversion factors for units in the English and metric systems.

ted, but lack of information on certain aspects of the biology of these animals led to failure in virtually all cases.

During the early 1970s speculators moved on to other areas, leaving the development of shrimp aquaculture to private and public researchers. The results of that research are beginning to pay dividends as the reality of profitable commercial shrimp farming approaches. At the same time, increased research activity into the aquacultural suitability of other species has increased. Salmonids and channel catfish are still the only species reared profitably in large numbers in the United States, but the potential for various other species in the future looks increasingly bright.

Table 1.2 lists some of the species that appear to hold potential for aquaculturists or currently are being reared profitably in the United States. Although some of these animals may not become viable aquaculture species for several years, we draw upon the literature on them for examples throughout this book.

Aquaculture is not a distinct discipline; rather it combines aspects of various sciences, crafts, arts, and several business fields. Aquaculturists most often are trained primarily as biologists, although much of their work is only remotely biological. Chemistry is as important, or possibly more important to them, since so much of their work involves the maintainance of water quality and as aquaculture develops in this country, there will be increasing need for persons with backgrounds not only in science but in economics and engineering.

Few colleges and universities in the United States have programs leading to degrees with emphasis on aquaculture, although this situation has begun to change in recent years. Most aquaculturists come out of general biology programs or programs in wildlife and fisheries. More directly relevant courses are often available at the graduate level; however the neccessary practical experience is still somewhat difficult to obtain. A mastery of the theory of aquaculture is important but cannot replace the knowledge gained from day-to-day experience obtained under the supervision of a practicing culturist.

In addition to obtaining as much formal education as possible, the aquaculturist must be familiar with sources of information that can assist in solving the problems encountered from time to time; these sources also serve to relate new developments in the field. Depending on the problems that arise and the area of the country in which the aquaculturist is working, the following agencies may be called on to supply verbal and written information:

State Agricultural Experiment Stations and Cooperative Extension Services

Table 1.2 Partial List of Aquatic Species that Have Been Researched as Food Animals in the United States[a]

Common Name	Scientific Name	Habitat
White shrimp	*Penaeus setiferus*	Marine
Brown shrimp	*Penaeus aztecus*	Marine
Pink shrimp	*Penaeus duorarum*	Marine
American lobster	*Homarus americanus*	Marine
Blue crab	*Callinectes sapidus*	Marine
Blue mussel	*Mytilus edulis*	Marine
Quahog clam	*Mercenaria mercenaria*	Marine
American oyster	*Crassostrea virginica*	Marine
Southern flounder	*Paralichthys lethostigma*	Marine
Summer flounder	*Paralichthys dentatus*	Marine
Striped mullet	*Mugil cephalus*	Marine
Spotted sea trout	*Cynoscion nebulosus*	Marine
Red drum	*Sciaenops ocellata*	Marine
Black drum	*Pogonias chromis*	Marine
Dolphin	*Coryphaena hippurus*	Marine
Pompano	*Trachinotus carolinus*	Marine
Channel catfish	*Ictalurus punctatus*	Freshwater
Blue catfish	*Ictalurus furcatus*	Freshwater
White catfish	*Ictalurus catus*	Freshwater
Buffalo fish	*Ictiobus* spp.	Freshwater
Yellow perch	*Perca flavescens*	Freshwater
Tilapia	*Tilapia* sp.[b]	Freshwater
Grass carp	*Ctenopharyngodon idella*	Freshwater
Freshwater shrimp	*Macrobrachium* spp.	Freshwater
Frog	*Rana* spp.	Freshwater
Green sea turtle	*Chelonia mydas*	Marine

[a] Native fish common and scientific names according to American Fisheries Society (1970).

[b] Trewavas (1973) suggested that the genus *Tilapia* be split into two genera depending on breeding habits: *Tilapia* (substrate spawners and guarders) and *Sarotherodon* (mouthbrooders). Virtually all the aquaculture species under study in the United States would be placed in the genus *Sarotherodon* if this change were accepted. Most aquaculturists in the United States and abroad continue to utilize the generic name *Tilapia* and "tilapia" has become well established as the common name for aquaculturally reared animals in the family Cichlidae around the world. The use of *Tilapia* has been retained throughout this text.

U.S. Department of Agriculture (Soil Conservation Service)
U.S. Department of the Interior (Fish and Wildlife Service)
U.S. Department of Commerce (National Marine Fisheries Service)
State Departments of Fish and Wildlife
Commercial feed, fish, and supply dealers
University libraries

Publications exclusively devoted to aquaculture are not abundant. Several magazines have been developed in recent years, but most have failed. A few scientific publications publish aquaculture-related articles either nearly exclusively or with a degree of frequency sufficient to merit attention. Among these are:

Aquaculture
Fish Farming International
Fishery Bulletin (U.S.)
FAO Fishery Bulletin
Journal of the Fisheries Research Board of Canada
Journal of Fish Biology
The Progressive Fish-Culturist
Proceedings of the Southeastern Association of Game and Fish Commissioners
Proceedings of the World Mariculture Society
Transactions of the American Fisheries Society

The culturist also must, insofar as is possible, identify with the culture species and attempt to perceive the world from the point of view of those organisms. The aquaculturist who establishes a rapport with the animals often is more successful than those who find it difficult to relate or remain aloof.

EXOTIC SPECIES IN UNITED STATES AQUACULTURE

At least some of the popular aquaculture species being commercially reared in foreign countries have been introduced into the United States. Among those that have received significant attention by researchers in recent years are freshwater shrimp (*Macrobrachium rosenbergii*), grass carp (*Ctenopharyngodon idella*), and tilapia (*Tilapia aurea, T. nilotica, T. mossambica,* and others). A great deal of controversy presently exists with respect to the introduction of exotic species into the United States because of the negative impacts on natural populations of aquatic organisms which have resulted from exotic introductions in the past. The common carp (*Cyprinus carpio*) and the walking catfish (*Clarius batrachus*) are two examples of exotic species that have caused significant problems.

Both grass carp and tilapia were originally introduced into the United States as weed control organisms. Introductions of these species into natural waters in this country have been widely criticized, and grass carp have been banned in more than 30 states, although they are probably present in most of them. Studies on the benefits and harm of these and other species of exotics are continuing.

Ctenopharyngodon idella was introduced into the United States in 1963 (Guillory and Gasaway, 1978) and is now present in at least 35 states. The species is native to eastern Asia from the Amur River Basin to the West River (Lin,1935), hence one of the common names, white amur. It was thought at first that grass carp could not spawn in the natural waters of the United States and that fish that escaped from captivity would not be able to reproduce successfully, thus would eventually die out. However it is now believed that certain bodies of water associated with the Mississippi, Missouri, and Arkansas rivers, as well as a few other locations in the country, may support conditions suitable for spawning of *C. idella* (Stanley *et al.*, 1978).

The ability of grass carp to control undesirable rooted aquatic vegetation is well documented (Stott and Robson, 1970; Kilgen and Smitherman, 1971); however the presence of grass carp in natural waters has also been shown to have detrimental effects on certain aspects of water quality (Lembi *et al.*, 1978). Grass carp appear to be strict herbivores, although they are somewhat selective in their food habits and may not remove all the rooted vegetation types that appear in a pond (Colle *et al.*, 1978).

Grass carp are an acceptable human food and have been reared by United States aquaculturists in combination with other species, as well as in ponds fertilized with terrestrial animal wastes (Buck *et al.*, 1978).

Tilapia are tropical species native to Africa and the Middle East. They have probably been cultured for as long as 2000 years in certain parts of the world (Hickling, 1963); but are relatively new arrivals to the United States, appearing during the 1960s. While able to tolerate high temperatures (Allanson and Noble, 1964; Gleastine, 1974), most species die when water temperatures drop below 7 to 10 C (Chimits, 1957; McBay, 1961; Avault and Shell, 1968; Gleastine, 1974). Thus in virtually all the continental United States (except in south Florida and south Texas), tilapia will be unable to overwinter except during unusually warm years unless they are provided with artificially heated or geothermal water. In Texas, for example, several power plant cooling reservoirs have well-established populations of *Tilapia aurea* which are able to survive by moving into the warm water of discharge canals during winter.

Tilapia do not represent a major threat to natural fish populations in most parts of the United States because of their inability to survive cold temperatures. However fish culturists should avoid controversy and criticism by attempting to avoid escapement of both tilapia and grass carp from hatcheries on which these species are maintained.

The most popular aquaculture species of freshwater shrimp, *Macrobrachium rosenbergii*, must also be maintained in warm water in most portions of the United States because its intolerance to cold temperatures is similar to that of tilapia. Native species such as *M. carcina* and *M. acanthurus* reach suitable size for sale as human food but have not been domesticated to the point achieved with *M. rosenbergii*.

Exotic species presently occurring in the United States will probably remain part of the aquatic fauna of this country. However, it is becoming increasingly difficult to import exotic fishes because of the problems they have caused in the past. The fear of introducing new diseases that might decimate native fish populations is one among many concerns of persons opposed to exotic introductions. Aquaculturists should plan to utilize only species currently available in the United States, although it is probable that other species will be introduced on an experimental basis in the future. Such introductions will be on a small scale compared to the past. In any event a number of exotics that appear to have potential as aquaculture candidates are present now.

EXTENSIVE AND INTENSIVE CULTURE

A key to distinguishing between aquaculture and more traditional aspects of fishery management is associated with the degree of control exerted by man over the environment. Fishery managers attempt to manage natural aquatic ecosystems, usually for the benefit of sport or commercial fishermen. Various techniques are utilized to maintain natural productivity and to promote the growth and survival of desired species. Some of the techniques include regulation, selective stocking, habitat manipulation, vegetation control, and drawdown. In general, the aquaculturist exercises much more control on the environment than does the fishery manager. The aquaculturist manages water systems for maximum production of one or a small number of species and attempts to eliminate insofar as possible all competition and sources of mortality.

Traditional fishery management can be thought of as lying on one end of a continuum, with high density tank aquaculture at the other end (approximately). Fishery management is generally accomplished in bodies of water where standing crop biomass is fairly low and the amount of water surface area required to rear more than a few hundred kilograms of fish exceeds a hectare. This type of situation is known as extensive culture and has been practiced around the world by pond aquaculturists. Catfish farmers who produce fish at or below the level of about 1500 kg/ha may be referred to as intensive culturists. As one moves along the continuum away from the arbitrarily assigned most extensive end (characterized by fishery management of natural populations), the level of culture becomes increasingly intensive.

Intensive pond culture is exemplified by the catfish farmer who may produce 3000 kg/ha or more. More intensive yet is aquaculture conducted in raceways, tanks, silos, and cages where standing crops exceeding 10^6 kg/ha have been achieved. Production systems of these types are discussed in detail in Chapter 2.

As the intensity of culture increases, problems associated with crowding tend to increase: disease epizootics, water quality deterioration, competition for food, and in certain cases, cannibalism. Ever greater control must be applied by the aquaculturist as the intensity of culture is elevated. This also means an increase in energy utilization and in the sophistication of monitoring and life support equipment. High intensity aquaculture systems have been developed and are always being improved. However most successful commercial warmwater aquaculturists continue to rely on the less intensive pond systems for their production. The precise type of culture system selected depends on such factors as the availability and costs of land, water, and energy; the physical location of the facility; the expertise of the culturist; and the species to be reared.

MONOCULTURE AND POLYCULTURE

The majority of aquaculturists in the United States produce a single species in a given culture unit, regardless of whether it is a pond, a raceway, a tank, a cage, or other type of culture chamber. This practice is known as monoculture. Polyculture, the rearing of two or more species in each culture unit, enjoys wide popularity throughout much of the world (Bardach *et al.*, 1972), since the technique allows species with different food habits to take advantage of each of the feeding niches in a pond environment. Polyculture becomes less practical as the intensity of culture increases.

In pond monoculture the animal being reared usually fails to utilize all portions of the environment. For example, the culture of a demersal species leaves most of the water column unoccupied. In an attempt to increase production, however, a pelagic animal may be cultured along with the demersal one. The two or more species that coexist in a polyculture pond must, of course, be compatible. The Chinese have developed polyculture to its highest currently practiced level, utilizing four or more species of fishes that feed on benthos, zooplankton, phytoplankton, and aquatic macrophytes, respectively (Bardach *et al.*, 1972).

In the United States a polyculturist might stock tilapia to feed on phytoplankton, common carp or buffalo fish to consume zooplankton and benthos, and catfish, which are offered a prepared diet. In regions where grass carp are not prohibited, this species could be added to feed on aquatic macrophytes. Fertilizer would be added to the water to enhance the growth of natural food organisms. In a polyculture scheme of the type described, the catfish would bring the highest price in the marketplace on the basis of unit weight. Thus it is important that the other species not be produced at the expense of catfish. Production of the target species either should remain approximately the same as it would have been in monoculture or should be enhanced by polyculture (the latter probably relating to an increased supply of food as a result of pond fertilization). Stocking densities of each species in a polyculture operation are generally similar to those appropriate for the animals when stocked in monoculture.

To modify the traditional polyculture technique, one might utilize a series of ponds linked by pipes through which water flows by gravity without allowing passage of the culture animals. If such a system were placed in a position to receive the wastes of terrestrial livestock, a species tolerant to poor water quality (e.g., *Tilapia* spp.) could be reared on the phytoplankton present in the first pond. Relatively small amounts of nutrient-rich water would be flowed into the second pond, where secondary productivity would be utilized by a species such as channel catfish. A third pond might be utilized for rearing freshwater shrimp or some other species. Water from the pond lowest in the series would be recirculated back to the uppermost pond, thus conserving water within the system. Such systems have been designed in the United States and utilized in experimental aquaculture (Stickney and Hesby, 1978), and their theoretical basis has been described in more detail by Stickney (1978).

FRESHWATER AQUACULTURE VERSUS MARICULTURE

Both stenohaline and euryhaline species have been successfully utilized by aquaculturists. Among the euryhaline organisms, most warmwater species are typically found throughout their lives in either salt or fresh water, although anadromous species such as salmon, and catadromous eels, are cultured in many nations (Brown, 1977). In most cases aquaculture is conducted within a fairly narrow range of salinities and the species utilized occur naturally under the available salinity regime. It is possible to rear euryhaline species under salinity conditions in which they are not generally found in nature, although before this is attempted studies should be conducted to determine that any reduction in growth rate or increase in mortality that occurs as a result of this practice is not significant.

In certain instances it may be necessary to provide a range of salinities in response to the stage of the life cycle in which the organisms are found. For example, *Macrobrachium* species require water of low salinity during spawning and larval development but may spend the remainder of their lives in fresh water (Bardach *et al.*, 1972).

Various estuarine species can be adapted to fresh water, but since they generally are unable to reproduce in a low salinity environment, spawning must be limited to salt water. The red drum, *Sciaenops ocellata*, and flounders of the genus *Paralichthys* have been released by the Texas Parks and Wildlife Department into freshwater reservoirs in the state. Newspaper reports of returns indicate that growth may exceed that of the same species in the marine environment in some instances. High mortalities have posed a problem with that program. Mullet, *Mugil cephalus*, are commonly found in freshwater, and it might be possible to move this species inland for rearing.

Certain freshwater species can also be introduced into saline waters. Various species of tilapia have been found to tolerate salinities that may exceed those found in the open ocean (Chervinsky, 1961, 1966; Chervinsky and Yashouv, 1971; Chervinsky and Zorn, 1974). Catfish, including channel catfish, can tolerate in excess of 10 parts per thousand (o/oo) salinity (Perry, 1967; Perry and Avault, 1968; Stickney and Simco, 1971) and have been reared in coastal lagoons of relatively low salinity. However in most cases aquaculture is conducted in water of the salinity in which the animals are most commonly found in nature.

As a general rule, the difficulty of maintaining an aquaculture facility increases with increasing salinity. A saline medium not only introduces new aspects of water quality into the culture system, it also places a tremendous stress on facilities. Salt water is extremely corrosive to metal; thus all equipment and structures associated with a mariculture facility must be protected from the water or they will rapidly deteriorate.

DOMESTIC VERSUS FOREIGN SITES FOR AQUACULTURE

Although this book emphasizes domestic aquaculture practices, the reader should be aware of some of the potential advantages and disadvantages associated with the establishment of facilities in a foreign country. Such ventures are most commonly operated in developing nations with the objective of producing aquaculture crops at relatively low cost and exporting the products back to the United States or some other developed country for sale. United States firms interested in establishing aquaculture facilities in foreign countries usually select Central America or northern South America because of climate and proximity to the North American market.

The following facts should be taken into consideration when the prospects of siting an aquaculture venture in a developing tropical country are weighed:

1. Labor costs tend to be lower than in developed countries, especially with respect to manual labor. It is often necessary to train or import professional technical and managerial staff, but there may be routine jobs that can be filled by local residents after only a minimum of training.

2. Suitable land is often abundant and inexpensive in developing nations. In addition, many foreign countries promote outside investments through tax credits and other forms of dispensation.

3. Many developing countries have available large areas of suitable land that are not subjected to pollution, although this situation may change because many developing nations lack regulations to limit the environmental degradation that will occur as the countries industrialize. Presumably, as industries in the United States reduce pollution levels and as safer pesticides are developed, conditions will improve in regions that are presently contaminated. Just the opposite may occur in certain developing countries.

4. Construction costs, like labor costs, are often lower in developing countries. The cost of having ponds and buildings constructed, for example, may be much less than in the United States. However finding contractors with experience in building such facilities may be difficult.

5. One of the greatest advantages of locating an aquaculture facility in a tropical country is the water temperature, which remains at or near the optimum for many warmwater animals throughout the year. Thus a 12 month growing season is the rule rather than the exception. In most of the continental United States the growing season for warmwater aquaculture is 6 months or less. Production in the tropics may easily exceed twice that of more temperate climates.

6. Many developing nations have histories of political unrest. Disregarding the physical danger in which the aquaculturist might be placed, it is possible that the business would be nationalized and the persons who established it asked to leave the host country.

7. The availability of feed, supplies, and equipment may be severely limited in developing countries. It may be difficult to have equipment repaired, and parts may have to be ordered from the United States.

8. Some means of transporting the crop to market at harvest is required. Transport must be available at the appropriate time of year, and facilities for processing and freezing the crop must be established.

Not all tropical areas have all the advantages and disadvantages just alluded to, and there are certainly other factors that should be considered as well. For instance, the cost and availability of energy vary greatly from one country to another.

Many regions of tropical America are rapidly developing, and most of the large United States ventures in aquaculture are being established in those countries or territories. Such places as Puerto Rico, Brazil, Costa Rica, and Panama, are becoming increasingly favorable for aquacultural investments. Indeed, large amounts of capital are going into aquaculture in each of these areas.

CONSIDERATIONS IN DOMESTIC SITE SELECTION

The site selection process depends largely on the circumstances in which the culturist is found. If the individual is not committed to either a particular location or species, the site can be selected on the basis of complimentation. Many prospective aquaculturists, however, begin with certain self-imposed limitations: for example, they already possess a piece of land they wish to convert to aquaculture, or they feel that they must produce a certain species and will not consider others. Sometimes the two conditions are not compatible. For example, an individual committed to the production of American lobsters would not be very successful if no growing area except the state of Kansas were under consideration.

Many United States warmwater aquaculturists, especially those who rear channel catfish, have established operations on existing farms and ranches. Thus the profits or losses that may result from aquaculture seldom in themselves spell success or failure for the entire business enterprise. Other aquaculturists specialize in underwater agriculture and do not operate traditional farms or ranches in conjunction with aquatic animal husbandry. Persons wishing to expand existing agricultural operations into aquaculture generally select an aquatic species that others have found to be well suited to the growing area in which the land is found.

The basic requirement of any piece of aquacultural property is a plentiful supply of water, readily available and of suitable quality to support the organisms to be stocked in it. If extensive culture is to be practiced, the soil should have good water-holding capacity, or the culturist must be willing to go to the expense of sealing ponds against leakage. Alternatively, sufficient water must be available to replace seepage losses until natural processes (largely the accumulation of organic material over the period of several growing seasons) tend to seal the ground. In sandy soils the expense of pumping large volumes of replacement water could be prohibitive.

Land on which intensive nonpond culture facilities are to be established need not have soil that will retain water, although the total annual utilization of water may be higher in an intensive tank or raceway water system than in the less intensive pond system. Certain types of water system may require less water than would be utilized for the same amount of animal production in ponds (Chapter 2), and others take advantage of plentiful surface waters rather than well water.

THE FUTURE

Interest in aquaculture in the United States has been growing rapidly in recent years, although the previously outlined constraints on expansion voiced by the National Academy of Sciences (1978) must not be overlooked. More than 1000 registrants attended a meeting in Atlanta in January 1978; the Fish Culture Section of the American Fisheries Society, the World Mariculture Society, and the Catfish Farmers of America were represented. Government interest and acknowledgment of aquaculture were apparent and seem to be increasing, though slowly. Many congressmen who were once unfamiliar with the term "aquaculture" have made it part of their day-to-day vocabularies. Aquaculture will not solve the world food shortage, nor will it support as fish farmers nearly as many people as make a living from traditional agriculture. Aquaculture will, however, provide a career for at least several thousand persons in the United States, and the techniques involved can be applied by many thousands of farmers and ranchers to increase farm pond production for their own food and for recreation. As research in aquaculture advances our knowledge concerning the needs of aquatic animals, aquaculture will begin to make an important contribution to the total amount of fishery produce sold in the United States, as it already has in various foreign nations.

The remainder of this book examines water systems, water chemistry, and the management of water systems in some detail. The nutritional requirements of various warmwater species are also discussed, along with the manner in which prepared and natural feeds are prepared and distributed to the animals. Disease and parasitism are considered, with emphasis on maladies affecting channel catfish and shellfish. Finally, we cover harvesting, processing, and economics.

The commercial aquaculturist must be dedicated to two basic goals if the particular fish farming venture is to become and remain profitable. First, the culturist must always be concerned for the well-being of the animals, taking great care to provide them with conditions for growth and survival as near to optimum as possible. Second, and just as important, the culturist must produce a crop that the consumer not only finds acceptable but will return to the next time fishery products are purchased. This book is written with those two concerns continuously in mind.

LITERATURE CITED

Allanson, B., and R. G. Noble. 1964. The tolerance of *Tilapia mossambica* Peters to high temperatures. *Trans. Am. Fish. Soc.* **94:** 323-332.

Allen, K. O., and J. Avault, Jr. 1969. Effects of salinity on growth and survival of channel catfish, *Ictalurus punctatus*. *Proc. Southeast. Assoc. Game Fish Comm.* **23:** 319-323.

American Fisheries Society. 1970. *A list of common and scientific names of fishes*. American Fisheries Society Special Publication No. 6. American Fisheries Society, Washington, D. C. 150 p.

Avault, J. W., Jr., and E. W. Shell. 1968. Preliminary studies with the hybrid tilapia *Tilapia nilotica* × *Tilapia mossambica*. *FAO Fish. Rep.* **44:** 237-242.

Bardach, J. E., J. H. Ryther, and W. O. McLarney. 1972. *Aquaculture*. Wiley-Interscience, New York. 868 p.

Brown, E. E. 1977. *World fish farming culture and economics*. Avi, Westport, Conn. 397 p.

Buck, D. H., R. J. Baur, and C. R. Rose. 1978. Utilization of swine manure in a polyculture of Asian and North American fishes. *Trans. Am. Fish. Soc.* **107:** 216-222.

Chervinsky, J. 1961. Study of the growth of *Tilapia galilaea* (Artedi) in various saline conditions. *Bamidgeh.* **13:** 71-74.

Chervinsky, J. 1966. Growth of *Tilapia aurea* in brackish water ponds. *Bamidgeh*, **18:** 81-83.

Chervinsky, J., and A. Yashouv. 1971. Preliminary experiments on the growth of *Tilapia aurea* Steindachner (Pisces, Cichlidae), in saltwater ponds. *Bamidgeh*, **23:** 125-129.

Chervinsky, J., and M. Zorn. 1974. Note on the growth of *Tilapia zilli* (Gervais) in sea water ponds. *Aquaculture*, **4:** 249-255.

Chimits, P. 1957. The tilapias and their culture. A second review and bibliography. *FAO Fish. Bull.* **10:** 1-24.

Colle, D. E., J. V. Shireman, and R. W. Rottman. 1978. Food selection by grass carp fingerlings in a vegetated pond. *Trans. Am. Fish. Soc.* **107:** 149-152.

Finch, R. 1969. The U.S. fish protein concentrate program. *Comm. Fish. Rev.*, January: 25-30.

Gleastine, B. W. 1974. A study of the cichlid *Tilapia aurea* (Steindachner) in a thermally modified Texas reservoir. M. S. thesis, Texas A&M University, College Station. 258 p.

Guillory, V., and R. D. Gasaway. 1978. Zoogeography of the grass carp in the United States. *Trans. Am. Fish. Soc.* **107:** 105-112.

Hickling, C. F. 1963. The cultivation of *Tilapia*. *Sci. Am.* **208:** 143-152.

Howard Community College and Foundation for Self-Sufficiency, Inc. 1977. *Essays on food and energy.* Foundation for Self-Sufficiency, Inc., Catonsville, Md. 184 p.

Idyll, C. P. 1970. *The sea against hunger.* Crowell, New York. 221 p.

Kilgen, R. H., and R. O. Smitherman. 1971. Food habits of the white amur stocked in ponds alone and in combination with other species. *Prog. Fish-Cult.* **33:** 123-127.

Lembi, C. A., B. G. Ritenour, E. M. Iverson, and E. C. Forss. 1978. The effects of vegetation removal by grass carp on water chemistry and phytoplankton in Indiana ponds. *Trans. Am. Fish. Soc.* **107:** 161-171.

Lin, S. 1935. Life history of Waan Ue (*Ctenopharyngodon idella* Cuv. and Val.). *Lingnan Sci. J.* **14:** 129-135.

Ling, S. W. 1974. Keynote address. *Proc. World Maricult. Soc.* **5:** 19-25.

McBay, L. G. 1961. The biology of *Tilapia nilotica* Linnaeus. *Proc. Southeast Assoc. Game Fish Comm.* **15:** 20⁹-218.

National Academy of Sciences. 1978. Aquaculture in the United States. Report to the Senate Committee on Agriculture, Nutrition, and Forestry. Washington, D.C. 93 p.

Perry, W. G., Jr. 1967. Distribution and relative abundance of blue catfish, *Ictalurus furcatus*, and channel catfish, *ictalurus punctatus*, with relation to salinity. *Proc. Southeast. Assoc. Game Fish Comm.* **21:** 436-444.

Perry, W. G., Jr., and J. W. Avault, Jr. 1968. Preliminary experiments on the culture of blue, channel, and white catfish in brackish water ponds. *Proc. Southeast. Assoc. Game Fish Comm.* **22:** 397-406.

Ryther, J. H. 1969. Photosynthesis and fish production in the sea. *Science,* **166:** 72-76.

San-Dun, G. 1975. Fish biology in China. *Copeia,* **1975:** 404-412.

Stanley, J. G., W. W. Miley II, and D. L. Sutton. 1978. Reproductive requirements and likelihood for naturalization of escaped grass carp in the United States. *Trans. Am. Fish. Soc.* **107:** 119-128.

Stickney, R. R. 1978. The polyculture alternative in aquatic food production. In P. N. Kaul (ed.), *Drugs and food from the sea, myth or reality?* University of Oklahoma Press, Norman, pp. 385-392.

Stickney, R. R., and J. H. Hesby. 1978. Tilapia culture in ponds receiving swine waste. In R. O. Smitherman, W. L. Shelton, and J. H. Grover (Eds.), *Culture of exotic fishes symposium proceedings.* Fish Culture Section, American Fisheries Society, Auburn, Ala., pp. 90-101.

Stickney, R. R., and B. A. Simco. 1971. Salinity tolerance of catfish hybrids. *Trans. Am. Fish Soc.* **100:** 790-792.

Stott, B., and T. O. Robson. 1970. Efficiency of grass carp (*Ctenopharyngodon idella* Val.) in controlling submerged water weeds. *Nature,* **226:** 870.

Swingle, H. S. 1957. Preliminary results on the commercial production of channel catfish in ponds. *Proc. Southeast. Assoc. Game Fish Comm.* **10:** 160-162.

Swingle, H. S. 1958. Experiments on growing fingerling channel catfish to marketable size in ponds. *Proc. Southeast. Assoc. Game Fish Comm.* **12:** 63-72.

Trewavas, E. 1973. On the cichlid fishes of the genus *Pelmatochromis* with proposal of a new genus for *P. congicus*; on the relationship between *Pelmatochromis* and *Tilapia* and the recognition of *Sarotherodon* as a distinct genus. *Bull. Brit. Mus. (Nat. Hist.) Zool.* **25:** 1-26.

SUGGESTED ADDITIONAL READING

Hickling, C. F. 1971. *Fish culture*. Faber & Faber, London. 317 p.

McNeil, W. J. (Ed.). 1970. *Marine aquiculture*. Oregon State University Press, Corvallis. 167 p.

Meyer, F. P., K. E. Sneed, and P. T. Eschmeyer (Eds.). 1973. *Second report to the fish farmers*. U.S. Bureau of Sport Fisheries and Wildlife Resource Publication 113. 123 p.

Pillay, T. V. R. (Ed.). 1972. *Coastal aquaculture in the Indo-Pacific region*. Fishing News (Books), London. 497 p.

CHAPTER 2

Water Systems

GENERAL CLASSIFICATIONS

Selection of the proper type of water system for any proposed aquaculture facility is a critical factor in the optimization of production and proper utilization of the available land and water supply. The aquaculturist may have some latitude in choosing a water system, but its design should provide the best possible conditions for maintenance of the culture organisms without being uneconomical. Various types of water system have been employed by aquaculturists in attempts to meet the optimum production goal while avoiding stressful conditions. A basic principle of aquaculture, which applies to all aspects of any culture strategy but is particularly critical with respect to water systems, is that a healthy population of animals can be maintained only if those animals remain free from stress.

With respect to water movement, culture systems may be either static (lentic) or flowing (lotic). The most widely used static systems employ ponds as culture units. Some of the more common flowing water systems utilize tanks, raceways, silos, and cages as rearing chambers. In static systems provision must be made for adding and removing water during filling and draining, and water must be available to replace losses due to evaporation and seepage and to help maintain water quality. In flowing systems water continuously enters and leaves the culture chambers, although the turnover rate (time required to replace the total volume in a culture tank) can vary considerably from one system to another.

Flowing culture systems may be either open or closed. Open systems are those in which the water exiting the culture chambers is not recycled back through the system, whereas closed systems employ recirculated water, which is usually passed through some type of device or series of devices that treat the effluent to restore water quality. A third type is the semi-closed system in which part of the water is recirculated and part is replaced with new water every day.

Generally, extensive aquaculture is undertaken in static waters and intensive culture is conducted in flowing water systems. Pond culture of channel catfish may produce 3000 kg/ha annually or more, whereas the equivalent of more than 10^6 kg/ha can be produced in some types of open and closed lotic systems. These figures are somewhat misleading, however, since they are based on available water surface area rather than volume. With either open or closed lotic systems, the total volume of water to which the animals are exposed during a growing season is much greater than that in the culture tank at any given instant. Viewed volumetrically, the apparent disparity in production figures between static and flowing systems is considerably reduced. In any case the amount of physical space required for extensive culture systems is much greater than that required for intensive systems capable of the same annual production.

The culturist must remember that in terms of initial cost and energy requirements, the more intensive the culture system, the more expensive it will be to operate (and possibly to construct). As recently as the early 1970s the trend was toward increasing sophistication in culture system design and promotion of high intensity operations. Each new design tended to require significantly more energy to operate than its predecessor. Present high costs of all commercial sources of energy have forced many aquaculturists to retreat from such systems and return to culture ponds.

This does not mean that intensive culture systems should not be further developed. In some instances intensive culture systems can be operated inexpensively (e.g., when heated water is readily available from such sources as power plant condensers, geothermal springs, or warm artesian wells). Intensive tank and raceway systems are almost a necessity in many types of aquacultural research because of the problem of replicating experimental conditions in ponds.

Under certain circumstances, closed recirculating water systems present the only reasonable approach to economical aquaculture. For example, in urban areas an excellent market may exist for certain aquatic species, but physical space is at a premium. A closed recirculating water system might be constructed indoors in an office building, warehouse, or some other structure. If the urban area is in a climate that features a short growing season for warmwater species, indoor culture could extend the effective growing season, even to 12 months. Indoor intensive culture systems may be economical if more traditional culture facilities for the species selected are so far away that the cost of importing the product is a significant percentage of the market price. Shipping costs for animals reared in an urban closed water system would be minimal even if other expenses of culture were higher than in less intensive operations.

The remainder of this chapter discusses water sources and specific types of water systems. Since many of the features of one system can be incorporated into others, when a feature or topic is introduced it is usually discussed in detail. Thereafter its utility in other types of situations may be mentioned, but with little elaboration. The general designs presented here are not meant to represent specific plans, but are provided as an aid to assist the culturist in selection and design based on individual needs, capabilities, and finances. A brief tour of aquaculture facilities in any part of the world will serve to quickly demonstrate that each is distinct, reflecting in part the personality of its designers. Each has flaws and most have strong points, but no two are exactly alike.

WATER SOURCES AND PRETREATMENT

The source of water for any aquaculture system is critical to the success of the enterprise. Some systems enjoy the advantage of utilizing only small volumes of new water after the culture chambers have been filled initially (e.g., well-sealed ponds in areas of low evaporation and closed recirculating water systems); however all the new water added to any culture system must be of suitable quality to ensure that the culture organisms are not stressed. The aquaculturist may be able to select from among several sources of water, or the alternatives may be extremely limited. If only a single water source is available, aquaculture must proceed within the constraints dictated by that water, whether its quality, volume, or both. As an example, a culturist may be obliged to utilize municipal water in a culture system. Since the levels of chlorine added to municipal water supplies are lethal to many aquatic organisms, it may be necessary to run the water through activated charcoal or to treat it chemically with sodium sulfite to remove the chlorine before allowing the water to enter the culture chambers. Either process would pose little difficulty or expense in closed recirculating culture systems but would be cumbersome and expensive in open or large static systems.

Well Water

Many aquaculture facilities are sited in locations that have good supplies of ground water, although the activities of industry and irrigation demands on the groundwater supply are significantly reducing its availability in many portions of the country. In regions of plentiful groundwater, various qualities may be available in different volumes from several strata. Well logs from the vicinity of proposed aquaculture facilities will provide an indication of potential flow rates and water quality.

The culturist who wishes to drill a well to supply any type of facility should carefully assess his needs with respect to maximum demand in liters per minute; then a well must be drilled to provide sufficient water to fill those needs. Though increasing well diameter adds significantly to drilling costs, a well that produces insufficient quantities can be both troublesome and contributory to crop failures. In many instances of water quality deterioration, the cure involves the addition of large amounts of new water. If water is unavailable, heavy losses of animals can occur. A well that produces more than currently anticipated needs will allow for future expansion of facilities, and it is less expensive to drill one large diameter well than two small ones capable of providing the same volume of flow.

The absolute requirements for volume of water flow per minute vary with the type of culture system being utilized and the species of animals maintained. As a general rule, sufficient water should be available to fill an individual static pond in 1 week and preferably within 24 to 48 hours. If numerous ponds are employed the culturist should be able to fill them all in no more than 3 to 4 weeks. If sufficient water is available to meet the criteria outlined, there should be plenty of excess on demand for emergencies during the growing season, since once filled, ponds do not generally have a continuous demand for large volumes of water.

An open water system must have sufficient water available to exchange its volume several times daily. Depending on the density and species of organisms involved, turnovers as frequently as once or twice per hour may be required. This water is required 24 hours per day, 7 days a week, throughout the growing season. Wells providing water for such systems should have the capacity for extra flow if required during mechanical failures or if the biomass in culture chambers exceeds the capacity of the system to maintain proper water quality at standard rates of flow.

Most wells are cased and screened to prevent inflow of water from strata where the quality is poor and to keep out particulate matter. Recommendations for the type of casing, well diameter, pumps, and other factors can be obtained from competent well drilling firms. A variety of pump types and delivery systems for water are available. The characteristics of some of these have been discussed in detail by Wheaton (1977). In most cases, an aquaculturist will require at least 1000 l/minute for even a modest facility. Large operations may require many times that volume.

In some parts of the country, especially agricultural and industrial areas, shallow wells may produce water contaminated by surface recharge containing sewage, pesticides, fertilizer nutrients, and other chemicals, although the water may be well oxygenated. Deep wells, on the other hand, are often free of contaminants but may be depleted in oxygen. As well depths reach and exceed a few hundred meters, the temperature of the water often begins to increase, and this property can be used to advantage by aquaculturists. The excess costs involved in drilling a deep well should be taken into consideration, however, when determining the economic benefits of increased water temperature. In certain areas, such as coastal regions of the southeastern United States, the water from deep wells may be too warm for aquaculture and must be cooled prior to use. Even deeper wells (ca. 700 m) in coastal South Carolina and Georgia may contain salt at levels suitable for mariculture.

Shallow saline wells may also be drilled in certain regions. In the Pecos River area of West Texas, salt water is readily available just below the surface in quantities sufficient to support large aquaculture enterprises. The salinity of this water is generally less than one-fourth ocean strength, but certain euryhaline marine forms and many freshwater organisms can be reared in it. In the case of any water of this type, the relative concentrations of the various salts contained should be determined and compared with seawater. Other standard chemical tests should also be run to verify that the water will support aquatic organisms. Perhaps the best way to test water for its suitability for aquaculture is to conduct a small-scale growth study utilizing animals of the species that will ultimately be stocked if the water is suitable. If the aquaculture animals are also to be reproduced at the facility, further studies to determine the suitability of the water during the breeding season may be required.

Not all deep wells are free from the influence of contamination that appears to be associated with surface water recharge. We have examined three wells (two several decades old and one drilled in 1977) that enter a water table approximately 500 m below the Brazos River flood plain in East Central Texas. In each case the total ammonia concentration was about 2 mg/l, whereas deep well water should be ammonia free. The source of the ammonia remains unconfirmed, but it has been theorized that the water table involved outcrops at or closely approaches the surface near a town several kilometers from the location of the wells, and that the contamination occurs at the point of outcrop. Surface recharge of contaminants into deep wells may also occur if the wells are improperly cased, or if the casing becomes damaged or corrodes. Surface recharge can be especially severe in areas of porous soils.

An increasingly common problem being encountered by aquaculturists and others is the lowering of existing water tables as a result of steadily increasing demands. As the level of a particular water table falls, wells begin to dry up and new, deeper wells are required, assuming sufficient water can be obtained at the new depths. Eventually a stratum may be unable to meet the requirements of all users and the well must be abandoned by consumers requiring large volumes on a continuous or frequent basis. The expense of having to drill new, deeper wells or find alternative sources of water may prohibit the continuation of otherwise successful aquaculture ventures. In some instances (especially in Arkansas), catfish farmers have sold out when water became scarce, relocating their facilities in states where supplies are more dependable (e.g., Louisiana). In coastal regions the removal of fresh groundwater may lead to saltwater intrusion. Land subsidence may also occur as subterranean water is removed and not replaced.

Mariculture operations often use surface sources of salt water, although water of excellent quality may also be available from shallow wells. If the quality of surface salt water is poor (e.g., if it is contaminated by pollutants, has high turbidity, or has a drastic salinity range), it may be possible to drill a shallow saltwater well and circumvent the problem. Such wells may be drilled directly beneath surface waters or on land adjacent to surface salt water. Many coastal river mouths, marshes, and beaches have large sand lenses beneath them through which surface water percolates and can then be pumped out. These sand lenses can usually be depended on to remove contaminants and often provide water of more constant salinity than that which is present at the surface.

Surface Water Sources

Trout culturists sometimes utilize mountain streams or, more frequently, springs that feature excellent water quality, in flow-through raceway production and hatchery operations. The water is typically allowed to reenter the stream at a point below where it was diverted. Warmwater aquaculturists are seldom fortunate enough to be located in areas having high volumes of clear, unpolluted stream or spring water of the proper temperature. Mountain streams and springs are much too cold to support rapid growth of warmwater culture species, and much of the surface water available in suitable warmwater areas of the United States must be treated in some fashion before it can be utilized for aquaculture. Even so, many warmwater culturists must utilize surface water as the sole source for their facilities or in combination with insufficient supplies of well water.

Catfish farming, which has developed in the southern United States, is practiced largely in agricultural areas. If river or creek water is selected for use in aquaculture in regions where crops are grown, it should be checked for contamination by agricultural chemicals that may have entered in runoff. Pesticides and herbicides pose the most immediate threats to the aquaculturist, although high levels of fertilizers can lead to problems of excessive noxious plant production in ponds. Surface waters may also be contaminated with a variety of industrial chemicals as well as by domestic sewage and feed lot wastes. Since the concentrations of chemicals of these types may change seasonally or following periods of heavy precipitation, a single chemical analysis should not be relied on. An additional type of contamination from surface waters is the presence of unwanted species of fishes and disease organisms.

When a river or stream is utilized to supply an aquaculture facility, the maximum and minimum annual flow rates should be ascertained to ensure the availability of sufficient water throughout the year (Milne, 1976). Reservoirs and natural lakes, though generally dependable as water supply sources, can also go dry under certain circumstances, or the water level may be reduced to the point that the culturist is unable or not allowed to remove any of the remaining water. Intake structures for an aquaculture facility should be placed in such a manner that anticipated changes in water level do not render them inaccessible.

Before the aquaculturist begins to pump water from a surface body, permission must be obtained from the appropriate state and, if necessary, federal authorities. Laws governing the use of surface waters for aquaculture and the return of such waters to their place of origin vary from state to state. The federal government becomes a party in certain instances, such as when navigable waters are involved. The issuance of permits may follow lengthy proceedings, and the payment of fees may be exacted.

Lake and reservoir waters are employed less frequently than are rivers and streams, but such large bodies of water can be effectively utilized in a variety of culture strategies. In flowing systems that return the water to its source after use, it is desirable to run the effluent from the culture facility into a portion of the lake or reservoir well removed from the intake so that degraded water will recover before being recycled through the system.

Foreign Species and Fouling Organisms

One problem that occurs more frequently in aquaculture systems employing surface water than in those utilizing water from wells is the contamination of facilities and culture animals by foreign organisms, both pathogenic and nonpathogenic. Disease organisms and parasites present in incoming water can be carried throughout the culture system and may lead to serious problems. It is difficult to remove bacteria, protozoans, and especially viruses from incoming water. Less difficult to control, but still sometimes a significant problem in some areas, are species of nonpathogenic organisms that enter the culture chambers and compete with the aquaculture crop.

In fresh water various centrarchids, such as the green sunfish, commonly invade culture systems. Small carp, buffalo fish, gizzard shad, and a variety of others can also enter culture systems by surviving passage through pumps and plumbing. Removing centrarchids and other undesirable fishes is not a significant problem in tank and raceway culture, but it can be difficult in ponds. If not removed, such as by passing incoming water over a Saran screen or other fine mesh material (Figure 2.1), such undesirables may

Figure 2.1 A simple wooden box with a nylon or metal screen bottom can be utilized to filter out nonpathogenic organisms, such as invading fishes, from incoming water.

grow rapidly at the expense of the cultured species. Although it is often possible to keep predators out of intensive culture systems and ponds by means of water filtration, there is little the culturist can do to prevent the invasion of cages and net enclosures placed in large bodies of water. In small freshwater lakes and reservoirs a fish toxicant may be used before the stocking of cages, but in large water bodies this approach is impractical.

Fouling organisms can cause considerable problems in marine and freshwater environments. Sponges and bryozoans have been known to colonize the plumbing of aquaculture systems in both situations. In salt water, barnacles, tunicates, and various other groups of animals often foul pipes and colonize in culture tanks so heavily that severe water quality deterioration occurs. Many mariculture systems employ dual plumbing between the source of water and the culture chambers. Incoming water is pumped through one side of the dual system while the other is filled with standing freshwater to kill fouling organisms. Depending on the rate at which fouling occurs, the two halves of the system are alternated at intervals of one to several weeks.

The use of net or wire enclosures and cages in the marine environment can result in serious fouling problems. Many types of netting and wire that have not been treated to prevent fouling or deterioration from continuous exposure to seawater may become badly impaired. Fouling can be so heavy as to restrict the passage of water through cages, resulting in water quality deterioration. In addition, the cage material may tear under the weight of fouling organisms. Experiences of this type have occurred in the Inland Sea of Japan (Milne, 1976). Studies conducted in Scotland (Milne, 1976) indicated that in terms of resistance to fouling, galvanized welded wire seems to be superior for use in enclosures to material such as nylon and polypropylene. Welded wire coated with rubber or plastic is also available and will not corrode as long as the coating is not disrupted; however coated wire does not resist fouling to any appreciable extent. Materials that are impregnated with copper, or alloys high in copper, have been used to reduce fouling. Copper alloy wire is available for cages, but caution must be exercised to avoid copper toxicity to the culture organisms. Wire that has a high copper content remains to be thoroughly tested on many aquaculture species.

In some instances mechanical removal of fouling organisms is the only practical method of control. This can be accomplished with a degree of difficulty by cage culturists, but persons who rear aquatic organisms in large nets or fenced areas in estuaries, natural lakes, or streams will find it very difficult to remove the barriers for cleaning without allowing the animals to escape. One solution is to construct enclosures with paired nets or fences, so that one remains in place while the other is being dried and cleaned. Alternatively, divers can be hired to scrape fouling organisms from the nets. Either solution can be extremely expensive.

Cages can be removed from the water for cleaning at intervals dictated by the fouling rate. The animals in the cage must, of course, be moved into another cage and placed back in the water while their former home is being renovated. This handling process could be coupled with weighing of the fish and adjusting feeding rates (Chapter 5), thus accomplishing two goals at one time. The original cage should be allowed to dry for several days, after which the fouling organisms can be removed with a stiff bristle brush. If frequent cleaning of all cages is required, it may be necessary to construct twice the number that will be in the water at any one time. Fouling of cages is usually not very severe in fresh water, and cleaning, if required, can be done between growing seasons.

Suspended Sediments

A frequent and often significant problem associated with the use of surface water in aquaculture is associated with the sometimes high levels of suspended sediments (clay, silt, and fine sand) that are present. In sufficient concentration, sedimentary materials will deposit in ponds and other types of culture chambers and can even fill them in with time. Such sedimentation can be especially troublesome in tanks and can lead to mortality in some instances. If some type of substrate is used over the surface of the tank bottom (e.g., it is sometimes necessary to provide oyster shell, sand, or gravel substrates for certain species of animals), that material may tend to collect fine suspended materials and keep them from exiting with the effluent water. Sedimentation can lead to the death of some benthic organisms (Chapter 4).

Dissolved Gases

Some water sources, particularly wells, are low in dissolved oxygen but may be high in other dissolved gases, particularly carbon dioxide and nitrogen. Oxygen-poor water must be aerated prior to use. Although mechanical aeration (discussed later in this chapter) is possible, passage of water over splashboards or spraying it into the culture chambers are also effective aeration techniques.

Well water containing high concentrations of carbon dioxide or nitrogen gas may be toxic when these gases are present at certain concentrations. The dissolution of carbon dioxide and nitrogen in well water is generally a function of pressure. When the water is brought to atmospheric pressure, these gases are usually quickly dissipated to safe levels. Aeration, which as previously noted is necessary in oxygen-depleted well water, will help rid the water of excess carbon dioxide and nitrogen.

A problem that has occurred in both saline and fresh well water and can be handled much like the problem of low dissolved oxygen, is the presence of high levels of hydrogen sulfide. Hydrogen sulfide is toxic to aquatic animals and is readily detected by the human nose as the smell of rotten eggs. If present, hydrogen sulfide should be eliminated before exposing the water to aquatic animals. Strong aeration may be an effective means of driving off this gas. Splashboards may also be effective, and both techniques also serve to aerate the water.

Iron

Well water often contains levels of iron higher than those found in most surface waters. Even when iron levels do not cause direct mortality of aquatic animals, the metal may be present in sufficiently high concentration to coat with a layer of rust any metals with which it comes in contact. Underground, iron is usually in the reduced form. When exposed to oxygen at the surface, iron is oxidized, forming ferric hydroxide. Passing water rich in iron over splashboards or aerating it strongly accelerates the oxidation process. The ferric hydroxide precipitates out and settles. This operation may best be undertaken in a sedimentation basin that precedes the culture chambers. Thus the iron-containing precipitate is not passed into rearing chambers. As in the case of high levels of unwanted gases in well water, the flocculation of ferric hydroxide by the techniques mentioned also serves to oxygenate the water.

Mechanical Filtration

Removal from incoming water of both suspended solids and foreign organisms (except for certain microorganisms) can be achieved by means of mechanical filtration. Sand and gravel filters have been used extensively and are available from commercial sources. Diatomaceous earth filters are also available and perform well but tend to become clogged more quickly than sand and gravel filters because of the very small pore size of the diatomaceous earth medium. When extremely high loading is applied to any mechanical filter, clogging becomes more frequent. All filters must be backflushed on occasion, and if this turns out to be necessary more than once daily, filters of larger capacity should be utilized.

In circumstances where silt and fine sand sediments predominate, it may be practical to flow incoming water through a settling basin prior to filtration and use in aquaculture chambers. Most suspended materials will settle within a few minutes to a few days if the water is allowed to stand. A reservoir large enough to hold a volume equal to or larger than a one week supply for the culture facility should be adequate. Wind turbulence may tend to resuspend the sediments, however, and this factor should be considered in the design of a settling reservoir. A long, narrow reservoir oriented with the short axis in the direction of prevailing winds will have a short fetch and will reduce turbulent mixing.

Water may be filtered by allowing it to percolate through the filter medium by gravity (Figure 2.2) or by forcing it through the filter bed under pressure (Figures 2.3 and 2.4). Both techniques are effective, although the former calls for pumping the water twice, whereas only one pump is required in the latter. If the proper plumbing is installed, the same pump can be used both to deliver the water into the filter and to backflush when the medium becomes clogged. Pressure gauges located at the point of inflow and in a location below the filter will indicate clogging by showing a significant pressure drop through the filter. Many water systems are operated at between approximately 2.1 and 3.5 kg/cm^2.

Very fine particles may pass through sand filters and later flocculate and settle out in culture chambers. In ponds this is no problem, but it can lead to turbid water or undesirable sediment accumulations in tanks and raceways. Cartridge filters of various types and sizes are available through plumbing outlets and scientific or engineering supply houses. Such filters

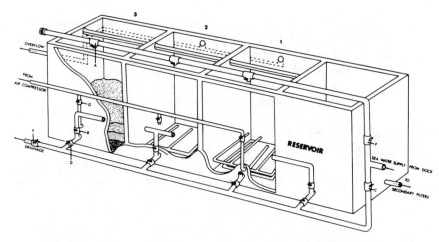

Figure 2.2 Gravel filter utilizing gravity flow. Filtration takes place in three separate chambers, allowing the culturist to backflush one or two without shutting off the water flow to the culture animals. Water enters through valves above the filter bays (*A*), and excess water is allowed to overflow (points 1, 2, and 3). Valves labeled *B* through *F* are associated with backflushing and were described by White *et al.* (1973). Original drawing by Daniel Perlmutter.

will remove particles as fine as 1 micron (μ); however they impede water flow significantly and must be operated under relatively high pressure. Thus cartridge filters may be effective for adding small amounts of water to a closed recirculating culture system but are not appropriate for large capacity lotic systems or ponds. Cartridges must be replaced frequently because they become clogged, adding to the expense of operating the water system. Such filters have application primarily in research and are generally impractical for commercial aquaculture.

CLOSED RECIRCULATING WATER SYSTEMS

Closed recirculating water systems are being used primarily for experimental work and for the rearing of larval organisms in commercial or research facilities. Many commercial aquaculturists have attempted to produce crops of marketable animals in closed water systems, but largely because of high costs for overhead, few commercial ventures have been successful. Development of closed recirculating water systems is continuing with emphasis on specialized uses for such systems.

Figure 2.3 Pressurized sand and gravel filter constructed from a steel aircraft engine shipping container. The tank was half-filled with pea gravel and sand. Water was pumped in at the top and forced down through the filter medium under pressure (approximately 2 kg/cm^2). Water exited through the bottom and entered the culture system located inside the building in the background. The vertical pipe at the right was used in backflushing. Water was shunted downward through that pipe and entered the filter at the bottom. When the backflush water exited the top of the filter, sediments were washed from the filter medium and diverted from the water system through the vertical pipe on the left and into a receiving area.

Basic Components

Some of the components of closed recirculating water systems are unique to this classification of aquaculture system, whereas others, such as the types of culture chamber and backup equipment, have wide applicability in semiclosed and open water systems. Closed systems are generally comprised of four components: the culture chambers, a primary settling chamber, a biological filter, and a final clarifier or secondary settling chamber (Figure 2.5). Each of these units is important to the system, although some closed recirculating water systems designs have eliminated one or more of the four components. The components may be separate units or they may be arranged in combinations that make the system appear to have only one or two compartments. Each component may be very large, or relatively small, but each must be in proper proportion to the others if the system is to perform properly. Closed systems small enough to be

Figure 2.4 Commercially available sand filters designed for use in swimming pools can be readily adapted for use in aquaculture systems. The filters shown are easily backflushed. Utilization of two or more filters allows for one to be backflushed while the others continue to provide water to the culture chambers located in the building in the background.

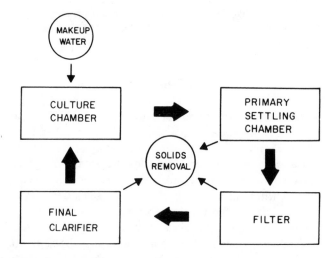

Figure 2.5 Generalized configuration of closed recirculating water systems. Large arrows indicate the direction of water flow, small arrows the direction of flow of incoming makeup water and solids effluent. At least one pump is required between any two chambers. Gravity flow moves water among the remaining chambers.

carried on the bed of a pickup truck have been designed, as have those capable of rearing several thousand kilograms of animals.

Culture Chambers

Closed, semiclosed, and open aquaculture systems usually employ relatively small culture chambers such as circular tanks, rectangular raceways, or silos. The material used for construction of the culture chambers varies widely, depending on such factors as the preference of the culturist, the availability of materials, costs, and in some cases the species of organism to be cultured. The most commonly used materials are concrete, wood, fiberglass, various types of molded plastic, and sheet metal (usually aluminum or stainless steel, and usually only in fresh water because of corrosion problems in the marine environment). Culture chamber dimensions are variable, although most circular and rectangular tanks are less than 10 m in diameter and seldom exceed 1 m in depth. Commercial tanks tend to be larger than those employed in research. The latter may be less than 1 m in diameter and the same depth or less. The material used for construction and any paint or other substance used to coat the inside of culture chambers should be tested to ensure that there is no toxicity.

Figure 2.6 presents a modification of the general closed recirculating water system featuring outdoor raceways and a holding reservoir (where water purification occurs). Raceways (linked in series or in parallel) can be constructed in the form of long, narrow earthen, concrete, or plastic-lined

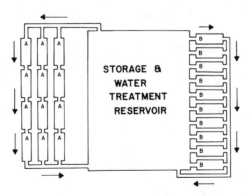

Figure 2.6 Schematic representation of two types of closed system raceway designs utilizing a storage and water purification reservoir. The scale for such systems is variable. *A*, raceways in series, with water running from one to another by gravity in the direction of the arrows; *B*, design in which each individual raceway receives reservoir water, which is then recycled in the direction of the arrows.

units. One pump is required either to lift water to the head of each individual raceway or series of raceways or to pick up water at the bottom of the raceways and return it to the reservoir.

Flow through the raceways is by gravity. The volume of water in the reservoir should be large relative to the combined volume of the raceways so that at least several days residence time is available for water treatment between passes of water through the culture system. Such systems must be designed to prevent escapement of culture animals into the reservoir. In most cases the reservoir should remain unstocked. Exceptions would be necessary if cage culture were conducted within the reservoir or if a natural water body or multipurpose reservoir were utilized.

If raceways in such a system are linked in series (situation A in Figure 2.6), care must be taken to avoid degradation of water quality down the series. If too many units are utilized in a series, or if the individual raceways are too heavily stocked, ammonia increases and dissolved oxygen depletions can occur in the raceways at the lower end of the series. As a rule the total length of the series of raceways should not exceed 200 to 300 m, if such problems are to be avoided.

Water systems using culture tanks of wood, metal, fiberglass, or plastic may be located outdoors; however indoor construction is preferred because better environmental control is possible and fouling of plumbing or degradation of water quality due to heavy algal growth can be eliminated when the tanks are not exposed to sunlight. In either case, each culture chamber in a typical closed recirculating tank system has its own water supply, although a common drain line may be shared by a large group of tanks.

Round tanks (often called circular raceways) have an advantage over rectangular raceways in that the former have no areas of dead water, such as corners, where waste products and feed can accumulate and foul the water. Circular tanks tend to concentrate solids in the middle, where they can be removed if the proper type of drain is installed. Corners can be eliminated in rectangular raceways by fitting the tanks with curved ends (Figure 2.7).

Water normally enters a rectangular raceway at one end from above and exits the other through a standpipe, which should be equipped with a venturi drain. The venturi drain system (Figure 2.8) is especially useful in conjunction with round tanks, since fecal material and waste feed collect between the two standpipes and can be flushed from the system by simply removing the inner standpipe. Such tanks tend to be self-cleaning and often require little routine maintenance. Another advantage of the venturi drain system is that water levels in the tanks are maintained in cases of pump or other mechanical failures leading to cessation of water flow.

A venturi drain cleans tanks most efficiently when some circular velocity

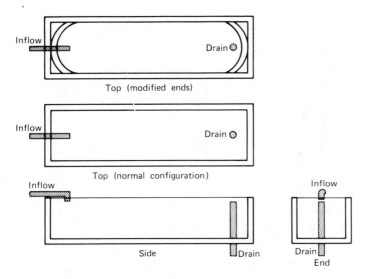

Figure 2.7 Configuration of rectangular culture tanks indicating the location of inflow plumbing and drains. To promote an even flow of water throughout the tank, the ends may be modified as indicated in the upper drawing. All dimensions are variable.

is maintained. This is readily accomplished by placing inflow water pipes tangential to the water surface. Water movement does not have to be particularly rapid to produce the desired results (a few centimeters per second is sufficient). In some cases circular velocity is helpful in providing a current with or against which culture animals can orient and obtain exercise. Juvenile and larval animals may be adversely affected by the spraying of water and by high circular velocity. Spray can be dampened by decreasing pressure, while maintaining flow through the use of gate valves or by introducing inflow water below the surface. Sufficient circular velocity may still be available for self-cleaning. If any appreciable water movement would force larvae against fine meshed netting or other materials placed over the drain to prevent escapement, very slow turnovers and large surface area drains may be necessary. The venturi system may be retained, but self-cleaning will be reduced. Large areas of the outer standpipe can be cut out and covered with bolting cloth of suitable size, through which exiting water can pass, but slowly enough that larvae are not swept along and trapped on the material.

Culture tanks or raceways should be small enough to allow the venturi drain to function properly. Although the drain diameter can be increased for larger culture chambers, this is effective only to a limited extent. Outside standpipes larger than about 30 cm are rarely used, and inside standpipes are normally not larger than approximately 10 cm. The inside stand-

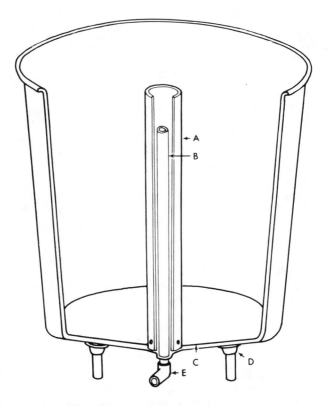

Figure 2.8 Cutaway of a circular raceway indicating the configuration of the venturi drain: *A*, outer standpipe or sleeve, with holes or slots at the bottom to allow waste to collect inside; *B*, inner standpipe, which controls water level and can be removed to allow waste to be flushed from the tank; *C*, bottom of tank, sloped toward the drain; *D*, legs screwed into flanges on the bottom of the tank keep the tank off the floor; allowing room for drain line; *E*, elbow connecting center standpipe to drain line. Original drawing by Daniel Perlmutter.

pipe must, of course, be large enough to accommodate anticipated maximum flow. Culture tanks larger than 20 m in diameter are probably impractical. Similarly, rectangular raceways should not be so long that water quality is severely impaired between the points of inflow and effluent. Indoor raceways are normally no wider than 3 to 5 m and no longer than 25 to 50 m. Most are significantly smaller.

It should be possible to drain and refill culture tanks rapidly during harvest and restocking. These processes are facilitated when oversize inflow and drain lines are used. During normal operation, inflow lines can be fitted with flow regulation devices to reduce the volume of water entering the tanks as compared with that available during filling when, if desirable, the flow regulation device can be removed. Apparatus of various types has been utilized to regulate the flow of incoming water.

Commercial flow regulators are available, or the aquaculturist may wish to fabricate them. In large tanks that receive high volumes of water, effective flow regulation can be achieved by reducing the diameter of the inflow pipes just before water is injected into the culture chambers. Combined with suitable pressure, this results in a jetting action that assists in tank aeration. Alternatively, water pressure may be regulated simply through use of gate valves, although the jetting action may be reduced to some extent. Commercial flow regulators often work by forcing the water through a constriction. This may also result in a jetting action that is beneficial in terms of aeration.

Rectangular tanks or raceways, even those fitted with venturi drains, tend to become fouled more frequently than do circular tanks, since much of the waste deposited on the bottom of the former is not carried to the drain unless extremely high rates of flow are utilized. Routine siphoning may be required to keep such tanks clean. One advantage that rectangular tanks do have is efficient utilization of floor space.

In the maintenance of larval animals and small concentrations of juveniles or adults, it may be desirable to hold animals for a period of time in a static or slowly flowing recirculating water system. To maintain a slow rate of circulation in such circumstances, airlifts can be utilized (Figure 2.9). Airlifts placed at intervals around the circumference of a culture chamber promote slow circulation, eliminating dead spaces where water quality might be impaired. In addition to water movement, airlifts provide oxygenation of the water. Air pumps are discussed in a later section of this chapter.

A specialized type of culture tank that has been used to a limited extent in pilot-scale commercial culture systems, primarily for rearing trout, is the silo. Silos are basically tall circular tanks (Figure 2.10). Because of the large water volume in silos relative to tanks of similar diameter, a rapid flushing rate may be required in the former to maintain water quality. Silos are of most utility in maintaining pelagic species, since the bottom area available for demersal forms is small relative to tank volume. Silos have the advantage of adding greatly to available water volume when surface area is limited but vertical space is not. Silos are often many meters high and may be constructed outdoors or inside a building.

To maintain water level in the event of pump failure, inflow should be at the bottom and effluent from the top of a silo (with a one-way valve on the inflow to prevent water from draining out through that line when the pressure is reduced). A standpipe drain would be feasible, but it could

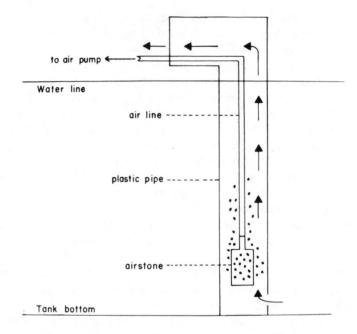

to air pump

Water line

air line

plastic pipe

airstone

Tank bottom

Figure 2.9 Airlift schematic indicating relative position of water line, tank bottom, air line, and airstone. Water is carried in the direction of the arrows by air bubbles rising through the plastic pipe (normally pipe diameter does not exceed 2 to 5 cm). Rate of flow can be adjusted by raising or lowering the airstone.

be difficult to manipulate in a tall silo. A second, bottom drain should be incorporated to permit emptying the tank without having to remove the standpipe. Supplemental aeration is important. In silos it may be necessary to introduce air at several levels to ensure even distribution.

Harvesting animals from a silo may present difficulties because of depth. Certain design features, such as the incorporation of a harvesting hatch near the bottom of the silo (Figure 2.10), might facilitate harvesting. The major problem encountered with silo culture is maintenance of water quality, especially when the silo is operated as part of a closed recirculating water system. In an open system the silo merely becomes a raceway standing on end and has no particular advantage if horizontal space is not limiting. The use of silos in trout culture has been discussed by Buss *et al.* (1970).

Each closed recirculating water system may employ one or more culture chambers in conjunction with each of the other three main components of the system. The effluent from several tanks could be collected in a common

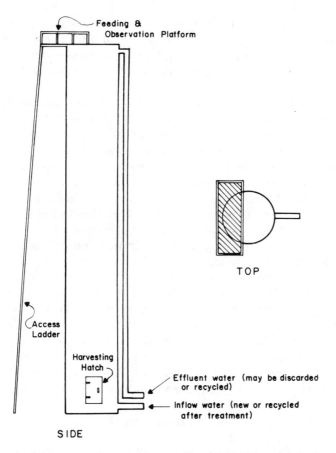

Figure 2.10 Schematic representation of a silo for the culture of pelagic fish. Scale is variable.

drain system and flowed through the settling chambers and the biofilter before reentering the tanks. In such systems each water treatment component must be large enough to handle the combined volume of the culture chambers. Advantages of building large settling chambers and biofilters include economical use of space, especially indoors; limitation on the number of backup components required; and reduction in the need to duplicate plumbing. In general the cost of constructing a few large chambers is less than that incurred for the building of several smaller units with the same combined capacity.

When the effluent from several culture chambers is pooled for treatment, an equipment malfunction could result in the loss of all animals in the system. Total power failures have serious consequences on any closed recirculating water system; however the failure of a single pump or aerator would be less crucial to the producer if each culture chamber had its own mechanical devices than if several culture chambers shared pumps and aerators.

A disease outbreak in a large water system rapidly spreads from one tank to another, but the spread can be prevented or better controlled if each culture tank has its own waste treatment system. Disease treatment often results in the damage or destruction of beneficial bacteria in a closed recirculating water system, leading to water quality deterioration and increasing the stress on the animals under culture.

From the standpoint of research, it is often undesirable to mix culture tank effluents. In studies where various levels of chemical or water quality conditions are being compared (e.g., temperature, dissolved oxygen, salinity, or ammonia), mixing of the water from various treatments can make readjustment of influent water quality impractical. In nutrition experiments diets with varying levels of a particular component are often utilized to allow the researcher to determine the animals' requirements of that dietary ingredient. If effluent water from various treatments is pooled before being treated and introduced once again into the culture tanks, the dietary ingredient being tested may become dissolved in the water; if this occurs the ingredient will be present even in tanks where the animals are not meant to be exposed to it. Under such conditions each culture chamber should be on a separate water treatment system.

Primary Settling Chamber

In most closed aquaculture system designs effluent water from the culture tank passes directly into a primary settling chamber. Water should enter and exit that chamber at or near the surface, to ensure that previously settled wastes are not resuspended. Inflow and effluent points should be as far apart as possible to allow maximum residence time in the tank for settling.

The volume of the primary settling tank should be sufficient to reduce the flow rate in that tank appreciably from that of the influent water. As the rate of flow is reduced, particulate matter will begin to settle. A drain at the bottom of the settling chamber is incorporated for the removal of settled material and for drainage when necessary.

Sedimentation of particulate matter is important because this step reduces the demand on the filter component of the system. In some designs the primary settling chamber is ignored or incorporated as a part of the biological filter. Sediments collected in the settling chamber or by an alternative method (e.g., by mechanical filtration) could be utilized as organic fertilizer. The material collected from the typical closed recirculating water system is composed largely of feces, waste feed, and bacterial floc.

Biofilters

Mechanical filters have been tried on experimental closed recirculating water systems, but clogging and resultant poor filtration efficiency can lead to serious water quality problems. Mechanical filters may become anaerobic and begin to produce toxic substances. Although when operating properly mechanical filters do remove particulate matter efficiently, they are not effective in the removal of dissolved metabolites. Removal of such substances is the primary purpose of the biological filter, or biofilter.

A major concern is the nitrification of ammonia (excreted by the kidneys and gills of aquatic organisms) to nitrate. The bacteria responsible for this process and the steps involved are discussed in Chapter 3. For the biological filter to be effective in reducing the levels of metabolites in a culture system, aerobic conditions must be maintained. If a biofilter becomes anaerobic, ammonia will be produced rather than detoxified, and the water may become lethal to the animals in the culture chambers. Aerobic conditions can be ensured by injecting air into the water entering the biofilter or by bubbling air through the biological filter itself.

Biofilters can be installed either indoors or out, but the internal portion should be protected from exposure to sunlight or bright artificial lights to prevent the growth of undesirable algae. Algal growth can lead to clogging of the biological filter and, if blue-green algae become established, there is the potential of off-flavors or even direct toxicity from certain metabolites produced by those organisms (Chapter 5).

Four basic types of biofilter have received attention from experimental aquaculturists. These are trickling, submerged, updraft, and rotating disc filters.

Trickling Biofilters. Water enters trickling filters from the top and is allowed to pass by gravity through the filter at a rate that does not allow the medium to become submerged, although all internal portions of the filter are continuously wetted. Municipal sewage treatment plants often employ trickling filters with rock media. Those units are much larger than units now utilized for experimental work except for one study undertaken in a municipal sewer plant that had been converted to a closed recirculating channel catfish rearing system (Davis, 1977). Whereas large trickling filters are equipped with rotating arms that spread incoming water evenly

over the filter medium, most research units developed to date have stationary water distribution systems. Figure 2.11 gives a generalized schematic
design of a small trickling filter.

Submerged Filters.   The design of submerged biofilters is often similar to
that of primary settling basins, except that the submerged filter contains a
medium on which bacteria become established. Water enters one end of
the filter and passes through the medium, exiting from the opposite end
(Figure 2.11).

Submerged biofilters can be operated by gravity flow or, with the incorporation of a watertight cover, water can be pushed through them under

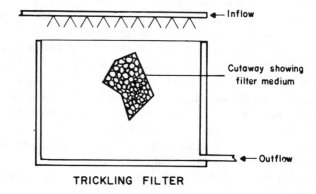

TRICKLING FILTER

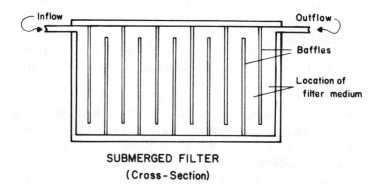

SUBMERGED FILTER
(Cross-Section)

Figure 2.11 Schematic representation of trickling and submerged biofilters indicating the
most common locations of inflow and effluent lines and of the filter media. The submerged
filter is fitted with a waterproof lid and contains baffles to route water throughout the
chamber.

pressure. If pressure is used, inflow and outflow pipes can be at any desired height without danger of losing water because of overflow. In a gravity system the pipes would logically be placed as in Figure 2.11.

Updraft Filters.   Water enters an updraft filter at the bottom and moves upward through the filter medium to exit at or near the top. A sedimentation chamber (primary settling chamber) can be incorporated into an updraft filter by designing the unit to allow solids to settle out below the level in influent (Figure 2.12). Solid material tends to accumulate in all biofilters as bacterial mats slough from the walls of the chamber and from the filter medium. Therefore it is recommended that a drain valve be incorporated at the bottom of the biofilter, regardless of type, so that settled material can be evacuated as necessary.

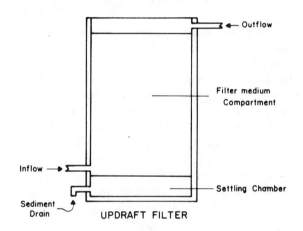

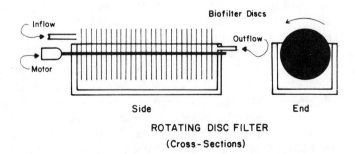

Figure 2.12   Schematic representations of updraft and rotating disc biofilters. Not all updraft filters incorporate a sedimentation chamber.

Rotating Disc Filters.   Rotating disc biofilters utilize a concept somewhat different from that of the filter types already described. In this case the medium is moved through the water, whereas trickling, submerged, and updraft filters utilize stationary media. Rotating biofilter media are composed of numerous circular plates placed on an axle and set in a trough with half of each disc submerged and half exposed to the atmosphere (Figure 2.12). The discs are rotated slowly (only a few revolutions per minute). Bacteria colonize the plates as in other types of biofilters. Alternating exposure to the metabolite-laden water in the trough and to the atmosphere provide the bacteria with a continuous supply of nutrients and oxygen. Experimental aquaculturists have tested the efficiency of rotating disc biofilters and some favorable results have been obtained (Lewis and Buynak, 1976). The other three filter types may be able to support higher loading rates and have been more widely utilized by aquaculturists.

Biofilter Size.   The size of the biofilter required for a particular culture system depends on many factors. Reliable formulas for calculating the relative size required as a function of the number, kind, and biomass of animals in the culture chambers, total water volume of the system, and flow rate remain to be adequately developed, although some work has been conducted on these aspects of closed recirculating water systems (Parker and Simco, 1973; Davis, 1977). When a closed water system is first put into operation the biofilter will not function until the medium has been colonized by the appropriate bacteria. This may take several weeks unless the filter is seeded with scrapings from an existing biofilter that is operating efficiently. Alternatively, an inorganic ammonia source (e.g., ammonium chloride) may be used to induce colonization of the biofilter by bacteria.

The total biomass of organisms present in newly stocked culture systems is usually small relative to the carrying capacity of the system; thus there is only a low level of metabolite production at the onset of rearing. It may take several days or even weeks for bacterial colonization to occur, but the microorganisms are present in the atmosphere and will invade the biofilter with time. Since proper biofilter colonization depends partly on the presence of a nutrient load, an unstocked or lightly stocked water system will develop the proper bacterial flora more slowly than a system that is more heavily stocked. This does not mean, however, that heavy stocking is mandatory in a new closed recirculating water system. The metabolite loading in such circumstances will be sufficient to bring about rapid colonization of the biofilter, but a great deal of stress may be placed on the animals from metabolite accumulation before the biofilter begins to operate at peak efficiency. In lightly stocked systems, even though bacterial colonization proceeds slowly, metabolite production often does not reach critical levels before the biofilter has begun to mediate the system.

Early in the growing season the size requirements for biofilters may be small, since there is little biomass in the culture chambers and the quantity of waste to be treated is not great. As the culture animals increase in size, the efficiency of the biofilter may be impaired to the point that water passing through that component of the system receives little or no improvement in quality. Water quality should be monitored routinely by sampling the effluent from the biofilter at least once daily and determining the concentrations of ammonia and dissolved oxygen. When an indication of loss of filter efficiency is detected, steps should be taken to correct the situation. This may involve altering the rate of flow through the biofilter, adding more air to that chamber, increasing the amount of filter medium in the system, or reducing the biomass.

The use of biofilters is presently more art than science in warmwater aquaculture and remains in a research phase of development. Design criteria for calculating water treatment requirements have been developed for salmonid hatcheries (Speece, 1973; Liao and Mayo, 1974) and should be applicable to warmwater hatcheries with appropriate modifications for temperature, ammonia, dissolved oxygen, and other factors regarding which differences between salmonids and many warmwater culture organisms exist. As a general rule, it is better to have too much filter capacity than not enough, although extensive overdesigning can be expensive. By utilizing units of proper size and design, portions of the biofilter component can be held in reserve early in the growing season and placed in operation as required.

Biofilter Media. The chemical transformations that occur in biofilters are accomplished by various groups of bacteria that require an aerobic environment and plenty of surface area on which to grow. Nearly any nontoxic material that provides abundant surface area is suitable as a medium for biofilters. The most commonly used media in the past were sand and gravel, since both provide abundant surface area per unit volume. However severe problems are often encountered when these media are utilized for filtration that is not mechanical.

In the first place, sand and gravel are dense materials, so the filter chamber must be strong enough to support them without rupturing. Second, biofilters containing sand and gravel have only a small percentage of the void space of lighter materials and frequently become clogged, thus requiring backflushing as often as once or twice daily. Frequent backflushing may be acceptable in open water systems where mechanical filtration is the primary purpose of the filters, but it can lead to problems in closed recirculating water systems if the biofilter must be removed from service even briefly while backflushing is accomplished. When sand or gravel filters are used in multiples, it is possible to shut down one or more filter while allowing flow to continue through the others. This is usually adequate for an open water system; in a closed recirculating system that is operating at maximum loading, however, removal of part of the biofiltration capacity may quickly lead to stress on the culture animals, even if backflushing can be accomplished within a few minutes. Direct mortality may not occur, but stress induced by a reduction in dissolved oxygen, coupled with an increased ammonia level, can lead to cessation of feeding, or disease and parasite outbreaks in the culture chambers.

Clogging of biofilter media such as sand and gravel occurs when feed particles, fecal material, and other unsettled wastes enter the filter or when bacterial films slough from the surface of the medium. What generally occurs is the phenomenon called channeling: water continues to pass through the filter, but it follows highly restricted pathways. The effect is to prevent the bulk of the filter medium from being exposed to the water. Filter efficiency declines rapidly and, as water trapped in the clogged areas stands, it becomes anaerobic. Under these conditions the biofilter may begin to produce ammonia rather than nitrify it (Chapter 3). When channeling occurs, backflushing becomes necessary.

Backflushing involves running high volumes of water backward through the filter. The plumbing must be designed to allow the sediments and sludge to be carried into a drain rather than back into the system. The rate of water flow necessary for backflushing is generally as high or higher than that used when the system is running normally. Backflush water should be expelled from the system rather than being allowed to move back into the primary settling chamber, where the flow rate may be so great that much of the suspended matter is carried on through and into the culture chambers.

Many aquaculturists have the impression that if the filter medium is densely packed, the volume of medium required can be greatly reduced. Because of clogging and channeling, this view is incorrect. In fact, a medium with less surface area and more open space between particles often is much more efficient than sand or gravel occupying the same amount of biofilter volume.

Charcoal, limestone, and oyster shell are employed (in addition to sand and gravel) in aquaria, either as filter media or in conjunction with sub-bottom filters. Because of the low density of animals in a typical aquarium, these media are often very efficient in maintaining excellent water quality. However high density, closed system aquaculture is a different matter. If the organisms inhabiting an aquarium are maintained at a density of up to several hundred grams per liter and fed a few percent of body weight daily, the filtering capacity of the system is quickly overwhelmed. In aquaculture these conditions are the rule rather than the exception; therefore high efficiency biofiltration is required.

Within the past several years a variety of filter media have been tested, and many have been found to be far superior to the standard sand and gravel media that preceded them. The most successful media are inert plastics. Polyvinyl chloride (PVC) has been used extensively as a biofilter medium. Scraps of PVC are often available around aquaculture facilities because this type of plumbing is widely utilized instead of metal. PVC will not corrode, nor is it toxic to aquaculture animals. Pieces of PVC pipe, PVC sheets, and other types of plastic have been successfully used as filter media. In addition, such materials as Teflon and Styrofoam have shown promise. Pieces of medium should be sized to present as much surface area as possible without leading to the kind of dense packing that might result in channeling of water. If buoyant media are chosen, a cover must be placed on the biofilter to keep pieces from floating out.

The plates in rotating disc filters are usually composed of plastic or fiberglass. Corrugated fiberglass sheets are readily available and seem to work well when cut into circles.

When lightweight filter media are utilized, the need for backflushing is often virtually eliminated, although some material may pass through the primary settling chamber and become deposited in the biofilter. In addition, pieces of bacterial mat sloughed from the filter medium may become deposited on the bottom of the biofilter. Thus a drain should be installed at or near the bottom to remove these materials.

Maintenance of pH.   The accumulation of dissolved chemicals in closed recirculating water systems leads to a depression in pH unless the system is buffered. As the water becomes more acid, stress is placed on the culture organisms, and if the pH becomes too low eventually death will occur. Microorganisms colonizing the biofilter may also be adversely affected by low pH. Organic acids and carbon dioxide are the primary causes of increased acidity. Chapter 3 treats the chemistry involved and the role of carbon dioxide with respect to water quality.

Wetzel (1975) noted that ammonia is strongly sorbed to particulate matter at high pH—another compelling reason for preventing the water from becoming acidic. Since bacteria are also associated with surfaces, the nitrification process may be enhanced when adsorbed ammonia and microorganisms are placed in close proximity on a waste particle or on the biofilter medium.

For most freshwater aquaculture systems, the pH should be in the vicinity of 7.0 (range of 6.5 to 8.5), whereas saltwater systems should be maintained at a pH in excess of 8.0 (reflecting the differences in normal pH of natural fresh and salt waters). To accomplish pH control, calcium carbonate is often used as a buffering agent. This material may be in the form of limestone, or more commonly, whole or crushed oyster shell. As pH decreases, calcium carbonate slowly dissolves and forces the pH back up (Chapter 3).

Oyster shells and limestone are relatively inexpensive and require little or no attention once incorporated into the water system. Bacterial mats build up on the surface of the buffering material and may interfere to some extent with dissolution of the calcium carbonate. Cleaning may be necessary at intervals of several weeks to ensure that the buffering capacity of the system is maintained.

The amount of buffer material present in the biofilter is not critical. Most culturists utilize a few kilograms per cubic meter of filter capacity, often placing the buffering agent at a location just before the point of effluent from the biological filter.

The Entire Water System as a Biofilter.   Although each component of a closed recirculating water system has its own particular function, all compartments work to a degree in biofiltration. This is because all the units in the system, as well as all the plumbing, have surfaces on which bacteria will grow. Theoretically this increases the total efficiency of the system, but it can also lead to problems. Bacterial sludge has been known to build up so heavily in pipes that water flow is impeded or even stopped. One means of decreasing the risk of such blockage is to use oversized pipes and design the system with the fewest constrictions possible. Access ports for cleaning should be provided at strategic locations.

Secondary Settling Chamber

Some closed recirculating water systems utilize a secondary settling chamber or final clarifier; such a compartment is useful for the collection of solid materials that may pass through the biofilter, but it is not a mandatory feature. The design of the final clarifier is virtually the same as that of the primary settling chamber and again, flow reduction is necessary to promote settling (Figure 2.13).

As the concentration of dissolved organic materials in the system increases, foam often begins to form, especially if water exiting the biofilter is allowed to splash into the secondary settling chamber. Foam can be produced from a variety of organic materials, including dissolved proteins. The removal of foam reduces the total demand on the biofilter by removing organic matter from the system. Foam removal can be accomplished most

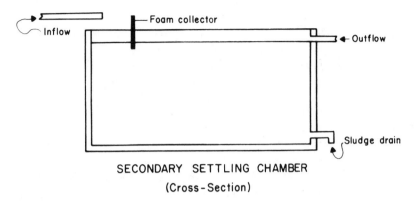

SECONDARY SETTLING CHAMBER
(Cross-Section)

Figure 2.13 Schematic representation of a typical secondary (or primary) settling chamber with foam stripping panel and sludge drain.

simply by skimming it off the surface of the secondary settling chamber or any other component of the system where it becomes a problem. Placement of a vertical board or sheet of metal across the surface of the final clarifier close to the location of foam formation concentrates the foam into a small area (Figure 2.13). The foam overflows the chamber and can be allowed to run down a drain or evaporate. A horizontal sheet of metal or board can be used in place of a vertical one, in which case billows of foam form on the horizontal surface and evaporate. Other, more elaborate foam strippers, including moving panels, have been designed and may be more appropriate on large water systems than on those described earlier, which are used primarily for research.

Moving Water

The fewer mechanical devices in a recirculating water system, the less chance exists for failure. One high quality, continuous duty water pump is often all that is required to make a water system operate efficiently. The pump can be placed between any two components, except when an updraft filter is utilized, in which case it is desirable to place the pump on the influent line to the filter. In all types of system units downstream from the pump can obtain water through gravity flow. When more than one pump is utilized, sophisticated electrical circuitry is sometimes required to ensure that water flow is balanced in all components. Increasing the number of electrical devices associated with a water system also increases the likelihood of equipment failure.

Pumps can be expected to run continuously for up to several years if they are properly maintained. If pumps are run intermittently, starters may occasionally fail. In saltwater systems pumps with metal impellers generally function well if allowed to run continuously but rapidly corrode when used intermittently. The use of metal pumps with PVC plumbing poses few trace metal toxicity problems in marine systems, since little metal is in contact with the water.

The amount of water to be moved through a water system is related to the size of the system and the optimum flow rate (the latter must often be determined experimentally and may change as the demand on the biofilter increases because of growth of the culture organisms). A certain amount of flexibility should be incorporated into any design. Several types of water pumps that are currently available and utilized by aquaculturists have been discussed in detail by Wheaton (1977).

Plants and Tertiary Water Treatment

Nitrates and phosphates accumulate in closed recirculating water systems even if a highly efficient biofilter is in operation. Although both these substances are required nutrients for plant growth and are not directly toxic to aquaculture organisms under normal circumstances, nitrates can be toxic at extremely high concentrations (Westin, 1974). Nitrates can be removed through denitrification to elemental nitrogen, and this technique has been applied to aquaculture systems (Meade, 1974; Balderston and Sieburth, 1976). However efficient removal of both nitrates and phosphates can be accomplished by incorporating aquatic plants in the system. The plants might be best placed in a position immediately following either the biofilter or the final clarifier, or they could be placed in the final clarifier.

Candidates for use in tertiary water treatment include water hyacinths (*Eichhornia crassipes*) and Chinese water chestnuts (*Eleocharis dulcis*). Each grows rapidly and is efficient in removing dissolved nutrients from the medium. Water chestnuts have potential economic value as human food. Water hyacinths, though considered a nuisance when they invade natural waters, hold some potential for livestock feed, as do a large number of other aquatic plant species (Boyd, 1968a, 1968b).

Plant growing chambers associated with closed water systems must be either placed outdoors or provided with artificial light of suitable quality and quantity. The cost of operating artificial lights should be considered in the overall economics of the water system, since a significant amount of energy may be required in a large culture operation.

Regardless of whether the plants are to be utilized as animal food, they must be harvested periodically. Extremely rapid growth of plants can be expected; thus it may be necessary to harvest at frequent intervals. Unmarketable plants might be spread on the ground to dry, and sufficient space should be available to accommodate the anticipated production. Because of the high moisture content in aquatic plants, those used for livestock feed must be dried before shipment or fed in the immediate vicinity of the aquaculture facility; otherwise the cost of handling them makes their use uneconomical.

If the animals in a culture system are not going to be maintained for more than a few months at high density, it may be better to avoid plant culture and allow the nutrients to accumulate in the system. At harvest the nutrient-rich water can be discarded and eventually replaced for the next growing season. If exceedingly high levels of nutrients have built up, a reasonable approach is to exchange all or part of the water in the system to reduce the nutrient concentrations, although this may not be possible if water is available in limited quantity. Another alternative, as previously mentioned, is to increase the filter volume by adding extra units to the system. Finally, the biomass in the culture system may be reduced.

Backup Systems and Auxiliary Apparatus

Each mechanical component of an intensive aquaculture system should have a backup, or some means of maintaining the system with an alternative piece of equipment should be furnished in case of pump failure or the breakdown of other mechanical apparatus. Since aeration is generally provided in a closed recirculating water system, an apparatus is usually already available to provide sufficient oxygen to maintain the system in case of pump failure, assuming that a backup pump can be placed on the line fairly quickly or that the problem is minor enough and can be rapidly repaired. Aeration devices such as blowers (Figure 2.14), air compressors (Figure 2.15), mechanical agitators, and bottles of compressed air or oxygen have all been used by aquaculturists. Except for compressed gas bottles, however, all require electricity for their operation, thus would not operate during a power failure. Many culturists maintain a gasoline or diesel generator (Figure 2.16) as a component of their backup system. Such a device can run both pumps and aerators.

Many culturists fail to provide adequate backup systems for their facilities and risk losing their crops because of their unwillingness to spend the amount necessary for protection in cases of emergency. Although not inexpensive, a backup system often more than pays for itself if even a portion of one crop is saved by its availability.

Backup devices are essential for the operation of any aquaculture facility, but various types of auxiliary equipment may be considered optional, though desirable. Some means of pretreating incoming water is indicated

Figure 2.14 Electrically operated air blower, which produces large volumes of air at low pressure. Such blowers operate continuously and can provide sufficient aeration for a number of culture chambers and biofilters.

Figure 2.15 Electrically operated air compressor capable of supplying fairly high volumes of air at high pressure. The motor runs intermittently as the storage tank is emptied.

in many cases, especially if the water is obtained from a surface source. Mechanical filtration has been discussed previously, but other forms of treatment also exist. If, for example, the water is extremely hard or alkaline, or if it contains undesirably high levels of various inorganic chemicals, it can be passed through an ion exchange column of the proper type to remove the contaminants. Ion exchange columns can be very expensive (depending on the type of ion exchange resin required and the volume of water to be treated), and the resin must be replaced or recharged whenever it becomes saturated. In a closed system it may be economical to pass incoming water through an ion exchange column, but in open systems the expense is almost always prohibitive.

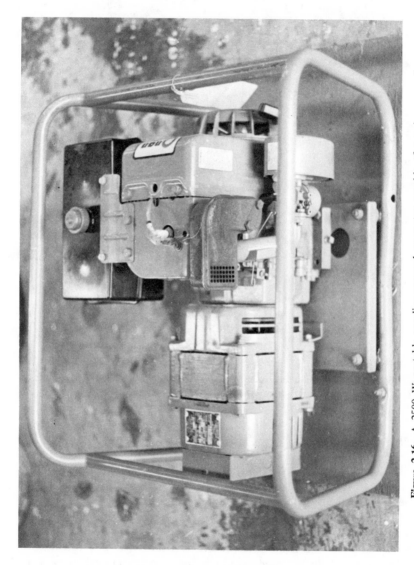

Figure 2.16 A 2500 W portable gasoline-powered generator capable of running a small blower, one or more pumps, or other critical apparatus during a power failure.

Ozone generators have been utilized in aquaculture to oxidize organic matter and kill bacteria in the circulating water. However ozone is extremely toxic, and a potential problem involves the necessity of giving the ozone sufficient time to convert back to molecular oxygen (from O_3 to O_2) before the water containing it enters either the culture chamber or the biofilter. Since the half-life of ozone is about 15 minutes (Layton, 1972), an ozone generator can be utilized only if the system has a slow flow rate or if part of the water can be shunted out of the mainstream of the system during treatment and until the ozone decays to molecular oxygen. Wheaton (1977) suggested that measurable quantities of ozone be avoided in aquaculture systems. Ozone can be reduced or eliminated from the water by passage through an activated charcoal filter, and this technique is recommended (Sander and Rosenthal, 1975).

Contact times of from one to several minutes with ozone levels of less than 2 mg/l result in a kill rate on bacteria, viruses, and protozoans of certain types in excess of 90% according to a review of several studies made by Kelly (1974). The use of ozone can eliminate circulating bacteria in the culture system without adversely affecting the bacterial activity of the biofilter, provided ozone is removed from the water following treatment.

Ultraviolet (UV) light has also been utilized as an effective bactericidal treatment. The impact of UV light on bacteria and other microorganisms is dependent on wavelength. UV light ranges from about 150 to 4000 angstroms (Å), but the most effective wavelength for bactericidal action is about 2600 Å (Koller, 1965).

UV lamps can be suspended over culture tanks or other system components, or submerged in the water. In suspended systems standard fluorescent light fixtures can be equipped with readily available bulbs that emit a high percentage of UV. These lights provide white light in addition to UV. Commercially available submerged UV systems capable of handling fairly large volumes of water can be readily incorporated into open or closed aquaculture systems. In submerged UV systems the water flows past UV bulbs placed in glass or quartz tubes. The layer of water passing the light should be as thin as possible for greatest effectiveness; thus many systems employ several closely spaced UV bulbs encased in a cylindrical waterjacket. Problems associated with such systems include fouling of the surface of the glass or quartz tubes (necessitating cleaning of the tubes to restore the efficiency of the system) and the need to replace the bulbs approximately once a year.

SEMICLOSED WATER SYSTEMS

All water systems, including closed recirculating ones, require at least small amounts of water continuously or intermittently to replace losses due to leakage, splashout, and evaporation. A system in which the water is replaced at regular intervals but is at least partly recirculated is referred to as a semiclosed water system. Since such systems are not much utilized by warmwater aquaculturists, discussion of them must be largely theoretical.

Semiclosed water systems may be derived from any closed recirculating water system merely by adding continuously significant amounts of new water and allowing a like amount to exit the system. In fact, early in the growing season when the stresses being placed on water quality are small, it makes economic sense to operate the system in the closed recirculating mode. As water quality begins to deteriorate, new water may be added, at a slow rate initially, and in larger volume as required. Near the end of the growing season the turnover rate may be as rapid as once daily. All treatment components of the system continue to operate, but the demands on them are reduced through dilution of metabolites and flushing out of solid wastes.

New water entering a semiclosed system should be flowed directly into the culture chambers. Preaeration may be required in cases of water that is drawn from an oxygen-depleted source. Excess water can be removed from either of the two settling chambers or from the biofilter effluent. If surface water is utilized it may be desirable or necessary to pretreat with ozone or UV light to avoid introducing disease organisms.

OPEN RUNNING WATER CULTURE SYSTEMS

The open running water culture technique allows the aquaculturist to dispense with virtually all the major components of closed and semiclosed systems except the culture tanks and pump. Mechanical filtration, UV or ozone sterilization, and backup or auxiliary apparatus of various kinds may be necessary, but fewer such items are required than in the water systems discussed previously. The primary requirement for open running water culture is an abundant supply of suitable water. Water is usually sprayed into culture chambers and allowed to exit by way of a venturi drain into a receiving stream or impoundment. The tanks or raceways utilized may be round, rectangular, or some other shape (Figures 2.17 and 2.18). Turnover rates in the culture chambers are a function of the species of organism under culture; its density, metabolic rate, and condition; and the feeding regime being employed. With animals at low density it may be possible to maintain water quality while changing the volume in each culture tank only once or twice daily; high density culture will require exchanges as frequently as every several minutes.

To conserve water, the flow rate should be low when the animals are

Figure 2.17 Flow-through aluminum raceway for holding small animals for experimental use or maintaining fry prior to stocking in other types of water systems. Water enters at the near end and exits by way of a venturi drain at the far end. Supplemental aeration is provided by a blower through airstones.

stocked and gradually increased as growth and degradation of water quality occur. The exact requirements vary for each system and must be determined through water quality monitoring. Because excellent water quality can be maintained in open flowing water systems, a pump failure or electrical power outage may not have an adverse effect on the culture animals immediately, although water quality will begin to deteriorate rapidly. The

Figure 2.18 Circular polypropylene tanks for experimental studies. Water enters through flow regulators and exits through venturi drains.

time between mechanical failure and onset of stressful conditions in the culture chambers may be sufficient to allow repairs to be completed or the backup system to be placed in operation.

Open running water culture is impractical for most warmwater aquaculturists because of the large quantities of water required. If such systems are not available at low cost, the economics of operating them are unfavorable. The amount of energy that must be expended to move water through open culture systems is often large unless artesian flow is available. The technique is impractical except on a small-scale research level if the water must be heated or cooled before passing through the system.

Open culture systems offer considerable potential for use in conjunction with electrical generating plants where large volumes of water at and above ambient temperature are available. To date, power plants have not been designed with the idea of incorporating aquaculture systems, but as electric utility companies become aware of the possibilities of utilizing waste heat for food production, new approaches may be seen. Currently experimental aquaculture (and even a small amount of commercial aquaculture) is being undertaken in conjunction with power plants. In virtually every case the aquaculture system has been established as an add-on following plant construction.

CAGE CULTURE

Certain invertebrates can be reared in cages, but nearly all warmwater cage culturists in the United States raise fish. Cages may be placed in any type of water body, lentic or lotic, although their use in ponds is not believed to be advantageous unless harvesting might otherwise be extremely difficult. In streams and large impoundments where a culturist could never recover fish that were stocked without some means of confinement, cages can provide a good alternative to other intensive culture methods. Fish in cages are contained in a known, relatively small volume of water; feeding, harvesting, and possibly disease treatment are facilitated; and management of the culture system is simplified (Figure 2.19).

Cages of various shapes and sizes have been utilized in aquaculture. Circular and rectangular cages appear to work equally well. Cages for research purposes, usually not more than 1 m³ in volume, often contain several hundred animals averaging 0.5 kg at harvest. Large cages are difficult to handle, and they may rupture if lifted from the water; moreover if one of them is invaded by disease, the financial loss will be heavier than would be true if a smaller unit were involved. The total cost of construction per unit volume of water impounded is higher when small cages are used. Careful consideration should be given to unit costs when evaluating the economics of cage culture.

Figure 2.19 Cages of catfish in a reservoir. Each cage has a volume of 1 m³. The float in the foreground indicates the location of a concrete block anchor to which the cages are tied with polypropylene line. A second anchor is placed at the far end of each string of cages.

Cages can be constructed from a variety of materials. Most culturists have selected steel frames covered with some type of coated or uncoated welded wire. Galvanized wire or that which has been coated with rubber or vinyl is desirable in fresh water and is necessary in salt water where rapid corrosion occurs. Uncoated materials will survive only a few months in salt water; thus cage frames should be constructed of an inert material or protected with paint, fiberglass resin, epoxy, or some other substance that will prevent direct contact of metal with water. Nylon and other types of synthetic material may be utilized as cage fabric but are subject to tearing.

Each cage must be provided with some type of flotation (Figure 2.20). Styrofoam is popular but subject to deterioration and should be protected from severe wave action and abrasion with durable, lightweight covering. Other synthetic flotation materials are available, but most are more expensive than Styrofoam.

If the cages are free floating (not firmly attached to a fixed or floating pier), they are usually tied together with cable or rope and maintained in position by cables anchored at the sediment surface or along the shore (Figure 2.19). A properly floated cage should have only a few centimeters of freeboard.

Cages must be fitted with tops to prevent fish from escaping (Figure 2.20).

Figure 2.20 Cages of 1 m^3 volume being fitted with tops prior to placement in a reservoir. Note the Styrofoam floats attached to each cage for flotation and the feeding ring incorporated into the cage lid. These cages were described by Kelley 1973).

Solid tops of wood or other materials as well as those made from the same fabric as the cage walls and floor have been used. Cages may be fitted with feeding rings (Figure 2.20) made of metal, wood, or plastic. The sides of the feeding ring should extend several centimeters into the water. Floating feed is retained inside the feeding ring rather than passing through the sides of the cage. Floating feed also can be retained in cages if a band of fine mesh netting or screen several centimeters wide is attached to the upper inside walls of each cage. The fine mesh material should also extend a few centimeters below the water surface. Pellets will be retained without impeding water circulation through the bulk of the cage. Sinking pellets cannot be utilized, since large amounts of feed would be lost from the cage.

Initial construction costs of cages can be very high. Depending on the types of material chosen and the size of individual units, cages may cost up to several hundred dollars each. Life expectancy varies depending on the materials used and the environment in which the cages will be placed. Estimates of cage life seldom exceed 5 years for the frames. Replacement of the wire may be required more frequently.

One of the best uses for cages in fish culture, at least in theory, is in conjunction with the heated water discharge of power plants. Both fossil fuel and thermonuclear facilities require large amounts of water for steam condensation. To ensure that cooling water is readily available, reservoirs are commonly constructed adjacent to the plant, or water is drawn into the plant from an estuary, natural lake, or river. Water flows through intake canals into the condensers and is returned to the original source or elsewhere through discharge canals. Because of potential negative environmental impact, the change in water temperature as a result of passage through the plant must be no more than a few degrees celsius. Regardless, temperatures in discharge canals in temperate and tropical regions may be too warm for aquaculture during the summer.

However, the warm water in discharge canals may significantly extend the growing season if cages are placed there after thermally unaltered waters in the vicinity cool below optimum for the culture species. In some instances year-round culture is made possible. Once the discharge temperature increases in the spring to higher than optimum for the culture species, the cages may be moved into a more thermally tolerable environment. Fish such as tilapia, which cannot be maintained during the winter at ambient temperatures in temperate areas, can be kept alive in a warmwater discharge canal. Channel catfish are presently being successfully cultured during the winter in heated water of this type. There is at least one commercial operation in Texas.

Power plants must periodically shut down one or all generators for routine maintenance, and equipment failure can cause the plants to suspend electrical generation, eliminating the heated water. If the whole plant is not shut down, there may be sufficient heat discharged to permit growth of the culture species to continue. If the generation of electricity is temporarily terminated during a period of low environmental temperature, species such as catfish can be expected to demonstrate an interruption or reduction in growth; intolerant species, such as tilapia, may die. From the perspective of economic production of electricity, maintenance shutdown during the winter—a period of reduced demand—is logical. Such winter shutdowns will probably continue until aquaculture becomes an integral part of power plant operation.

Placement of cages in flowing water, such as the intake and discharge canals of power plants, takes advantage of the rapid water exchange rate that is available. The intensity of a particular cage culture operation is limited only by economics and the carrying capacity of the water body in which the culture work is attempted. Studies on water quality associated with cage culture at a power plant near Dallas, Texas, indicated that water quality was not greatly affected even in the immediate vicinity of fish cages placed in the intake and discharge canals (Pennington, 1977), though the native fish population was highly developed (Germany, 1977).

A problem unique to power plants has plagued some cage culturists. During the winter water passing through a power plant may enter the condensers saturated with dissolved gases. When this water is rapidly heated, it may become supersaturated. Supersaturation, especially with nitrogen, can cause a malady known as gas-bubble disease. This condition is similar to the bends experienced by divers who ascend from depth too rapidly. Affected culture animals may experience the formation of bubbles in the gills, fins, and behind the eyes. Fish subjected to gas-bubble disease often exhibit exophthalmia (or "popeye"). This symptom can also occur in conjunction with certain nutritional deficiencies (Chapter 5) and pathogenic diseases (Chapter 7); however its presence during winter in the heated water below a power plant is generally a sign of gas-bubble disease.

To avoid gas-bubble disease, the levels of gases in the water should be monitored with a saturometer. Since not all the fish will exhibit symptoms simultaneously, a culturist who does not have access to a saturometer, or is willing to take a chance, can examine the crop daily during the winter and check for signs of the disease. If an outbreak occurs or is imminent, the fish should be moved into cooler water where the degree of saturation of dissolved gases is lower.

An alternative to moving the cages laterally is to move them vertically— that is, to sink them, since increased pressure will also reduce the level of supersaturation (Chamberlain and Strawn, 1977). Sinking the cages just a few meters usually is sufficient. The major disadvantage of this technique is that feeding becomes difficult unless the cages are raised to the surface for a few minutes daily while the animals feed. If sufficient natural food is available in the water column, however, the cages can remain at the bottom of the supersaturated canal until the danger of gas-bubble disease passes. Another alternative is the use of cages that reach from surface to bottom. The fish can come up to feed but are able to compensate for supersaturation by moving to the cage bottom at other times.

Treatment of pathogenic diseases in caged fishes is facilitated by the use of dip tanks. The desired solution for treating diseased animals can be mixed in the tank; then cages containing affected fish are submerged in the treatment solution. Large cages cannot be treated in this manner unless they are sufficiently reinforced to withstand the stresses that occur when they are lifted from the water. Other methods of disease and parasite treatment for caged fish include removal of the animals for treatment on shore or in a boat-mounted treatment tank, construction of a treatment tank in the side of the bank, with gates or locks through which cages can be floated (eliminating the need to lift the cages from the water), and surrounding the cages to be treated with plastic bags into which chemicals are poured.

Cage culturists seem to be affected by poachers to a larger extent than other aquaculturists, although everyone in the profession should be aware of the problem. Cage culture is often conducted in association with public waters, and many people do not hesitate to dip fish out of a cage that is untended. Some states allow cage culture in leased areas, which can be fenced or at least posted. Power plant intake and discharge canals are normally off-limits to recreational fishermen and are better protected from poaching than the main body of their receiving waters. Area lights, alarms, or security guards may be required at some locations.

POND CULTURE

The majority of aquaculture throughout the world is conducted in ponds. The site selected for a pond should have soil with good water-holding capacity and a dependable supply of high quality water. If one of these features must be sacrificed in any instance, it should be the former. Ponds can be sealed against leakage, but nonexistent water cannot be created in arid areas; and although poor water can be treated to make it suitable for aquaculture use, the expense involved may be prohibitive.

Sealing Ponds Against Leakage

If a site is selected because it has a good supply of suitable water but sandy soil, ponds can still be constructed and sealed against leakage. The best soils for ponds are those that have a high percentage of clay. Even in regions that feature soils of the proper quality, however, lenses of sand may be present. If one of these is struck during construction of ponds, leakage may occur in the affected area. Leakage may be acceptable if large amounts of new water are available at little or no cost, but it must be avoided when water conservation is required (as is usually the case).

If water rapidly seeps from a pond, some means of overcoming the problem must be devised. Polyethylene lining sheets have been used effectively, but these must be replaced periodically and are subject to tearing when people enter the ponds. Small holes in a plastic pond liner may not be significant, but eventually they become enlarged and allow a great deal of seepage to occur. Other, more durable, but also more expensive pond liners may be purchased, including heavy duty plastic, rubber, and concrete. Thin polyethylene and other materials, which are sometimes affected by sunlight, may require replacement after a few years. Clear polyethylene is subject to photodegradation and may be effective for only a single growing season, especially in shallow, clear water.

In some areas it may be economical to haul clay into ponds. Clay should be spread evenly over the sides and bottom, then packed with a sheepsfoot or similar device before the pond is filled. Bentonite, an expandable clay used in drilling operations by the petroleum industry, has been widely used as a pond sealant. This material is available from various commercial sources and is not prohibitively expensive if long-distance hauling to the aquaculture site is not required. For ponds that do not have severe leakage problems, a rate of about 20,000 kg/ha is often recommended by suppliers of bentonite. Serious leakage problems may require levels of 125,000 kg/ha or more, which may become prohibitively expensive. Bentonite is generally spread over the pond bottom and then mixed into the sediments with a disc. When wetted, bentonite expands and forms a seal.

Newly constructed leaky ponds often seal themselves to a large extent within a few years of production as organic material and reworking of the sediments by the aquaculture animals occur. Species such as carp, goldfish, tilapia, and shrimp are often helpful in reworking the sediments and may aid pond sealing more than do pelagic species. Since the time required for natural sealing to take place may be in the order of several growing seasons, high rates of water utilization can be anticipated while natural processes are at work.

Saltwater ponds are often constructed in sandy soils because of the absence of other sediment types in coastal regions. If such ponds are constructed below rather than above existing ground level, hydrostatic pressure from the nearby estuary or open sea is often sufficient to maintain pond water level. In many instances saltwater ponds refill themselves after being pumped out because of the hydrostatic pressure in the area.

Pond Size

There is no absolute best size for aquaculture ponds, but extremely large or small ones should be avoided. Ponds should not be so small that they become uneconomical to construct, nor should they be so large that harvesting, draining, filling, and feeding become difficult. Ponds on the order of 0.05 to 0.25 ha are often useful for spawning and larval or fingerling production, but grow-out ponds should be larger.

A typical grow-out pond for catfish, tilapia, shrimp, and other species should be able to accommodate several thousand animals. Such ponds are generally not larger than 5 to 10 ha and may be as small as 0.5 ha for commercial production. If the animals in an individual pond are lost as a result of degradation of water quality or an outbreak of disease, it is better to have the failure in a small pond than a large one, assuming that only one pond is involved. On the other hand, the cost of construction per unit area of water impounded increases as the size of the ponds is reduced. Not only will there be more dikework, but each pond must have its own inflow and effluent plumbing. Because of the space required for levees, the amount of usable water surface for any given plot of land increases as the size of the ponds increases. All factors should be carefully weighed when pond designs are being considered. Ponds in the vicinity of 1 ha each may be a good compromise size for the rearing of most species.

It should be possible to fill a pond of any size from the existing water supply within a few days. In addition, it should be possible to drain the pond completely and relatively quickly (e.g., within 24 to 48 hours). Ponds that depend for water on land runoff after rains and have no means of draining are in use by some aquaculturists, but these are not recommended because of the undependability of the water supply and the difficulty involved in managing and harvesting. Wild fish introductions are more common in ponds that receive land runoff than those constructed such that their only source of inflow is controlled by the aquaculturist.

The time required to fill an individual pond is a function of the amount of water available per minute and the diameter of the inflow plumbing. Small pipes handle considerably more water when a pump is used than when flow occurs by means of gravity alone. Ponds are rarely pumped out but usually drain by gravity. As the diameter of the inflow and effluent lines increases, the cost may increase almost logarithmically. Currently metal and PVC valves, fittings, and pipes are among the most expensive items associated with pond construction.

Although some pelagic animals may be suitable for aquaculture, most species now being cultured in warmwater are benthic invertebrates or demersal fishes. Therefore most aquaculture ponds can be fairly shallow.

Catfish ponds, for example, are usually from 0.9 to 1.8 m deep in the southern United States, but may be on the order of twice those depths in the north. The difference with respect to latitude is dictated by whether the ponds can be expected to freeze over for long periods during the winter. Winterkill due to oxygen depletion can occur under heavy ice cover; thus greater depth is important in the north as a means of increasing the supply of dissolved oxygen available to the culture animals. In the south, ponds shallower than 0.8 m may be appropriate for such animals as shrimp, crayfish, and flounders, which remain on the bottom most of the time. Such bodies of water have the disadvantages of sometimes becoming too warm during the summer to support aquatic life and becoming weed choked because unless the water is very turbid, their bottoms are completely exposed to the sun.

The whole pond bottom is generally not of uniform depth. In most cases the depth increases toward the drain. A full pond may be less than 1 m deep at the shallow end and nearly 2 m deep near the drain, with the average being within the prescribed depths. Variable depth is important for complete drainage and can be of help in harvesting, since all the animals can be confined in a fairly small area but considerable volume when part of the pond is drained.

Pond Configuration and Levee Design

Most ponds are rectangular, although some are square, round, or irregular. Square or rectangular ponds are preferred, since space utilization is more efficient and such ponds are less difficult to harvest than are other shapes.

Actual pond construction is a highly technical and expensive process. With few exceptions, aquaculturists contract construction to experienced companies. The culturist may instruct the construction engineer or architect with respect to the number, surface area, and depth of the desired ponds, but normally design and construction are left to experts. The culturist should be certain that all debris is removed from the ponds following construction and prior to filling. The presence of stumps, brush, rocks, and snags of other types will make harvesting difficult.

Ponds are usually constructed on level or gently sloping land. They may be cut into the surrounding terrain, with the earth hauled off or used for fill in low places; earth may be hauled in and the levees built above the original land elevation; or the earth removed from the excavation may be used to build levees above ground, resulting in ponds that are partly above and partly below the original ground elevation.

The levee along at least one side of each pond should be wide enough to drive on, thus facilitating access for harvesting and feeding (Figure 2.21).

Figure 2.21 Levees between ponds should be wide enough to allow a vehicle to pass. This feature is helpful in maintenance, harvesting, and feeding.

Ideally, it should be possible to drive completely around each pond. All the levees should be wide enough to allow the passage of a mower, even if a truck cannot be driven on them. Wide levees are required if a tractor is used for mowing (Figure 2.22).

The recommended slope for pond levees is 2:1 (for every unit of depth, the width increases by 2 units). Slopes of 1:1 and 3:1 are often effectively utilized, although the former results in a 45° angle, which may leave the banks prone to erosion. Slopes of 3:1 are desirable at the deep end of ponds to facilitate entry and exit during harvesting, but banks with such a shallow slope also leave significantly more bottom area for invasion by

Figure 2.22 A tractor has many uses around an aquaculture facility, not least of which is maintenance of the grounds and pond levees by frequent mowing.

aquatic macrophytes within the optimum depth range than do ponds with levee slopes of 1:1 or 2:1.

Drains and Inflow Lines

Pond drains are normally located at one end of the pond, with the bottom sloping toward them. The configuration of the drains in a particular set of ponds depends on the preference of the culturist and the amount of money available. The simplest drain is a standpipe protruding from the pond bottom. The lower end of the standpipe is screwed into an elbow, which connects to the main drain. The upper end of the standpipe controls the level of water in the pond (Figure 2.23). When the water level is to be raised or lowered, the angle of the standpipe is changed by rotating the elbow. This is the least expensive drain available. Gate valve drains are the most desirable from the standpoint of being able to control them from the pond bank (Figure 2.24). Some means of screening the water before it goes through a gate valve is necessary to keep the animals from being lost (or the flushing of animals out of the pond may be utilized in harvesting, as indicated below).

Figure 2.23 Pond with turnover standpipe drain. The standpipe can be moved laterally to control water level.

The drains from several ponds on an aquaculture facility may feed into a common pipe of large diameter that carries the effluent water to a receiving area. Alternatively, drains may be plumbed through levees and allowed to empty into a canal that carries the water to a receiving area. In the latter case it is sometimes possible to place a net over the downstream end of the drain pipe and collect animals as they are washed from the pond. This does not work very well with species that orient themselves into currents, but it may be at least partially effective for most species if the drain line is large, since most individuals will not be able to overcome so swift a current after they approach the drain area.

Harvesting basins are sometimes constructed in ponds. These are usually concrete-lined depressions in the vicinity of the drain in which some water is retained after the remainder has flowed out (Figure 2.24). Animals not removed by seining or other means collect in the harvest basin and can be removed with dip nets. If a harvest basin is not available, the deep end of the pond can be contoured with earthmoving equipment during construction to provide a depression in which the animals will collect during harvest. This is much less expensive than the construction of a concrete harvesting basin and is generally more practical for the commercial aquaculturist. More of the crop can generally be saved for live-hauling if harvest basins or harvest depressions are incorporated into rearing ponds.

Valves to control the flow rate of water should be attached to inflow

Figure 2.24 Pond with concrete kettle drain system featuring a stairway for access during harvesting, a harvest basin into which new water may be flowed during harvesting, a gate valve drain that can be opened from above (wheel on top of concrete structure), and slots to hold screens for filtering the water during draining operations. Photograph by Meryl Broussard.

pipes to each pond. This allows the culturist to fill one or more ponds while excluding water or adding small amounts to other ponds. If the source of water is from a river or nonartesian well, pumping may be required whenever water is needed in the ponds, since the culture chambers must be built above the level of the inflow source. In ponds constructed below a reservoir, or where artesian water is available, flow may be obtained without pumping. Although a pump may be required when large volumes of water are needed to fill ponds initially, maintenance of water level may be achieved through gravity or artesian flow.

Since water under pressure is often utilized in inflow lines, the pipes can be of smaller diameter than drain lines. Small diameter pipes are less expensive than large sizes, and the fittings, especially valves, are a great deal cheaper.

An alternative to pipes for water distribution is a system of canals leading from the water source. Simple wooden gates may be removed from canals entering each pond to allow inflow of water as required. Such a system can provide high volumes of water at low pressure and necessitates no gate valves except at the wellhead when artesian water is being utilized.

Other Considerations

Additional features that can be designed into ponds include splashboards at the point of inflow to help aerate water coming from wells or other sources where dissolved oxygen is low. The same technique can be utilized to remove iron from water when that metal is present in high concentration. Ferrous iron (Fe^{2+}) forms ferric hydroxide ($FeOH_3$) when oxidized, and this compound precipitates from the water (Ruttner, 1953). Splashboards also assist in dissipating the force of water entering ponds and can curtail erosion of the banks during the early stages of filling.

Windbreaks may be required in association with the levees of large ponds to prevent erosion by wind and waves. Trees and hedges are not recommended because they need time for growth, as well as a certain amount of trimming. In addition, periodic problems with falling leaves entering the ponds can occur. The major consideration is the addition of unwanted organic matter, with resultant demand on dissolved oxygen as the leaves decay. Windbreaks of snow fence or similar material may be best. Alternatively, riprap can be placed along the sides of the banks that are subject to be eroded by action of the prevailing winds. Also, ponds can be designed and oriented with the narrowest aspect at right angles to the prevailing winds, to cut down on fetch and reduce erosion.

Pond levees should be seeded or sprigged to grass. In the southern United States, coastal bermuda grass is popular. It spreads rapidly and holds the soil well. The tops of levees that will be regularly driven on can be graveled to prevent wear. Gravel roadways are also important to allow access to the ponds after heavy precipitation.

Winter draining of ponds is often practiced. Following fall harvest, any ponds that have been drained are allowed to dry out, and they remain dry until spring. Pond bottoms can be disced to aerate the soil, smoothed or reshaped as necessary, and limed to adjust water chemistry (Chapter 4) and sterilize the sediments. In the spring the ponds may be treated with herbicide either before or after filling, as a means of limiting the establishment of unwanted vegetation and to kill terrestrial plants that have become established while the pond was dry. Some aquaculturists prefer to refill ponds after harvest and keep water in them throughout the winter. At least some period of drying should be scheduled after draining to allow oxidation of the sediments, and ponds should be disced every 3 to 5 years.

SPECIALIZED AQUACULTURE SYSTEMS

Various systems of aquaculture that have been utilized as commercial or research facilities but do not fit the general schemes outlined above can be found throughout the world. These specialized systems have been developed as modifications of more general designs or for rearing species that do not adapt well to ponds, raceways, tanks, or cages. One modification of the typical raceway system that utilizes a reservoir through which water is recirculated has already been described (Figure 2.6). Such a system, incorporating two series of raceways, has been used for channel catfish and trout research in Georgia with encouraging results (Hill *et al*. 1974). There are, however, aquaculture systems in use that are basically different from those discussed thus far.

Modified Pond Systems

Human Sewage Effluent for Aquaculture Pond Fertilization. A virtually unlimited supply of nutrients is released annually from domestic sewage plants in the United States. With proper treatment, pathogenic organisms can be virtually eliminated and the nutrients utilized for phytoplankton production. Such a system has been developed by researchers at the Woods Hole Oceanographic Institution (Huguenin and Ryther, 1974; Huguenin, 1975; De Boer *et al*. 1977). The system employs the nutrients in the effluent from a domestic sewage treatment plant in the nutrient media of phytoplankton and macroscopic algae. Plants that cannot be directly utilized for human or animal consumption are fed continuously to oysters, clams, mussels, and scallops. The feces and pseudofeces (mollusks that have an abundance of food often produce fecal pellets composed almost completely of undigested algal cells) produced by these invertebrates are consumed by marine polychaetes, which can be sold for bait. Fishes have also been introduced into the system to feed on invertebrate herbivores.

Plants and animals produced in systems such as this may be suitable for human consumption once such sale has been approved by the appropriate federal agencies. The U.S. Food and Drug Administration and the Public Health Service are the agencies most closely associated with ruling on the future of such practices. Currently not only illegal, but also impractical on aesthetic grounds, such systems may become important in the future both as a means of providing human food inexpensively and as a means of disposing of nutrient-rich wastes. The use of human and livestock excrement for land and water fertilization is widely practiced in Asia and other parts of the world. Organic waste research appears to be timely in the United States, although application may be delayed for several years.

Domestic Animal Wastes in Aquaculture. A significant problem associated with the rearing of terrestrial livestock is the disposal of liquid and solid wastes. In addition, severe odor and fly problems exist around live-stock production operations such as poultry houses and cattle and swine feed lots. As is true in the case of human sewage treatment, the wastes from terrestrial animals can be deposited in sewage lagoons, where degra-dation occurs. The high levels of nutrients released from animal waste oxidation can support the growth of extensive phytoplankton blooms and lead to high levels of secondary productivity in the zooplankton and ben-thos communities. In the United States little thought has gone into pro-ducing higher forms of animal life in such systems, although fish and other organisms have been cultured in sewage lagoons in other parts of the world (notably in Asia) for many years. Recently the practice has been researched to some extent in Israel and the United States (Schroeder, 1974; Stickney et al., 1977a, 1977b; Buck et al. 1978; Stickney and Hesby, 1978).

Many species of fish and invertebrates are unable to tolerate the low dissolved oxygen and high ammonia levels associated with sewage lagoons, but tilapia appear to be tolerant of the conditions that exist when animal wastes are added to ponds at controlled rates. The optimum rate of ma-nuring for maintenance of the desired plankton bloom and water quality has not been determined for poultry, swine, and cattle manure, but these areas are being researched. Our experience has indicated that the rate of manure addition and quality of the resultant phytoplankton bloom is best controlled when poultry serve as the waste source (Figures 2.25 and 2.26).

One problem with rearing cold-tolerant species like those in the genus *Tilapia* is that brood stock must be overwintered in warm water. Thus fish production ceases during the winter, although manure production does not. For this type of aquaculture system to operate effectively, a species that is cold tolerant must be incorporated, the water must be heated, or during the winter the manure must be treated in a manner different from that characteristic of the fish growing season. Of these choices, water heating is the least desirable because of economics.

Fish produced in ponds fertilized by domestic animal wastes could conceivably be marketed for human consumption if it could be demon-strated that there are no pathogenic organisms associated with the product. As a means of assuring a sanitary product and also to reduce the aesthetic stigma surrounding such products, the fish could be removed from the ponds and held in tanks that receive manure-free water for a period during which commercial feed is offered. This treatment may be as short as a few days or as long as several weeks. Microbiological testing should demon-strate the effectiveness of the technique in cases where pathogenic bacteria are present in the ponds.

Of possibly more immediate utility is the concept of rendering the fish

Figure 2.25 Poultry wastes deposited in a pond can promote a phytoplankton bloom that will support a herbivorous fish like *Tilapia aurea*.

Figure 2.26 Chicken cages placed over plastic-lined wading pools for determination of energetics of poultry-*Tilapia* production systems. Each round tank contains 50 fish receiving the waste from zero, one, two, or four laying hens.

produced into fish meal for use as livestock feed. Studies utilizing tilapia meal indicate that broiler growth was as good or better than when chickens were fed either commercial anchovy meal or an all-grain diet (Rowland et al., 1977). Recycling domestic livestock wastes, if it can be done economically, would solve the odor and fly problems and conserve nutrients. In the long run, the cost of domestic livestock production may also be decreased.

Alternative Aquaculture Schemes

String Culture. Though not widely practiced in the United States, string or hanging culture of oysters and mussels is common in other parts of the world. Strong ropes are suspended into the water from rafts or fixed platforms. Some type of cultch material (the most popular being mollusk shells) is strung onto the ropes and serves as substrate for the attachment of the aquaculture animals. The cultch may be colonized *in situ* by placing the strings in an area where natural oyster or mussel spawning occurs, or colonization may take place in the laboratory (Chapter 6). In areas of high production of the appropriate species of mollusks, the settling larvae (called spat in the case of oysters) may attach to the cultch material in such numbers that a great deal of thinning is necessary. This must be done by hand.

Hanging culture effectively utilizes the whole water column, whereas bottom production takes advantage of only the two-dimensional surface available on the substrate. In addition, hanging culture keeps the mollusks above the reach of predators such as oyster drills and starfish as long as the strings are not allowed to come in contact with the sediments. Strings must be placed in areas with high phytoplankton productivity to ensure that all the animals receive sufficient food for rapid growth. Strings can be lifted one at a time for selective harvest, checking on growth rates, disease treatment, and removal of fouling organisms.

With the increase in offshore drilling, interest in adapting string culture to such structures as oil drilling platforms has been expressed in recent years. However drilling platforms have not, to date, been designed to carry the extra weight associated with strings of mollusks. If it appears to be in the interest of drilling companies to construct platforms capable of supporting strings of shellfish, there is little doubt that this will be done.

Crayfish Culture. The culture of crayfish, largely restricted to the state of Louisiana, employs techniques that are somewhat different from those utilized by persons involved in more traditional types of pond aquaculture. At present United States production is based largely on two species: the red swamp crayfish (*Procambarus clarkii*) and the white river crayfish (*P. acutus*), with the former representing the bulk of the crop of most producers (LaCaze, 1976). Other species may be utilized in other parts of the country, when and if crayfish farming is expanded.

Commercial fishing for crayfish (or "crawfish" as they are called in Louisiana) has centered historically in the Atchafalaya Basin in the south-central portion of the state (de la Bretonne and Fowler, 1976). The harvest has ranged from just over 90,000 to 4.5 million kg annually between 1959 and 1965 (Broom, 1963; LaCaze, 1966). Because of flucuations in natural supplies, pond production was introduced to ensure sufficient supplies during the harvest season. At present, more than 300 crayfish ponds comprising over 18,000 ha are in production in Louisiana (Gary, 1973).

Three types of production ponds are currently in use (LaCaze, 1976; de la Bretonne, 1977). The most desirable from the standpoint of maximum utilization of land involves the rotation of crayfish with rice. Because of different water depth requirements for the two crops, it is often necessary to increase the height of rice pond levees to accommodate between 30 and 75 cm of water for crayfish, whereas rice may require only 15 cm (LaCaze, 1976). Open ponds that are employed only for crayfish production are the second method. The third involves the impoundment of wooded or swampy land characterized by the presence of numerous shrubs and trees in the ponds.

Brood crayfish are stocked in the flooded ponds during late May or June. LaCaze (1976) recommended that the numbers of males and females among the brood stock be similar. Brood crayfish should number approximately 20 to 30 to the kilogram. Depending on the amount of cover available in the ponds, brood animals should be stocked at between 20 and 100 kg/ha (LaCaze, 1976); the highest rate applies to open ponds with little or no vegetation.

Reproduction occurs during the late summer and fall (the details are presented in Chapter 6), with harvest the following January through March (de la Bretonne and Fowler, 1976). Harvesting is often accomplished with baited traps of various configurations rather than by draining of the ponds or seining. Since crayfish ponds contain various types of vegetation, seining is not generally a practical means of harvest.

In rice fields, the growing crayfish, as well as the original brood stock, are able to feed on the rice stubble left after harvest. In open ponds, wild rice, millet, or sorghum may be planted to supply the crayfish with food. Other plants often found in crayfish ponds include filamentous algae (various species), cattail (*Typha* sp.), and pondweed (*Potamogeton* sp.). Each may be utilized by crayfish for food, but each can also lead to water quality or management problems (LaCaze, 1976).

Crayfish farming holds potential in states other than Louisiana but has not been widely developed. The applicability of crayfish farming to marginal lands and as a second crop in rice farming areas should create increasing interest in crayfish aquaculture in certain portions of the United States, especially with current prices. Although prices are variable, relating to some extent to the availability of natural stocks coupled with the short harvest period, the low is generally about 40¢/kg and the high slightly in excess of $1/kg in Louisiana (de la Bretonne and Fowler, 1976). This is somewhat less than the price paid for live weights of other aquaculture products; in contrast, however, to most shrimps that dress out at 40% or better, crayfish yield only about 10% edible portion upon processing.

Theoretical Aquaculture Systems. A few interesting concepts in open ocean aquaculture have been advanced in discussions among aquaculturists, and some of them may eventually be put into practice. The use of trained porpoises to herd aquaculture animals is only one of those ideas. Another involves the use of sound waves to contain animals that have been trained to stay within hearing range. Such ideas have received little support to date, but modifications of them may be practical in the future, and certainly they are more realistic now than some other suggestions that have been made, including rearing aquatic animals in space platforms or on distant planets.

At present aquaculturists either fence large areas (primarily in embayments) for highly extensive mariculture or practice ocean ranching, as described briefly in Chapter 1. Heavy losses to predation are problems encountered with both techniques, whereas maintenance of enclosures is a significant problem in connection with the former. Ocean ranching is limited, of course, to anadromous species, of which there are no warmwater aquaculture candidates.

CONCLUDING REMARKS

The diversity of water systems that have been utilized for aquaculture is rather remarkable. Successes and failures have occurred in virtually every type of system designed to date, and the choice is largely dependent on economics, site constraints, and to some extent, personal experience and preference of the culturist. Of course the requirements of the species to be reared must also be a primary factor in selection of a water system. Each water system is unique, reflecting the particular water source being utilized, land configuration, available materials for construction, and the personality of the designer. When two culturists take the same design and implement it into their respective facilities, the final results are often difficult to recognize as arising from a single source.

Although the systems described in this chapter may be applied for the most part in either freshwater or marine aquaculture, the corrosion problem in the saltwater environment must always be kept in mind, and materials selected accordingly. Metal must be avoided to the extent possible in mariculture systems. PVC plumbing is recommended for systems of both types, but the presence of exposed metal in buildings, culture chambers, or auxiliary equipment quickly leads to corrosion and failure in salt water, whereas metal withstands association with fresh water for long periods. All the goals that can be accomplished, as well as the problems that can occur in fresh water, apply to the marine environment. The choice of species for culture is potentially greater in salt water, but the problems may be more serious and challenging.

LITERATURE CITED

Balderston, W. L., and J. McN. Sieburth. 1976. Nitrate removal in closed-system aquaculture by columnar denitrification. *Appl. Environ. Microbiol.* **1976:** 808-818.

Boyd, C. E. 1968a. Evaluation of some common aquatic weeds as possible foodstuffs. *Hyacinth Control J.* **7:** 26-27.

Boyd, C. E. 1968b. Fresh-water plants: A potential source of protein. *Econ. Bot.* **22:** 359-368.

Broom, J. G. 1963. Natural and domestic production of crawfish. *La. Conserv.* **15** (3-4): 14-15.

Buck, D. H., R. J. Bauer, and C. R. Rose. 1978. Utilization of swine manure in a polyculture of Asian and North American fishes. *Trans. Am. Fish. Soc.* **107:** 216-222.

Buss, K., D. R. Graff, and E. R. Miller. 1970. Trout culture in vertical units. *Prog. Fish-Cult.* **32:** 187-191.

Chamberlain, G., and K. Strawn. 1977. Submerged cage culture of fish in supersaturated thermal effluent. *Proc. World Maricult. Soc.* **8:** 625-645.

Davis, J. T. 1977. Design of water reuse facilities for warm water fish culture. Ph.D. dissertation, Texas A&M University, College Station. 100 p.

DeBoer, J. A., B. E. Lapointe, and J. H. Ryther. 1977. Preliminary studies on a combined seaweed mariculture—Tertiary waste treatment system. *Proc. World Maricult. Soc.* **8:** 401-406.

de la Bretonne, L., Jr. 1977. A review of crawfish culture in Louisiana. *Proc. World Maricult. Soc.* **8:** 265-269.

de la Bretonne, L., Jr., and J. F. Fowler. 1976. The Louisiana crawfish industry—Its problems and solutions. *Proc. Southeast. Assoc. Fish Wildl. Agencies*, **30:** 251-256.

Gary, D. L. 1973. A geographical systems analysis of the commercial crawfish industry in South Louisiana. M.S. thesis, Oregon State University, Corvallis. 123 p.

Germany, R. D. 1977. Population dynamics of the blue tilapia and its effects on the fish populations of Trinidad Lake, Texas. Ph.D. dissertation, Texas A&M University, College Station. 85 p.

Hill, T. K., J. L. Chesness, and E. E. Brown. 1974. Growing channel catfish, *Ictalurus punctatus* (Rafinesque), in raceways. *Proc. Southeast. Assoc. Game Fish Comm.* **27:** 488-499.

Huguenin, J. E. 1975. Development of a marine aquaculture research complex. *Aquaculture*, **5:** 135-150.

Huguenin, J. E., and J. H. Ryther. 1974. Experiences with a marine aquaculture-tertiary sewage treatment complex. *Proceedings of the conference on waste water use in the production of food and fiber.* U.S. Environmental Protection Agency, EPA-660/2-74-041.

Kelley, J. R., Jr. 1973. An improved cage design for use in culturing channel catfish. *Prog. Fish-Cult.* **35:** 167-169.

Kelly, C. B. 1974. The toxicity of chlorinated waste effluents to fish and considerations of alternative processes for the disinfection of waste effluents. Virginia State Water Control Board.

Koller, L. R. 1965. *Ultraviolet radiation.* Wiley, New York. 312 p.

LaCaze, C. G. 1966. More about crawfish. *La. Conserv.* **18** (5-6): 2-7.

LaCaze, C. G. 1976. *Crawfish farming* (rev. ed.). Louisiana Wildlife and Fish Commission Fishery Bulletin 7. 27 p.

Layton, R. F. 1972. Analytical methods for ozone in water and wastewater applications. In F. L. Evans (Ed.), *Ozone in water and wastewater treatment*, Ann Arbor Science Publishers, Ann Arbor, Mich., pp. 15-18.

Lewis, W. M., and G. L. Buynak. 1976. Evaluation of a revolving plate type biofilter for use in recirculated fish production and holding units. *Trans. Am. Fish. Soc.* **105:** 704-708.

Liao, P. B., and R. D. Mayo. 1974. Intensified fish culture combining water reconditioning with pollution abatement. *Aquaculture*, **3:** 61-85.

Meade, T. L. 1974. *The technology of closed system culture of salmonids.* Marine Technology Report 30. University of Rhode Island Sea Grant Publications, Narragansett. 30 p.

Milne, P. H. 1976. Engineering and the economics of aquaculture. *J. Fish. Res. Bd. Can.* **33:** 888-898.

Parker, N. C., and B. A. Simco. 1973. Evaluation of recirculating systems for the culture of channel catfish. *Proc. Southeast. Assoc. Game Fish Comm.* **27:** 474-487.

Pennington, C. H. 1977. Cage culture of channel catfish, *Ictalurus punctatus* (Rafinesque), in a thermally modified Texas reservoir. Ph.D. dissertation, Texas A&M University, College Station. 111 p.

Rowland, L. O., D. M. Hooge, and R. R. Stickney. 1977. Evaluation of tilapia meal as a protein source for broilers. *Poult. Sci.* **56:** 1752.

Ruttner, F. 1953. *Fundamentals of limnology.* University of Toronto Press, Toronto. 295 p.

Sander, E., and H. Rosenthal. 1975. Application of ozone in water treatment for home aquaria, public aquaria and for agriculture purposes. In W. J. Blogoslawski and R. G. Rice (Eds.), *Aquatic applications of ozone.* International Ozone Institute, Syracuse, N.Y., pp. 103-114.

Schroeder, G. 1974. Use of cowshed manure in fish ponds. *Bamidgeh*, **26:** 84-96.

Speece, R. 1973. Trout metabolism characteristics and the rational design of nitrification facilities for water reuse in hatcheries. *Trans. Am. Fish. Soc.* **102:** 323-334.

Stickney, R. R., and J. H. Hesby. 1978. Tilapia culture in ponds receiving swine waste. In R. O. Smitherman, W. L. Shelton, and J. H. Grover (Eds.), *Culture of exotic fishes symposium proceedings,* Fish Culture Section, American Fisheries Society, Auburn, Ala., pp. 90-101.

Stickney, R. R., H. B. Simmons, and L. O. Rowland. 1977a. Growth responses of *Tilapia aurea* to feed supplemented with dried poultry waste. *Tex. J. Sci.* **29:** 93-99.

Stickney, R. R., L. O. Rowland, and J. H. Hesby. 1977b. Water quality—*Tilapia aurea* interactions in ponds receiving swine and poultry wastes. *Proc. World Maricult. Soc.* **8:** 55-71.

Westin, D. J. 1974. Nitrate and nitrite toxicity to salmonid fishes. *Prog. Fish-Cult.* **34:** 100-102.

Wetzel, R. G. 1975. *Limnology.* Saunders, Philadelphia. 743 p.

Wheaton, F. W. 1977. *Aquaculture engineering.* Wiley-Interscience, New York. 708 p.

White, D. B., R. R. Stickney, D. Miller, and L. H. Knight. 1973. *Sea water system for aquaculture of estuarine organisms at the Skidaway Institute of Oceanography.* Georgia Marine Science Center, Technical Report Series No. 73-1. 18 p.

SUGGESTED ADDITIONAL READING

Andrews, J. W., L. H. Knight, J. W. Page, Y. Matsuda, and E. E. Brown. 1971. Interactions of stocking density and water turnover on growth and food conversion of channel catfish reared in intensively stocked tanks. *Prog. Fish-Cult.* **33:** 197-203.

Bonn, E. W., and B. J. Follis. 1966. Effects of hydrogen sulfide on channel catfish (*Ictalurus punctatus*). *Proc. Southeast. Assoc. Game Fish Comm.* **20:** 424-431.

Broussard, M. C. 1975. High density culture of channel catfish in a recirculating system. M.S. thesis, Memphis State University. 43 p.

Collins, R. A. 1971. Cage culture of catfish in reservoir lakes. *Proc. Southeast. Assoc. Game Fish Comm.* **24:** 489-496.

Davis, W. P. 1970. Closed systems and the rearing of fish larvae. *Helg. Wiss. Meeresunters.* **20:** 691-696.

DeManche, J. M., P. L. Donaghay, W. P. Breese, and L. F. Small. 1975. *Residual toxicity of ozonated seawater to oyster larvae.* Oregon State University Sea Grant College Program Publication ORESU-T-75-003. Oregon State University, Corvallis. 7 p.

Guidice, J. J. 1966. An inexpensive recirculating water system. *Prog. Fish-Cult.* **28:** 28.

Hettler, W. F., Jr., R. W. Lichtenheld, and H. R. Gordy. 1971. Open seawater system with controlled temperature and salinity. *Prog. Fish-Cult.* **33:** 3-12.

Hirayama, K. 1974. Water control by filtration in closed culture systems. *Aquaculture,* **4:** 369-385.

Honn, K. V., and W. Chavin. 1976. Utility of ozone treatment in the maintenance of water quality in a closed marine system. *Mar. Biol.* **34:** 201-209.

Honn, K. V., G. M. Glezman, and W. Chavin. 1976. A high capacity ozone generator for use in aquaculture and water processing. *Mar. Biol.* **34:** 211-216.

Huguenin, J. E. 1976. Heat exchangers for use in the culturing of marine organisms. *Ches. Sci.* **17:** 61-64.

Knepp, G. L., and G. R. Arkin. 1973. Ammonia toxicity levels and nitrate tolerance of channel catfish. *Prog. Fish-Cult.* **35:** 2-8.

Liao, P. B., and R. D. Mayo. 1972. Salmonid hatchery water reuse systems. *Aquaculture,* **1:** 317-335.

Loyacano, H. A., Jr., and R. B. Grosvenor. Undated. *Effects of Chinese waterchestnut in floating rafts on production of channel catfish in plastic pools.* Tech. Cont. No. 1105, South Carolina Experimental Station, Clemson University, Clemson, S.C. 8 p.

Marvin, K. T. 1964. Construction of fiberglass water tanks. *Prog. Fish-Cult.* **26:** 91-92.

McGregor, D. 1973. An inexpensive cool recirculating system for the maintenance of marine flat fish. *Lab. Anim.* **7:** 13-17.

Meade, T. L. 1976. *Closed system salmonid culture in the United States.* Marine Memorandum 40, University of Rhode Island, Narragansett. 16 p.

Olla, B. L., W. W. Marchioni, and H. M. Katz. 1967. A large experimental aquarium system for marine pelagic fishes. *Trans. Am. Fish. Soc.* **96:** 143-150.

Palmer, F. E., K. A. Ballard, and F. B. Taub. 1975. A continuous culture apparatus for the mass production of algae. *Aquaculture,* **6:** 319-331.

Parisot, T. J. 1967. A closed recirculating seawater system. *Prog. Fish-Cult.* **29:** 133-139.

Parker, N. C. 1972. The culture of channel catfish, *Ictalurus punctatus,* in an indoor recirculating raceway system. M.S. thesis, Memphis State University. 29 p.

Pruder, G., C. Epifanio, and R. Malout. 1973. *The design and construction of the University of Delaware mariculture laboratory.* Sea Grant Report DEL-SG-7-73, University of Delaware. 96 p.

Robinette, H. 1973. The effect of selected sublethal levels of ammonia on the growth of channel catfish (*I. punctatus*). *Prog. Fish-Cult.* **38:** 26-29.

Russo, C. E., C. E. Smith, and R. V. Thurston. 1974. Acute toxicity of nitrate to rainbow trout. *J. Fish. Res. Bd. Can.* **31:** 1653-1655.

Schmittou, H. R. 1970. The culture of channel catfish, *Ictalurus punctatus* (Rafinesque), in cages suspended in ponds. *Proc. Southeast. Assoc. Game Fish Comm.* **23:** 226-244.

Smith, C. E., and W. G. Williams. 1974. Nitrite toxicity in rainbow trout and chinook salmon. *Trans. Am. Fish. Soc.* **103:** 389-390.

Spotte, S. H. 1970. *Fish and invertebrate culture.* Wiley, New York. 145 p.

Stickney, R. R. 1977. The polyculture alternative in aquatic food production. In P. N. Kaul and C. J. Sindermann (Eds.). Drugs and food from the sea myth or reality? University of Oklahoma Press, Norman, pp. 385-392.

Stickney, R. R., T. Murai, and G. O. Gibbons. 1972. Rearing channel catfish fingerlings under intensive culture conditions. *Prog. Fish-Cult.* **34:** 100-102.

CHAPTER 3

Nonconservative Aspects
of Water Quality

GENERAL CONSIDERATIONS

Many prospective aquaculturists express the fear that their particular source of ground or surface water may not be suitable for supporting culture organisms, but in most cases the concern is without foundation. As long as the basic salinity requirements of the culture animals are met, most nonchlorinated water sources will support desired species of aquatic life, assuming that the water is not contaminated with pollutants such as acid strip-mine drainage, industrial or domestic wastes in high concentrations, or high levels of agricultural chemicals. In general, if a source of water is inhabited by a healthy community of plants and animals, it will support aquacultural activity. Surface waters can easily be checked by observing the communities that exist under natural conditions. Most well water will also support aquaculture, although unusually hot or cold wells and those contaminated with pollutants, hydrogen sulfide, and other naturally occurring toxicants do exist. Well logs, available from drilling firms and the U.S. Department of Agriculture Soil Conservation Service, and the experience of other aquaculturists can provide valuable information on quality, depth, and potential flow rates.

Once an acceptable source of water has been found, maintaining its quality must be considered as one of the most important challenges facing the aquaculturist. A simple approach, and often one of the best responses to water quality deterioration, is the addition of copious amounts of new water to the culture system. This procedure can be somewhat expensive, depending on the source of water and means of distribution, but the cost is usually justified when mass mortality or disease is avoided. Water quality can also be preserved through sound management, and it should be the goal of every aquaculturist to utilize management to avoid water quality problems. When unavoidable problems arise, however, it is helpful, and in some cases mandatory, to have a plentiful supply of new water readily available.

Knowledge of water chemistry is extremely important to the aquaculturist. Nearly every problem that arises in an aquaculture system is the result of, or leads to, degradation of water quality. For example, outbreaks of disease are often triggered by low levels of dissolved oxygen, although the pathology may not be manifested until some time (usually 3 days to 2 weeks) after an oxygen depletion has occurred. Conversely, a disease that results in the death of more than a few individual animals may contribute to an oxygen depletion because oxygen demands on the system increase as these animals decompose. The latter problem can be avoided if dead animals float to the surface and are removed, but some species (e.g., shrimp) may remain on the bottom after death and are not always recovered before decomposition takes place.

No aquaculturist can have a complete knowledge of water chemistry, since virtually every chemical compound known to man has been recovered from water. Understanding how each element or compound can affect water quality and the subsequent potential effects of those waters on aquatic life is a monumental task, made even more difficult when the effects of the interrelationships of various chemicals on water quality are considered. Very little is known with respect to synergisms that exist among various water quality parameters, and the complexity of that topic indicates that understanding will be slow in coming.

Fortunately for the aquaculturist, a relatively few, fairly easily measured parameters of water quality provide the basic information required for sound management. Assuming that an aquaculture facility is appropriately sited and that the source of water is free from pollutants, concern for water quality is limited to the routine monitoring of a small number of parameters, and intermittent monitoring of several others.

Direct contamination of aquaculture systems by agricultural land runoff and aerial sprayers can occur when facilities are sited in farming areas, and precautions must be taken to avoid these problems. Crop dusters should be informed of the location of aquaculture facilities and should not spray upwind (or at all if the wind velocity exceeds a few kilometers per hour), nor should they ever fly over aquaculture ponds: even when they are not spraying, if a drop or two of pesticide from a leaky nozzle enters a culture chamber, high levels of mortality can result.

Assuming that contamination beyond the control of the aquaculturist does not occur, the major concern with respect to water quality involves determination of a variety of characteristics of the incoming water and being aware of natural fluctuations that may occur with respect to water chemistry. The remainder of this chapter discusses nonconservative aspects of water quality—the parameters that are altered by biological activity. Chapter 4 considers "conservative parameters"—several aspects of water quality that are independent of biological activity.

PHOTOSYNTHESIS AND PRIMARY PRODUCTION

The Photosynthetic Process

Many of the nonconservative properties of water are related to variations that result from the activity of photoautotrophic organisms, including aquatic macrophytes, benthic algae (periphyton), phytoplankton, filamentous algae, and certain types of bacteria. Photosynthesis, the process by which carbon dioxide is converted to organic compounds, is described most simply by the equation:

$$CO_2 + H_2O \xrightarrow{\text{light}} CH_2O + O_2 \tag{1}$$

Photosynthesis occurs in the presence of light, chlorophyll, and certain enzymes. Since various intermediates are formed through a series of complex reactions between the two sides of equation 1, the formula presented does not indicate the complexity of the process. The process of photosynthesis is carried out in the electron transport system and involves the substances ATP (adenosine triphosphate), ADP (adenosine diphosphate), and NADP (nicotinamide adenine dinucleotide phosphate). According to Steemann Nielsen (1975), the changes that occur in these substances as a result of carbon fixation during photosynthesis can be outlined as follows:

$$12NADPH_2 + 18ATP + 6CO_2 \rightarrow$$
$$C_6H_{12}O_6 + 18ADP + 12NADP + 6H_2O \tag{2}$$

Most photoautotrophs absorb light of wavelengths ranging from about 350 to 700 nm, whereas photoautotrophic bacteria often absorb light at up to 900 nm. This appears to be related to the capacity of the latter also to utilize chemical energy for the elaboration of new biomass (Steemann Nielsen, 1975), whereas plants are unable to grow except by photosynthesis. The photoautotrophic mechanisms have apparently evolved differently in the two types of organism.

For the majority of plants, photosynthesis occurs as a response to light wavelengths approximately within the visible range. When light enters water it becomes dispersed or attenuated with increasing depth as a result of absorption and scattering. The vertical attenuation coefficient of light in water as presented by Steemann Nielsen (1975) can be determined by the formula:

$$E_z = E_0 e^{-Kz} \qquad (3)$$

where E_z is the irradiance on a horizontal plane at a depth of z meters, E_0 is the irradiance just below the surface of the water, e is the natural logarithm, and K is the attenuation coefficient. From the form of equation 3 it is apparent that light intensity decreases rapidly with increasing depth.

As light penetrates water, both quality and quantity are affected. Long wavelengths are absorbed first, with shorter ones finally being absorbed as depth increases. Underwater color photographs taken in visible light are generally bluish green because of the absence of the longer wavelengths even at only a few meters depth. Many marine animals are red, which would be very conspicuous at the surface but renders them invisible in fairly shallow waters because red is the first color in the visible spectrum to disappear underwater. The depth to which light of any wavelength penetrates in water depends in part on the amount of suspended particulate matter present. Dissolved organic material can also exert a great influence on light penetration. Certain coastal waters are high in organic acids, which give the water a color similar to that of tea. The same situation occurs in bog lakes, most commonly found in temperate regions.

Light penetration is generally much greater in the open ocean than in estuaries, streams, or even lakes. Seawater in offshore regions is generally low in suspended solids, organic materials, and other substances that might impart color or turbidity. The presence of plankton affects light penetration, with the highest concentrations likely to occur inshore. Dense surface plankton blooms, especially those involving phytoplankton, are not uncommon in the open sea, however. Oligotrophic lakes may allow light to penetrate to relatively great depths, although by definition such lakes are of low productivity and would not be found in most areas where warmwater fish and invertebrate culture is practiced. Eutrophic lakes of the type common in the southern United States and in the tropics are excellent in many respects for aquaculture, although they are often highly turbid and allow little light penetration.

Since aquaculture is usually practiced in relatively shallow water, there is often little or no concern on the part of the culturist for ensuring that light will penetrate to appreciable depth. It is generally accepted that the compensation depth for aquatic plants lies at a level where light is about 1% of incident intensity. Even in the turbid waters of aquaculture ponds, light intensities at the bottom commonly exceed 1% of incident, and in ponds where this is not the case, light generally penetrates sufficiently to promote at least some photosynthetic activity. Most pond waters circulate to some extent, although a thermocline may be present. In any case, even limited circulation will carry phytoplankton cells into the light long enough each day to promote their continued growth. If light is somewhat limiting on the bottom of an aquaculture pond, the result may be curtailment of the growth of aquatic macrophytes, and this is generally beneficial to the culturist.

The rate at which photosynthesis occurs is dependent on the amount of chlorophyll available and the presence of light of the proper quality and quantity. In addition, photosynthesis is affected by temperature. In general, the rate of chemical reactions increases with elevated temperature, at least within the normal range of temperature tolerance of the species involved. Thus the rate of photosynthesis is expected to be higher in warm weather than during the winter. There are, of course, certain species of plants as well as animals that are more adaptable to cold water, and some seasonal alteration in species composition can be expected in the phytoplankton community. Regardless, highest rates of primary productivity occur during the warm months in temperature climates, provided appropriate nutrients are available to the plants.

At least three forms of chlorophyll occur in green plants: chlorophylls a, b, and c. Chlorophyll a appears to be the most important in terms of energy absorption. Other photosensitive plant pigments include the carotenoids (carotene, lutein, fucoxanthin, and peridinin) and the biliproteins (phycoerythrin and phycocyanin). Steemann Nielsen (1975) stated that some pigments other than chlorophyll appear to be able to transfer the energy they absorb to the type of chlorophyll best able to utilize that energy in carbon fixation. Carotenoids other than those listed above appear to have little or no function in photosynthesis.

The Importance of Primary Production in Aquaculture

Assuming that the primary culture species is an animal rather than a plant, primary productivity may or may not be important to the aquaculturist. Certainly it is seldom a factor in closed recirculating water systems unless the system is located outdoors where invading species occur or a plant rearing component is incorporated into the system. Plant growth of certain kinds is often desirable in ponds, but problems will arise if the wrong types of plant develop.

For most aquaculture strategies, macrophytes and filamentous algae should be avoided and phytoplankton encouraged. This can be accomplished through management and fertilization as discussed below. Phytoplankters provide food for the larvae and fry of certain aquaculture animals and are fed upon by juveniles in some cases. Maintenance of a phytoplankton bloom results in the shading out of undesirable plants and is thus a good management practice.

Aquatic macrophytes are common both to fresh water and to the marine environment; however the vast majority of all aquaculture problems associated with an overabundance of aquatic vegetation occur in freshwater ponds. The consequences of failure to control undesirable types of aquatic vegetation go far beyond their unsightliness and their removal of nutrients that might otherwise support phytoplankton.

Water temperature may be influenced to some extent by plants in culture ponds, especially diurnally, when clear water is present as a result of limited phytoplankton productivity. Water tends to warm more slowly and lose heat more rapidly when it is clear. Clarification of the deep water of ponds as a result of aquatic macrophytes removing the nutrients from the water allows sunlight to penetrate more deeply and will allow the rooted plants to spread from the margins of the pond into the middle. If not stopped, such plants can completely take over a culture pond in a matter of days or weeks.

Heavy growths of rooted aquatic macrophytes and filamentous algae of certain types can entrap swimming organisms, leading to their strangulation or increasing their vulnerability to attack by aquatic insects, snakes, turtles, and other predators. Access to prepared diets can be severely restricted in weed-choked ponds, and much of the supplemental feed provided may end up as fertilizer, worsening a bad situation.

An additional and significant problem involves the difficulty involved in harvesting and sampling ponds that contain even moderate amounts of undesirable vegetation. Seines are very difficult to pull through vegetation and tend to roll up. As a consequence, the animals being harvested have little difficulty in avoiding the seine in the first place, and once caught, many manage to escape. Those that are captured must be found among the masses of vegetation that clog the net.

The death and decay of aquatic vegetation of all types consumes oxygen, which is required for the health of the culture animals, and may lead to critically low levels of dissolved oxygen with consequent stress, resulting in disease epizootics or direct mortality.

The potential that aquatic macrophytes hold as livestock feeds has been investigated (Baily, 1965; Lange, 1965; Boyd, 1968a, 1968b; Culley and Epps, 1973), and the potential incorporation of macrophytes for tertiary treatment in intensive water systems was addressed in Chapter 2. The extremely high moisture content of aquatic plants not only leaves little matter remaining after desiccation, but makes transportation of the fresh material unfeasible. Any enterprise that intends to feed aquatic vegetation to livestock must provide facilities for rearing aquatic and terrestrial organisms in close proximity, or inexpensive drying and hauling capabilities for the plants must be present. Under most circumstances, vegetation-choked lakes and ponds are better sources of aquatic plants for animal feeding than are aquaculture ponds, since the growth of aquatic macrophytes at the concentrations required for livestock feed would severely limit the ability of those bodies of water to produce aquatic animals. It is, of course, possible to manage aquaculture ponds for the production of aquatic vegetation only, but this is not economical if the plants are to be utilized as animal feed.

Measuring Primary Productivity

The average commercial aquaculturist does not need to measure primary productivity and is, in reality, concerned with this subject only in terms of controlling nuisance vegetation and maintaining a suitable level of dissolved oxygen through photosynthesis and other means. However it is important from a conceptual standpoint to understand the biological process of photosynthesis and to further understand the ways in which primary productivity of aquatic ecosystems can be measured. As pond fertilization with organic and inorganic fertilizers becomes more widely accepted and practiced by aquaculturists, the need for making determinations of primary productivity may increase.

For annual, rooted aquatic macrophytes and floating plants, primary productivity can easily be measured by determination of the dry biomass per unit area at the end of the growing season. It can generally be assumed that the initial biomass (except for roots in some cases) is insignificant. This technique does not work well on phytoplankton, filamentous algae, or periphyton, all of which are present in natural waters throughout the year. These plants often follow patterns of erratic growth and may be dominated by different species from one period to another. An indication of periphyton activity can be obtained by placing clean substrates (e.g., glass microscope slides, glass plates, or glass rods) in the water at selected depths and determining the rate at which plant material accumulates with time. The determination is usually based on ash-free dry weight (the difference in weight of the material after being dried at between 80 and 105 C, and then ashed at 550 to 600 C). One problem with this technique is that the substrate seldom attracts plants alone. Animals of various kinds may also be present, and though it is possible to physically remove the larger insects in fresh water, or tunicates, barnacles, bryozoans, and so forth in salt water, it is not possible to remove small larval stages and such organisms as protozoans. Thus such measurements are actually of the *Aufwuchs* community rather than just the periphyton component of that community.

Phytoplankton productivity has been measured indirectly in a number of ways. One of the first was the use of the light and dark bottle technique (Gaarder and Gran, 1927) in which identical samples of water from a natural water body are placed in paired bottles: one transparent and the other opaque. The dissolved oxygen (DO) level in the initial water sample is determined; then each bottle is suspended at a preselected depth in the water body from which the samples came or in a chamber designed to simulate natural conditions. The bottles are allowed to incubate during daylight for a period of up to several hours, after which the DO concentration is again determined. Following incubation, DO in the transparent (light) bottle should be higher than at the onset of the experiment as a result of photosynthetic activity during incubation (as long as the bottles are suspended above the compensation depth), whereas DO in the opaque (dark) bottle should be reduced because of the respiration of organisms in the sample. The amount of oxygen consumed in the dark bottle is added to that produced in the light bottle to give an indication of gross primary production. The theoretical amount of carbon fixed during the incubation period can be calculated using the relationships in equation 1.

The light and dark bottle oxygen technique is considered to be crude by many investigators, and other methods have been developed. One of these is the measurement of primary productivity by determining the rate of carbon fixation. The carbon-14 (^{14}C) technique is essentially a modification of the light and dark bottle procedure except that the amount of radioactive ^{14}C fixed during incubation of a water sample is measured rather than the amount of dissolved oxygen produced or consumed. Measurement of radioisotope uptake is generally conducted by means of a liquid scintillation counter. This technique was developed by Steemann Nielsen (1951) and has been modified to some extent by a number of investigators who have introduced correction factors for various interferences or enhancements of apparent carbon uptake. For example, it has been learned that ^{12}C is absorbed by plants at a slower rate than ^{14}C (Steemann Nielsen, 1952); therefore a correction factor is required when the total amount of carbon fixed is calculated. In many cases the formula presented by Saunders *et al.* (1962) is utilized for determining photosynthetic production by the ^{14}C technique.   ⋅

The ^{14}C procedure involves the use of both light and dark bottles. Apparent carbon fixation in the dark bottle may be the result of absorption or adsorption, or it may indicate that chemoautotrophic bacteria are present in the sample. Despite the various problems associated with the ^{14}C technique, many investigators still employ it for primary productivity studies and are able to provide good indications of phytoplanktonic activity in various aquatic systems.

The total amount of chlorophyll in a water sample or in any other specimen containing plant material can be determined and utilized as a means of estimating primary productivity. Procedures for extracting chlorophyll from plant material as well as formulas for calculating the levels of various pigments through spectrophotometric analysis of the extract were presented by Richards and Thompson (1952), and have been reviewed and expanded on by Odum *et al.* (1958). Ryther (1956) found that although both chlorophyll content and rate of photosynthesis are variable, the rate of photosynthesis per unit of chlorophyll remains relatively constant, permitting some indication of primary productivity to be obtained through chlorophyll analysis.

Chlorophyll fluoresces when exposed to certain wavelengths of light, and the degree of fluorescence relates to the amount of chlorophyll present, thus to primary productivity. Fluorometers are available that provide a continuous reading of the extent of fluorescence in water pumped through them. Discrete readings from small water samples can also be obtained with the same instrument in most cases. The readout is nearly instantaneous, and no prior treatment of the samples is required. Thus it is feasible for up to several thousand individual water samples to be analyzed in a single day with the technique, whereas the oxygen production and ^{14}C uptake methods are generally limited to a few samples daily. Fluorometry provides at least a relative indication of primary productivity and is being utilized increasingly by limnologists and oceanographers as a means of obtaining rapid and numerous indications of phytoplankton activity. The technique appears to have merit in aquaculture research because the relative levels of primary productivity in numerous ponds, raceways, or other types of culture chamber can be compared quickly and often. The technique might be especially useful in determining when a culture chamber requires fertilization.

PLANT NUTRIENTS

Various nutrients are required by plants for photosynthesis to proceed at a rapid rate. Particularly important in this regard are phosphorus and nitrogen (primarily as PO_4^{-3} and NO_3^-, respectively). Silicon is critical for the proper growth of diatoms but is rarely limiting in aquacultural systems. When nitrogen or phosphorus becomes limiting, plant growth is curtailed. Addition of the limiting nutrient will stimulate primary productivity until some other nutrient becomes limiting or the supply of the nutrient just added is exhausted.

Nitrogen

Nitrogen is required by all living organisms, being an important component of protein and other biochemical substances. Nitrogen is removed from water primarily as nitrate (NO_3^-) by plants. Animals satisfy their nitrogen requirements by eating plants or other animals. Nitrogenous wastes are excreted by animals in several forms (Table 3.1), and nitrogenous compounds are released during the bacteriological decomposition of plant and

Table 3.1   Nitrogenous Excretory Products Released by Aquatic Organisms

Product	Formula
Ammonia	NH_3
Creatine	$NH \parallel$ ... $CH_3 \mid$ ... $NH_2 - C - N - CH_2 - COOH$
Creatinine	$HN \longrightarrow C - NH$, $O = C$, N, CH_3, CH_2
Urea	$O \parallel$ ... $NH_2 - C - NH_2$
Amino acids	$NH_2 \mid$ ... $R - CH - COOH$
Uric acid	OH, C, N, $CH - NH$, $OH - C$, C, $C = O$, N, N

Source: White *et al.* (1964).

animal matter. Nitrogenous wastes are eventually transformed into ammonia, which undergoes nitrification to nitrate through a nitrite (NO_2^-) intermediate:

$$NH_3 \rightarrow NO_2^- \rightarrow NO_3^- \qquad\qquad (4)$$

This process is conducted by aerobic bacteria. The bacterium *Nitrosomonas* is largely responsible for the step from ammonia to nitrite; *Nitrobacter* carries the nitrification process from nitrite to nitrate. Certain bacteria may then convert nitrate to elemental nitrogen (N_2), which may escape from the system as a gas. These reactions, found in both biofilters and the natural environment, are responsible for maintaining ammonia concentrations within acceptable levels (Meade, 1976).

Denitrification of nitrate to elemental nitrogen can be carried out by a variety of bacteria including *Pseudomonas, Achromobacter, Bacillus, Micrococcus,* and *Corynebacterium* (Meade, 1976). Energy for the reduction reactions involved may come from certain carbohydrates and alcohols. Some closed recirculating aquaculture systems employ special chambers that promote these reactions by supplying the proper organic substrate.

In pond environments there is little concern over the accumulation of nitrate because primary producers generally remove this ion from the water nearly as rapidly as it becomes available. However in closed recirculating water systems with no photosynthetic organisms present and in other types of water system receiving high levels of organic or inorganic fertilization, nitrate can become quite concentrated. Even if not directly toxic itself, the degree of nitrogen concentration might be indicative of deterioration by other aspects of water quality to the point of causing stress among the culture animals. Colorimetric tests have been devised to measure the level of nitrate in water samples (APHA, 1975). High precision in the determination of nitrate in water requires expensive equipment, however some manufacturers have developed reasonably priced colorimetric apparatus for nitrate and other nutrients. Such equipment does not provide a high degree of precision, but the aquaculturist is generally interested in knowing whether nitrate levels in the water are high or low, not whether they have changed 0.01 mg/l since the previous determination was made.

The level at which nitrate becomes toxic to aquatic animals is largely unknown, and according to Spotte (1970) nitrate may not be directly toxic under most circumstances. Hart *et al.* (1945) indicated that in 95% of the cases investigated, natural waters containing healthy fish populations had nitrate levels of less than 4.2 mg/l. Spotte (1970) stated that when culture systems are properly managed, nitrate levels above 20 mg/l are seldom observed.

Although nitrogen is available to plants in the form of nitrate, it apparently must be reduced to ammonia once again before it can be absorbed into the plant tissues (Fogg, 1972). The reaction appears to be light catalyzed (as is photosynthesis) and, according to Fogg, proceeds as follows:

$$NO_3^- + H_3O^+ \xrightarrow{\text{light}} NH_3 + 2O_2 \qquad (5)$$

This reaction also releases divalent oxygen, but it is probably insignificant in comparison with photosynthetic oxygen production.

After ammonia, the least desirable form of nitrogen in aquaculture systems is nitrite. Rare in natural waters because of the presence of suitable numbers of *Nitrobacter* to oxidize nitrite to nitrate as it is formed, nitrite sometimes does occur in high concentration in aquaculture systems, especially in closed recirculating systems that have just been filled with water for the first time. Even though nitrite is an intermediate, it can be present in high concentration if the bacteria have not colonized the biofilter, or if they are present in insufficient numbers to convert the toxic nitrite ions as they are formed. *Nitrosomonas* often colonizes a filter much more rapidly than does *Nitrobacter*; thus tests of the water may indicate no buildup of nitrate or ammonia. Many culturists fail to test for nitrite (also a colorimetric test), but assume that if ammonia and nitrate levels are not elevated, nitrite will also be under control. In time, *Nitrobacter* will enter the filter and attain sufficient density to remove nitrite from the system, but this may not occur for several days or even weeks, and by that time the highly toxic nitrite may have killed all the culture animals.

Lees (1952) reported that since the growth of *Nitrobacter* is inhibited in the presence of ammonia, nitrite may become concentrated in biofilters until the ammonia concentration is greatly reduced. Once the biofilter has begun to operate efficiently, nitrite usually ceases to be a problem unless some chemical enters the system and destroys the culture of *Nitrobacter*. This can happen during disease treatment, for example. Nitrite has not been found to cause problems in open systems or ponds.

Konikoff (1975) examined the acute toxicity of nitrite to channel catfish and found the 24, 48, 72, and 96 hour TLm's (median tolerance limits: the concentration that results in 50% mortality at the times indicated) to be 33.8, 28.8, 27.3, and 24.8 mg/l, respectively, for 40 g fish at 21 C.* These levels are extremely high compared with those normally observed in aquaculture systems except under extraordinary conditions as previously discussed. Chronic exposure to levels far below the acute toxicity levels just

---

*Appendix 2 gives equivalent temperatures between 0 and 40 C on the celsius and Fahrenheit scales.

given may, of course, lead to mortality, thus the values obtained by
Konikoff (1975) should not be used in the establishment of safe levels of
nitrite for the culture of channel catfish. In most aquaculture systems the
level of nitrite is below the detection limit of all but the most sensitive tests
(even crude tests can indicate the presence of nitrite at less than 1 mg/l).

Suspected cases of nitrite toxicity in fish can be quickly confirmed by
sacrificing an individual and examining the blood. Hemoglobin reacts with
nitrite to produce methhemoglobin, turning the blood a chocolate brown
color. Catfish subjected to lethal levels of nitrite die with open mouths and
closed opercles. The poisoned fish become quiet and lie resting on the
bottom, but they may begin to swim erratically for up to one minute
immediately before dying (Konikoff, 1975).

Fishes excrete most of their nitrogenous waste through the gills in the
form of ammonium, NH_4^+ (Hochachka, 1969). Ammonium ion accounts
for as much as 60 to 90% of the total nitrogen excreted (Smith, 1929;
Wood, 1958; Fromm, 1963). Other forms of nitrogenous wastes (Table
3.1), then, are relatively unimportant in most cases. Long-term exposure
by aquatic animals to elevated ammonia levels can result in reduced
growth, impaired stamina (Burrows, 1964), or death.

Ammonia occurs in water in both the ionized (NH_4^+) and un-ionized
(NH_3) forms. The toxicity of ammonia to aquatic organisms has been
shown to involve the un-ionized form (Chipman, 1934; Wuhrmann et al.,
1947; Wuhrmann and Woker, 1948), whereas the ionized form of the
chemical appears to be relatively harmless (Tabata, 1962). The state in
which ammonia occurs in water is related to temperature and pH, and is
also affected by such other factors as ionic strength of the solution. Un-
ionized ammonia decreases relative to ionized ammonia in hard and saline
waters (Emerson et al., 1975). Carbon dioxide levels in the water can also
affect the form in which ammonia occurs (Tabata, 1962). Mayo (1971)
indicated how the percentages of ionized and un-ionized ammonia change
with respect to pH at a few typical environmental temperatures, and
Emerson et al. (1975) calculated values for the percentage of un-ionized
ammonia present in water from 0 to 30 C in the pH range 6 through 10.
As an example, at 26 C, less than 1% of the ammonia present in fresh
water is in the un-ionized form at a pH of 7.0, whereas at pH 8.5 more
than 15% is un-ionized at the same temperature (Table 3.2).

Aquaculturists, especially those working in intensive culture systems,
should routinely measure ammonia levels. Many culturists do obtain indications of total ammonia, but few distinguish between total and un-ionized
ammonia concentrations. In closed recirculating water systems the addition of such buffering agents as oyster shell or crushed limestone leads
to an increase, or at least the stabilization of pH, resulting in higher levels

of un-ionized ammonia than would be present if the system were allowed to become more acidic. Since within the normal operating temperature and pH ranges of warmwater systems the percentage of un-ionized ammonia will not become unacceptable if the biofilter continues to operate properly, and since acidification of the water can lead to serious stress on the aquaculture animals, buffering is recommended in all cases. In a properly operating system the total ammonia concentration should remain below 1.0 mg/l (preferably less than 0.5 mg/l), which will result in the un-ionized ammonia level being within acceptable limits (Table 3.2).

Some species of animals appear to have relatively high tolerances for ammonia, whereas others succumb to low concentrations of that substance. The 24 hour lethal level for channel catfish is 2.766 mg/l if the chemical is present in the un-ionized form (Robinette, 1976). Channel catfish growth is affected at un-ionized ammonia levels as low as 0.12 mg/l (Robinette, 1976), and this is very important. The effects of ammonia on various coldwater fishes have been reviewed by Hampson (1976), but there is very little information available on the toxicity of this chemical on other fishes and invertebrates of aquaculture interest.

An exception is *Tilapia aurea,* for which the acute toxicity level has been studied and found to be similar to that reported for channel catfish (Redner, 1978). Of even greater interest, that study also indicated that *T. aurea* are able to develop a resistance to ammonia when exposed to sublethal levels in advance of acute bioassay tests. Thus this species may be able to survive a slow increase in ammonia concentration such as that which often occurs in aquaculture, whereas tilapia lacking previous ex-

Table 3.2   Percentage of Total Ammonia in the Un-ionized (NH$_3$) Form for Selected Temperatures and pH Values Typical of Warmwater Aquaculture Systems

Temperature (C)	pH				
	6.5	7.0	7.5	8.0	8.5
16	0.1	0.3	0.9	2.9	8.5
18	0.1	0.3	1.1	3.3	9.8
20	0.1	0.4	1.2	3.8	11.2
22	0.1	0.5	1.4	4.4	12.7
24	0.2	0.5	1.7	5.0	14.4
26	0.2	0.6	1.9	5.8	16.2
28	0.2	0.7	2.2	6.6	18.2
30	0.3	0.8	2.5	7.5	20.3

Source: Emerson *et al.* (1975).

posure to elevated ammonia may succumb. This pattern may be particularly important when fertilization is utilized for intensive tilapia production.

Ammonia is determined by colorimetric techniques (APHA, 1975) utilizing, most commonly, direct nesslerization or an ammonia electrode in conjunction with a digital pH meter. Direct nesslerization is susceptible to various interferences, especially in seawater; thus the samples may have to be distilled or otherwise treated before the determination can be made. Distillation does not affect the concentration of ammonia in water, since ammonia passes over with the condensing steam; however other interfering substances are left behind during the procedure.

Certain types of photosynthetic bacteria and blue-green algae are able to fix atmospheric nitrogen. According to Fogg (1972), the reaction proceeds in the dark with blue-green algae but is accelerated in the light according to the formula:

$$2N_2 + 6H_2O \rightarrow 4NH_3 + 3O_2 \tag{6}$$

Nitrogen fixation is not an important contributor to the total availability of nitrogen in the forms of ammonia, nitrate, or nitrite to plants in aquaculture systems. In intensive water systems there is sufficient nitrogen available to support high levels of primary productivity if plants are added to the system, and in most cases it is necessary to employ biofiltration to maintain the levels of the various nitrogen-containing chemicals within safe limits. The contribution of nitrogen-fixing bacteria and blue-green algae to the total ammonia level in closed recirculating water systems has not been reported to be of significance under normal operating conditions. Most extensive culture systems also have sufficient inputs of nitrogen to negate any great need for nitrogen fixation. Figure 3.1 is a schematic representation of the nitrogen cycle in natural waters.

Phosphorus

Phosphorus is a required nutrient for plant growth and is abundant in the bones and teeth of animals. Phosphorus is usually present only in minute concentrations in natural waters because of its high mobility, although the total phosphate concentration in natural aquatic systems may range from 0.01 to more than 200 mg/l (Wetzel, 1975). Unless present in extreme abundance or in instances of another limiting factor operating, phosphorus is rapidly removed by primary producers, and that which is not immediately utilized for growth can be stored. When a phytoplankton bloom becomes senescent and declines, phosphate is often released in high concentration, but is usually rapidly incorporated into other primary producer species, which begin to take over for the senescent one.

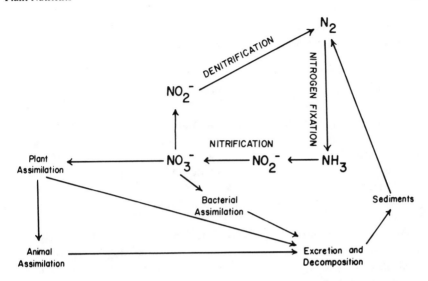

Figure 3.1 Schematic representation of the nitrogen cycle in natural waters.

In flowing systems phosphorus levels do not generally become concentrated because of the rapid exchange of water, although the continuous presence of slightly elevated phosphorus levels as a result of feed and animal wastes dissolved in the water may be sufficient to support luxuriant growths of primary producers. Ponds are excellent reservoirs for nutrient accumulations, and as in the case of nitrogenous compounds, phosphorus levels may become elevated, although the extent of elevation is usually not obvious in the water itself, but only indirectly in the large concentrations of plants that are supported in the system.

The ratio of carbon to nitrogen to phosphorus required by most species of phytoplankton is near 106-16-1, indicating the potential of even trace levels of phosphorus to influence primary productivity. Since phytoplankton blooms are often encouraged by aquaculturists, especially persons engaged in pond culture, it is important to maintain sufficient levels of phosphorus to encourage the desired type of productivity. If, on the other hand, improper management leads to a loss of the phytoplankton bloom, phosphate will just as rapidly go to support filamentous algae and aquatic macrophytes.

Phosphorus (measured as phosphate) is determined by colorimetric techniques (APHA, 1975). The determination is not difficult and will adequately detect differences in phosphate concentration within the range important to aquaculture (lower detection limit should be at or below 1 mg/l). Since various other water quality measurements often are utilized to provide the aquaculturist with an assessment of conditions in the culture system, frequently phosphate analyses are not a part of routine water chemistry.

POND FERTILIZATION

Fertilization of aquaculture ponds, especially with nitrates and phosphates, is often initiated in the spring to provide food for certain life stages of aquaculture animals and for the production of zooplankton and benthic organisms that can be preyed on by the cultured species. The presence of a phytoplankton bloom increases water turbidity and blocks sunlight penetration to rooted aquatic plants and filamentous algae on the pond bottom.

Fertilization is not always recommended in warmwater aquaculture ponds. Some soils contain relatively high levels of nutrients, which may leach into the pond water and support a sufficient phytoplankton bloom. Incoming water, especially that from a surface source, may contain enough nutrients. Also, the addition of feed to ponds increases the nutrient level, as do the excretory products of the culture animals. If measurements indicate very low levels of nitrates and phosphates, or if this is indicated by the presence of clear water of low primary productivity, fertilization may be appropriate. Fertilization has been used in both freshwater and saltwater ponds, although most of the experience with the technique in the United States is associated with inland aquaculture.

Most fresh waters respond well to approximately 50 kg/ha of 16-20-4 (percentages of nitrogen, phosphorus, and potassium, or N-P-K, respectively) fertilizer, although in soils where potassium levels are sufficiently high, good results can be obtained with 16-20-0. The optimum rate of fertilization varies from one region to another because of innate differences in soil chemistry and the concentrations of nutrients dissolved in the water. The application of fertilizer should be repeated at intervals of 10 to 14 days until a Secchi disc reading of approximately 30 cm is obtained. A Secchi disc is a circular plate painted white or having alternating pie-shaped wedges of black and white on its surface; the diameter varies (often 20 cm in fresh water and 40 cm in the marine environment). Secchi disc transparency is defined as the depth at which the disc just disappears from view when it is lowered into the water. The depth of the reading varies to some extent with time of day and the amount of cloud cover. If a Secchi disc is not available, the measurement may be made by determining the depth at which the culturist's hand disappears when lowered into the water at right angles to the forearm. Proper fertilization has been achieved if the hand disappears when the arm is submerged to the elbow.

In the spring, if an aquaculture pond is nutrient limited and will support only a fairly low density of phytoplanktonic organisms, it may be beneficial to fertilize. As indicated in the theoretical growth curve for algae in Figure 3.2, the phytoplankton community may respond to fertilization (indicated at point t_0 in Figure 3.2) by logarithmic growth after a lag phase, which generally lasts only a few days or less. Fertilization may be continued at intervals as specified earlier until the cell density reaches the desired level (t_m in Figure 3.2), which in the case of an aquaculture pond corresponds to a Secchi disc reading of 30 cm. Desired cell concentration can be maintained around the upper asymptote of the curve by adding small amounts of whichever nutrient becomes limiting.

Maintenance of a proper phytoplankton bloom is not as simple as it may appear. Often filamentous algae or rooted aquatic macrophytes bloom instead of phytoplankton, and the pond must be treated with herbicide and the procedure initiated again. Overfertilization may lead to impairment of water quality. In the latter case the high level of photosynthesis that occurs with any phytoplankton bloom may become excessive, with oxygen depletions resulting because of concomitantly high respiratory demands.

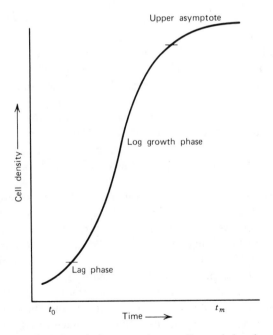

Figure 3.2 Theoretical phytoplankton growth curve. For symbol explanation, see text.

Literally thousands of species of phytoplanktonic algae are potentially able to bloom in a culture pond, although in practice each algal community is generally dominated by a single species and a variety of others occur in low concentrations. It is virtually impossible to predict which species will be dominant or when the bloom will become senescent and die out, whereupon the pond is taken over by another species of phytoplankton or becomes choked with undesirable species of vegetation.

In general, a well-managed culture pond does not have rooted aquatic plants or large mats of filamentous algae in it during any portion of the year. Exceptions do occur, however. For example, millet and rice have been grown in association with ponds utilized for shrimp and crayfish production. In Asia, milkfish (*Chanos chanos*) production is accomplished in ponds where *Aufwuchs* production (known as lab-lab) is encouraged (Pillay, 1972). In most cases present United States aquacultural strategy is to discourage nonagricultural rooted aquatic macrophytes in culture ponds. This trend may change to some extent if culturists here move toward the production of herbivorous species such as grass carp and as innovative methods of harvesting fishes from ponds containing large amounts of rooted aquatic vegetation are developed.

Algae culturists are often able to produce large volumes of unicellular, unispecific phytoplankters under controlled conditions, but virtually no aquaculturist attempts to seed ponds with a particular species of these organisms. Algae of the type that bloom in culture systems are cosmopolitan and will be introduced into any body of water as it is filled, even if sterile water is used. Algal cells are transported by birds and other animals and have spores and other types of reproductive bodies that can be carried by the wind or washed in with precipitation and runoff.

In water containing high levels of suspended sediments the establishment of a phytoplankton bloom is sometimes difficult because of natural shading. In such cases it may be necessary to reduce the extent of turbidity artificially before fertilization. Chapter 4 describes ways in which this can be accomplished.

The establishment of a proper phytoplankton bloom sometimes requires three to five applications of fertilizer. If a good phytoplankton bloom is not apparent after five applications, it is quite likely that other, less desirable plants are beginning to take over the system. When macrophytes or filamentous algae bloom instead of phytoplankton, these plants should be eliminated and the procedure started again. Fertilization with organic fertilizers such as livestock manures often results in excellent phytoplankton blooms (along with blooms of zooplankton), and these materials may be used more extensively by United States aquaculturists in the future.

The utilization of organic fertilizer in other parts of the world has been accepted practice for hundreds and probably thousands of years. The rates of organic fertilizer application are not precisely established for United States warmwater aquaculture, but most culturists agree that several kilograms per hectare can be added weekly. Research in this area is being initiated at several institutions.

Fertilization is generally initiated in the spring when the water temperature reaches about 15 to 18 C. Thereafter, filamentous algae blooms are more commonly encountered when a pond is fertilized. Attempts should be made to retain the phytoplankton bloom until the aquaculture animals are well established on prepared feeds, and maintenance of such blooms throughout the growing season is desirable to inhibit growth of undesirable plants.

Herbivores, such as certain species of tilapia and grass carp, feed on filamentous algae and aquatic macrophytes, although some species of the latter seem to be ignored by most aquatic organisms, including grass carp (Colle et al., 1978). Some species of tilapia prefer phytoplankton but will resort to the consumption of other types of aquatic vegetation when their preferred food becomes scarce. I have observed *Tilapia aurea* feeding on terrestrial grass seeds when the fish were present at high density in ponds that had low standing crops of aquatic plants.

CONTROL OF AQUATIC VEGETATION

Depending on the control method selected, it may be necessary to determine the species of plant or plants causing a particular problem in advance of treatment. Various publications have been drafted to help the aquaculturist, as well as the sportsman, farmer, and rancher determine what species of plant is occurring so that the proper control procedures may be undertaken. A few such publications are those by de Gruchy (1938), Eyles and Robertson (1944), Muenscher (1944), Smith (1950), Fassett (1960), Weldon et al. (1969), and Klussmann and Lowman (1975).

Filamentous algae may grow as mats associated with the bottom, in conjunction with other types of aquatic vegetation, on the surface of the water, or suspended in the water column. In addition, various types of aquatic macrophyte may occur. Some species remain completely submerged, and others are partly or mostly emergent. Emergent vegetation is most commonly associated with the shallow water at or near the shoreline of ponds. Finally, some species of aquatic macrophytes float at the surface.

Three types of basic control measures may be employed to rid aquaculture ponds of aquatic vegetation: mechanical (including proper pond construction), biological, and chemical. Although the latter has become the method of choice by many aquaculturists since the development of selective herbicides over the past several years, increasing interest has been demonstrated in the other two control measures because they are ecologically sound in most cases and generally are less likely to affect the culture species adversely than are chemicals.

Mechanical Control

The objective of mechanical vegetation control is either to physically remove the offending plants or to alter the environment to create conditions that will discourage the growth of the unwanted vegetation. When objectionable plants become established in a pond, harvesting the plants by hand is one of the simplest—and depending on the extent of the problem, one of the most expedient means of controlling certain aquatic macrophytes. This approach is not practical in large bodies of water where extensive areas of vegetation have become established, but often mechanical harvesting devices designed for lakes and reservoirs cannot be accommodated in an aquaculture pond. If an aquatic vegetation problem is detected early enough, the objectionable plants may be removed by hand. For example, arrowhead (*Sagittaria* sp.) may lend itself to hand harvesting (Figure 3.3) if the growth is not too extensive to make this method of harvest uneconomical. Certain floating plants such as the water hyacinth (*Eichhornia crassipes*) can also be removed from culture ponds manually, but care must be taken to remove all the offensive plants or they may quickly become reestablished. Duckweed (*Lemna* sp.) can be skimmed from the surface of ponds, but it is virtually impossible to remove it entirely, and rapid recolonization can be expected unless some environmental limiting factor is imposed on the growth of that plant.

Cattails (*Typha latifolia*) can also be controlled by hand harvesting. Cattails spread by sending out rhizomes, in addition to proliferation by seeds. During the spring and summer the spread of this plant by rhizomes can lead to the colonization of extensive areas in shallow water within a few weeks. Thus it is often necessary to return to the same area frequently to ensure that new plants have not sprouted from rhizomes that were inadvertently left from the previous mechanical harvest. An application of chemicals following mechanical removal may also be appropriate in the control of cattails. Cattails become rather large (up to nearly 3 m in height) and rapidly spread around the margins of a pond if not controlled (Figure 3.4). Given sufficient time, this plant will also extend into the middle of

Figure 3.3 Arrowhead (*Sagittaria* sp.), a rooted aquatic macrophyte that sends leaves to and often above the surface of the water.

Figure 3.4 A small patch of cattails (*Typha latifolia*) can be relatively easily removed by hand, but frequent checks should be made to ensure that unharvested rhizomes do not lead to recolonization.

most ponds if the water is not too deep to exclude it. In addition to being unsightly, cattails can severely limit access to culture ponds, and they interfere with the mobility and harvest of animals being reared.

Some submerged plants that appear to be easy to remove by mechanical harvesting techniques are, in reality, virtually impossible to control by this means. Bushy pondweed (*Najas* sp.) can be gathered by hand (Figure 3.5), but a considerable amount is always missed, and within a few days or weeks the pond again becomes choked by this type of vegetation. To prevent increases in oxygen demand as a result of destroying large biomasses of such plants with herbicides, it is often a good idea to remove as much of the plant material as possible through mechanical harvesting prior to

Figure 3.5 Bushy pondweed (*Najas* sp.) and other submerged aquatic macrophytes can be removed from culture ponds by hand, but reinfestation occurs very rapidly, since complete removal by that method is impractical.

chemical treatment, especially in cases where heavy infestations occur. Similar precautions should be applied to pondweed (*Potamogeton* sp.) and other submerged plant species.

Muskgrass or chara (*Chara* sp.) resembles bushy pondweed in some respects, although muskgrass is an alga. It grows in clumps, usually in association with the substrate in ponds (Figure 3.6). Mechanical removal of chara is possible, but it is difficult to completely control the plant in that manner and frequently some form of chemical control, often coupled with mechanical removal, is necessary. Chara can be distinguished from bushy pondweed by the musty odor of the former and the roughness of

Figure 3.6 Muskgrass or chara (*Chara* sp.), an alga that resembles bushy pondweed and other aquatic macrophytes, can be removed to some extent by mechanical means but is usually treated and controlled with chemicals.

the plant when rubbed between the fingers. Some authors report that chara has a "crunchy" feel (Klussmann and Lowman, 1975).

One of the best ways to limit emergent vegetation is to avoid having extensive shallow areas in culture ponds. Proper construction design should ensure that adequate slopes are built to prevent shoal areas where such vegetation can develop. If gently sloping banks are present, it may be necessary to deepen the edges of the pond to discourage colonization by macrophytic vegetation (Lawrence, 1968). Culture species that require primarily benthic habitats with large areas but little water volume (e.g., shrimp and crayfish) may require or at least adapt to ponds of less than normal depth; however invasion by aquatic macrophytes may cause a considerable problem in such ponds. In some cases the animals may create sufficient turbidity to limit the growth of rooted plants, but usually light penetrates to the bottom. Therefore it is often more desirable to construct relatively deep ponds even for animals that do not utilize the pelagic zone.

Although many species of submerged aquatic macrophytes can grow in relatively deep water, the shading effect of suspended sediments or phytoplankton (the latter is discussed below as a method of biological control) may be sufficient to limit their growth. Certain species of fishes, such as various species of carp and tilapia, tend to increase the turbidity of ponds as a result of their feeding activities. The same is true of shrimp in ponds with clay substrate, and some producers utilize a species exhibiting that habit as one of the species in polyculture. To date, this technique has not been widely accepted in the United States.

Warmwater aquaculture ponds are often drained during the winter and allowed to remain empty until just before restocking the following spring. Desiccation of pond bottoms results in the destruction of both living plants and at least some of the reproductive bodies of those plants, although many seeds and spores may be expected to survive to reinfest the pond the following year. In any case reintroduction of plants from other water bodies often occurs very rapidly following the filling of a pond. Many aquaculturists disc their ponds during the winter or spring to destroy any terrestrial vegetation that might have been established while the pond was empty. Such vegetation can lead to increased oxygen demand as a result of its decay after the pond is filled, and discing will reduce this problem by burying at least some of the material. If the pond is not filled immediately after being disced, most of the plant material that has been destroyed will decay and release nutrients into the soil, where they may become bound to the sediments. These nutrients will still be available to primary producers in the water over the long run, but they may leach from the sediments into the water more slowly than they would from the plants directly. More important, the demands on oxygen as a result of decomposition will be made on the atmosphere rather than on the dissolved oxygen of the water.

Biological Control

Biological plant control involves the use of phytoplankton blooms to shade pond bottoms, or reliance on animals to consume aquatic vegetation at a rate as fast as, or faster than the plants can grow to become a nuisance. Phytoplankton shading is effective if initiated early in the spring before aquatic macrophytes and filamentous algae have an opportunity to take over the ponds. According to Swingle (1947), a good phytoplankton bloom will shade out many of the rooted aquatic macrophytes in a pond deeper than 45 cm. The techniques involved in establishing and maintaining a phytoplankton bloom were described earlier in this chapter.

Most animal candidates for the role of biological plant control organisms have been fishes. However other, more exotic creatures have also been tried. For example, the sea cow or manatee (*Trichechus manatus latirostris*) has been stocked in weed-infested waters of Florida (Sguros et al., 1965). Large freshwater snails (Blackburn and Weldon, 1965) and the South American flea beetle, *Agasicles* sp. (Anderson, 1965), have been introduced into aquaculture systems to control undesirable vegetation. No outstanding successes have been achieved however. In most cases the stocking of such animals in ponds experiencing severe plant infestations accomplishes little, although their use to prevent the onset of plant colonization and takeover may be appropriate.

The grass carp, *Ctenopharyngodon idella,* was introduced into the United States from Malaysia in 1963 (Stevenson, 1965) because of its reported usefulness in aquatic weed control. Now banned in more than 30 states because of fears that they may compete to the disadvantage of native fishes, invade the estuaries, and consume vast quantities of marshlands, or reproduce in captivity and become a proliferating pest like the common carp, *Cyprinus carpio,* grass carp are of limited utility in most warmwater producing areas, although they are still widely used in Alabama and Arkansas in conjunction with aquaculture. Their suitability in vegetation control has been demonstrated (Avault, 1965a, 1965b; Mitzner, 1978; Colle et al., 1978), and the species can be of great benefit in control of aquatic vegetation in aquaculture ponds if stocked at the rate of several per hectare.

The Israeli strain of the common carp, *Cyprinus carpio,* have also been used in aquatic vegetation control but are not widely available or accepted by United States culturists because of their low market value. Various species of tilapia have been stocked into natural waters and aquaculture ponds with the purpose of controlling aquatic plants. *Tilapia aurea* and other species are primarily herbivorous after attaining certain sizes. The feeding habits of various species of tilapia have been elaborated by such authors as LeRoux (1956), McBay (1961), Yashouv and Chervinsky (1960, 1961), and Moriarty (1973).

Few native North American fishes of commercial interest feed exclusively on plant material, and even those that are available (notably minnows) feed on phytoplankton during most or all of their lives and are of little use to the aquaculturist with a weed-choked pond. Goldfish, though of no value as human food, are readily marketable at certain sizes and will assist in the control of nuisance vegetation. Crayfish may also assist in vegetation control and are a popular food item, particularly in Louisiana and other portions of the southern United States. Species such as the paddlefish feed extensively on plant material, but again, this is primarily phytoplankton and the techniques for rearing that fish have not been well developed. In addition, the market potential may not be great. Because of increasing bans on exotic introductions into the United States, the possibilities of bringing in new species of aquatic organisms for the purpose of plant control appear remote.

Chemical Control

The use of herbicides to kill aquatic vegetation is generally an effective means of control, but it also involves certain risks. In many cases a massive amount of plant material is killed as a result of the application of chemicals. This decaying vegetation can place serious demands on the oxygen in a culture pond, and dissolved oxygen depletions may result. Following the application of any type of chemical, the culturist should keep a close watch on DO for several days to ensure that depletions do not occur. As in any other case of oxygen depletion, the problem can be dealt with by adding large amounts of new, highly oxygenated water, or an aeration device may be utilized.

A second consideration when applying chemicals is the direct effect of herbicides on aquaculture animals, or, in some cases, the natural food organisms on which the culture species may be feeding. All herbicidal compounds are toxic to fishes and invertebrates if present in sufficiently high concentration. In some cases the concentration required to eliminate a plant problem may be similar to that which will kill the culture animals. A minor error in calculating the treatment level or mixing the chemical could lead to disaster. Another important consideration is that the lethal level of herbicides varies not only from compound to compound, but among species of culture animals. For example, crustaceans are killed at lower concentrations of many chemicals than are fishes.

Chemical control of filamentous algae and aquatic macrophytes will also result in the destruction of phytoplankton. Once a herbicide has dissipated sufficiently to permit the return of aquatic plants, it may be difficult to establish a new phytoplankton bloom without initiating another outbreak of unwanted plants, especially if the herbicide is applied late in the spring or in the summer.

All herbicides must be cleared for use in aquaculture by the federal government if they are to be used in the United States. Those that had been found safe as of February 1976 are presented in Table 3.3. This list is subject to change as new chemicals are cleared and as some already on the list are reevaluated. Other chemicals have been used with a great deal of success in aquaculture ponds but have not as yet received official approval from the federal government.

Table 3.3   Herbicides Registered by the U.S. Government for Use in Water Containing Food Fish (Meyer *et al.*, 1976)

Chemical	Registered Trade Name Examples
Copper sulfate	Cutrine, Algimycin, etc.
2,4-D	
Diquat dibromide	Diquat
Endothall	Aquathol, Hydrothol, etc.
Simazine	Aquazine, Princeps

The chemical of choice and the rate of application vary depending on the type of aquatic vegetation being treated (Table 3.4). In general, aquatic vegetation problems can be treated with those herbicides that have been cleared (Table 3.2). The clearance of simazine in 1976 completed an array of treatment chemicals legally available and necessary for the elimination of most kinds of vegetation.

Klussmann and Lowman (1975) provide a good presentation of the types of vegetation and control methods to use in conjunction with problems arising when those plants colonize ponds. Their discussion is aimed primarily at the control of aquatic vegetation in Texas; however the rates of application and the species of plants they consider are valid throughout the southern United States and in many cases apply over the whole country. Since treatment rates vary to some extent as a function of water quality (especially temperature, hardness, and alkalinity) the directions on the label of any herbicide should be carefully read and followed.

Simazine is generally effective against aquatic plants of all types, including algae. The value of copper sulfate to control filamentous algae

**Table 3.4   Partial List of Types of Aquatic Vegetation and Chemicals that
Can Be Used to Control Them: For Application Rates,
Consult the Manufacturer's Directions**

Plant Common Name and Genus	Herbicide
Filamentous algae (various species), muskgrass (*Chara*), nitella (*Nitella*)	Copper sulfate Diquat Endothall Simazine
Elodea (*Elodea*), coontail (*Ceratophyllum*)	Diquat Endothall 2,4-D
Naiad (*Najas*), pondweed (*Potamogeton*), water milfoil (*Myriophyllum*)	Diquat Endothall Simazine 2,4-D
Fanwort (*Cabomba*)	Endothall Simazine
Arrowhead (*Sagittaria*)	Endothall Silvex 2,4-D
Bulrush (*Scirpus*), rush (*Juncus*), water smartweed (*Polygonum*)	2,4-D
Cattail (*Typha*)	Diquat
Duckweed (*Lemna*)	Diquat Simazine 2,4-D
Water hyacinth (*Eichhornia*)	2,4-D

has long been recognized, and it remains among the chemicals of choice, although the results vary as a function of water hardness. Copper sulfate should not be used in waters of low hardness (Chapter 4) because of increased toxicity to fish under those circumstances (Davis, 1961; Inglis and Davis, 1972).

As of October 1977 persons applying certain chemicals within the United States are required to register with a government agency (the agency designation varies from state to state, as does the procedure for registration). In most cases registration is approved if the applicant can demonstrate awareness of the proper techniques of application and of the safety precautions that must be taken when pesticides and herbicides are used.

The new federal law limits the sale of many chemicals to persons who are registered to apply them.

According to Klussmann and Lowman (1975), many submerged aquatic macrophytes can be controlled with endothall, whereas rooted plants with floating leaves may be better controlled by applying such chemicals as 2,4-D. Emergent vegetation such as cattails, rushes, and arrowhead can be controlled with various chemicals including simazine, endothall, and 2,4-D, depending on the species involved (Table 3.4). Duckweed responds to applications of simazine or diquat, and water hyacinths have been effectively treated with 2,4-D.

Avoiding Problems with Aquatic Macrophytes

The inoculation of aquaculture ponds with seeds, cysts, and spores of aquatic macrophytes and filamentous algae cannot be controlled, since these reproductive bodies may fall into ponds from the atmosphere, having been picked up from neighboring or even remote bodies of water by the wind. The translocation of plants and their reproductive bodies on the feet and feathers of birds, as well as on the bodies of mammals, is also possible. Some plant seeds can pass through the digestive tract of vertebrates without damage.

The best means of preventing the invasion of ponds by aquatic vegetation are proper pond construction and provision of an established phytoplankton bloom before the other plant species have a chance to develop. However there is no way to ensure that unwanted plants will not eventually dominate. Even the best pond managers run into problems on occasion. Early identification and treatment of the problem is often critical to protection of the aquaculture crop. Mechanical and biological control measures are generally preferable to chemical control insofar as avoiding toxicity to the culture animals is concerned; most frequently, however, some combination of treatments is necessary. A significant amount of capital may have to be spent in attempts to avoid and/or control aquatic vegetation problems. This aspect should be considered in the anticipated overhead for any aquaculture venture, except for most mariculture operations, where plant problems seem to be somewhat less frequent.

THE CARBONATE BUFFER SYSTEM

Mixtures of weak acids and their salts are called buffers. Several buffer systems may exist in water, including those associated with phosphate, borate (important only in seawater), and carbonate. The latter is usually the most significant buffer system in natural waters and is primarily responsible for the maintenance of pH in aquatic ecosystems.

The carbonate buffer system involves a series of chemical reactions. Carbon dioxide (existing in the atmosphere at a level of about 0.03%) is one source of carbon for the system, and calcium carbonate (found in the sediments and in the exoskeletons of certain invertebrates) is another. When carbon dioxide is scrubbed from the atmosphere by rain, it goes into solution as carbonic acid:

$$H_2O + CO_2 \rightleftharpoons H_2CO_3 \tag{7}$$

A series of reversible reactions, each with its own dissociation constant, occurs in water, leading to the presence of bicarbonate ion (HCO_3^-) and carbonate ion (CO_3^{2-}) in chemical equilibrium:

$$H_2CO_3 \rightleftharpoons H^+ + HCO_3^- \rightleftharpoons H^+ + CO_3^{2-} \tag{8}$$

Buffer systems of this type resist changes in pH. If hydrogen ions (H^+) are added to the system, calcium carbonate will dissolve as follows:

$$CaCO_3 + H^+ \rightleftharpoons Ca^+ + HCO_3^- \tag{9}$$

and the hydrogen ions will be removed from solution. On the other hand, the equilibrium described in equation 9 will prevent an accumulation of OH^- ions by shifting to the left as H^+ ions combine with OH^- to form water. As long as the pool of carbonate and bicarbonate does not become exhausted, the solution will resist pH changes. Sometimes buffer systems are overcome and drastic diurnal fluctuations in pH occur, with potentially lethal effects on aquatic animals.

In fresh water pH is generally maintained between 6.5 and 8.5 by the carbonate buffer system, whereas the abundance of calcium carbonate and other buffering agents in seawater leads to pH levels nearly always greater than 8.0 (except in coastal waters under certain circumstances and in a few other specialized cases). The presence of acidic conditions in the sea would lead to the dissolution of mollusk shells, coral reefs, and carbonate sediments. Life in the sea as we know it would change dramatically.

Photosynthesis and respiration both place demands on the carbonate buffer system and may in some cases result in significant changes in pH when the carbonate-bicarbonate pool is exhausted. Photosynthesis removes carbon dioxide from the water. For every two bicarbonate ions present, the removal of one carbon dioxide molecule will result in the formation of one molecule of water and a carbonate ion. In addition, a replacement carbon dioxide molecule will be made available:

$$2HCO_3^- \rightarrow H_2O + CO_3^{2-} + CO_2 \tag{10}$$

Thus as long as bicarbonate is present in the system, there will be little or no change in pH. More bicarbonate is produced as previously shown by the dissolution of calcium carbonate when free hydrogen ions are being added to the system, but during photosynthesis this does not occur; thus the bicarbonate pool can be depleted. Once bicarbonate has been depleted, the removal of additional carbon dioxide by photosynthesis leads to a rapid increase in pH. At night respiration will add carbon dioxide and the bicarbonate pool will be restored. Figure 3.7 plots the daily responses of pH to photosynthesis and respiration.

Bicarbonate ion appears to be an important element in the evolution of oxygen during photosynthesis (Stemler and Govindjee, 1973) and sometimes can be substituted for carbon dioxide in the photosynthetic process. Bicarbonate usually serves as the carbon source in photosynthesis only when the concentration of that ion is at least 10 times that of free carbon

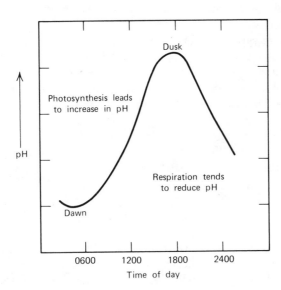

Figure 3.7 Potential pattern of diurnal pH fluctuation in ponds that have high levels of primary productivity, high respiratory demand, and a limited supply of carbonate and bicarbonate.

dioxide (Wetzel, 1975). This occurs frequently, since the carbon dioxide concentration in most natural waters is only about 10 micromoles (μmole) (Wetzel, 1975). Steemann Nielsen (1975) indicated that about 1% of the total inorganic carbon in seawater is in the form of carbon dioxide, 90% is bicarbonate, and the remainder carbonate. When bicarbonate is utilized in photosynthesis, it is exchanged for hydroxide ions:

$$HCO_3^- \rightarrow CO_2 + OH^- \tag{11}$$

At sufficiently high pH, calcium carbonate crystals form on the leaves of aquatic macrophytes:

$$OH^- + HCO_3^- \rightarrow H_2O + CO_3^{2-} \tag{12}$$

$$CO_3^{2-} + Ca^{2+} \rightarrow CaCO_3 \tag{13}$$

In alkaline fresh waters it is often possible to create a white cloud of calcium carbonate by disturbing rooted macrophytes that have been photosynthesizing in this manner. The utilization of bicarbonate ions in photosynthesis is probably more advantageous to macrophytic vegetation than to algae (Wetzel, 1975), and in aquaculture the growth of macrophytes is usually discouraged.

The buffering capacity of any aquatic system is related to such conservative properties of water as alkalinity and hardness (Chapter 4) and, of course to photosynthesis, which is nonconservative. Maintenance of pH is more or less under the control of the aquaculturist, who can make certain adjustments in the rate of carbon dioxide uptake and evolution by controlling photosynthesis and the biomass of animals in the system. Alkalinity and hardness can also be controlled within limits (Chapter 4) and are important in providing a sufficient pool of carbonate and bicarbonate ions to allow the system to resist changes in pH.

DISSOLVED OXYGEN

The level of dissolved oxygen (DO) present in the aquaculture environment is among the most important factors in water quality. If a sufficient level of DO is not maintained, animals will be stressed, becoming vulnerable to disease and parasite outbreaks, or they will die. At the very least, the animals may refuse to eat for a period during and after an oxygen depletion. Thus growth rate and food conversion efficiency will suffer, and feed will be wasted if prepared diets are provided. The reasons for oxygen problems and manner in which they can be avoided are of primary concern to aquaculturists. Of all the problems with water quality that can plague a warmwater culturist, oxygen depletion is the most frequent.

Aquatic animals possessing lungs (e.g., frogs and sea turtles), as well as a few species of fish (e.g., the walking catfish, *Clarias batrachus*), are able to exist in oxygen-depleted water, but most aquaculture species quickly succumb to certain minimum levels of DO. Many factors affect the DO level in water, including, but not restricted to, the activities of animals. If DO falls to a concentration that places stress on animals or threatens their lives, other water quality parameters tend to become insignificant until DO is reestablished at a safe concentration. The level at which DO begins to induce stress varies with species and undoubtedly is associated to some extent with other stress-causing factors, although the synergistic relationships between DO and un-ionized ammonia concentration, nitrite, temperature, salinity, hardness, alkalinity, and so forth, have not been evaluated in any detail.

Sources, Sinks, and Acceptable Levels

Oxygen is dissolved in water by diffusion from the atmosphere and photosynthesis and can be induced to dissolve through mechanical aeration. Diffusion from the atmosphere is aided when turbulence occurs, such as during stormy or windy weather. The creation of turbulence through artificial means, as well as placing large volumes of air or pure oxygen in contact with water, are the principles behind mechanical aeration.

Oxygen is removed from water primarily by respiration, which is essentially the reverse of photosynthesis:

$$CH_2O + O_2 \rightarrow H_2O + CO_2 \qquad (14)$$

All aquaculturists recognize that respiration is an important function in animals and that animal respiration can have a significant impact on the DO level in a given water system, but they do not often realize that autotrophic organisms also respire. Plants, like animals, respire continuously. During daylight photosynthesis usually produces more oxygen than is removed from the water by the combined respiratory demand of plants and animals. The net result is an increase in DO during the daylight hours. At night both plants and animals continue to respire while new oxygen is not being added to the water. In most instances the respiratory demand during darkness is not sufficient to deplete the water of oxygen; however under certain circumstances such a depletion may occur. Figure 3.8 presents a typical diurnal pattern of DO fluctuation in aquatic systems.

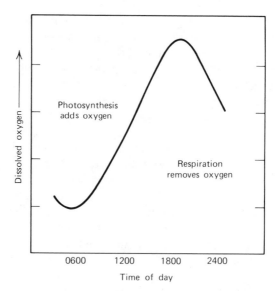

Figure 3.8 Typical pattern of DO fluctuation in an aquaculture pond with high rates of photosynthesis and respiration.

Oxygen is also removed from water as a result of certain inorganic chemical reactions (referred to as the chemical oxygen demand and having little importance in most aquaculture systems) and through the decomposition of organic matter by microorganisms. The requirement for oxygen by the latter process plus that associated with respiration comprise the biochemical oxygen demand (BOD). The BOD test is an empirical one that has been standardized for use in many types of water (APHA, 1975). The BOD in aquaculture systems can become very important when large amounts of aquatic vegetation are decaying (e.g., after a pond has been treated with herbicide) or when dead animals are allowed to decompose in the water system. Under certain circumstances high BOD may trigger oxygen depletions.

As a general rule, if DO is equal to, or in excess of 5 mg/l, no stress will be placed on aquaculture animals (Wheaton, 1977). Some species can tolerate DO concentrations well below 5 mg/l with little or no resultant stress, but no aquatic species must have higher concentrations to survive. Several species of tilapia appear to be able to survive well at DO concentrations as low as 1 mg/l (Uchida and King, 1962; Denzer, 1968) and continue to grow rapidly if exposed to such low levels for fairly brief times daily (Stickney *et al.*, 1977). Other species may die quickly if exposed to such a low DO level. Channel catfish appear to tolerate fairly low DO for brief periods, but it is generally conceded that some restorative action should be taken when the DO falls below 2 or 3 mg/l. The experience of many channel catfish farmers has been that the fish grow well following low DO if the level becomes high later in the day (as a result of photosynthesis, Figure 3.8), but they may not do as well if a chronic low level is present or if the daily maximum does not exceed several milligrams per liter.

The amount of oxygen that can be dissolved in water under ambient conditions is called the saturation oxygen concentration. DO saturation is dependent on temperature, salinity, and altitude, with saturation occurring at lower DO concentrations as each of these parameters increases (Table 3.5). Supersaturation of oxygen does occur both under natural conditions (as a result of high levels of primary productivity) and as a consequence of human activity (e.g., in the cooling water of power plants during winter). Even at fairly high temperatures and salinities, the DO saturation level exceeds 5 mg/l at sea level; thus unless some other factor leads to an oxygen depletion, the water utilized in warmwater aquaculture should

Table 3.5   Solubility of Oxygen in Water (mg/l[a] at 1 atm) with Varying Temperature and Salinity

Temperature (C)	Salinity (%o)				
	0	10	20	30	35
10	11.3	10.6	9.9	9.3	9.0
12	10.7	10.1	9.5	8.9	8.7
14	10.3	9.7	9.1	8.6	8.3
16	9.9	9.3	8.7	8.2	8.0
18	9.5	8.9	8.4	7.9	7.7
20	9.1	8.6	8.1	7.6	7.4
22	8.7	8.2	7.8	7.3	7.1
24	8.4	7.9	7.5	7.1	6.9
26	8.1	7.7	7.2	6.8	6.6
28	7.8	7.4	7.0	6.6	6.4
30	7.6	7.1	6.8	6.4	6.2
32	7.3	6.9	6.5	6.2	6.0
34	7.0	6.7	6.2	6.0	5.8

Source: Weiss (1970).

[a] Modified from milliliters per liter of oxygen by means of the following relationship: mg/l = 0.6998 ml/l.

have the capacity to hold sufficient oxygen to support any species being reared. Under hypersaline conditions this stricture may not apply, but few aquaculture operations have had to deal with salinities above those found in the open ocean (about 35°/oo).

Measuring Dissolved Oxygen

The aquaculturist should routinely monitor DO to ensure that depletions in the water system are not occurring. Measurements should be made daily, preferably at dawn, particularly in culture chambers where potential oxygen depletions are anticipated. Details regarding the causes and prevention of oxygen depletions are presented in the following section of this chapter.

DO concentration is most commonly measured by Winkler titration (APHA, 1975) or with an oxygen meter (Figure 3.9). Both methods can be utilized to determine DO to within 0.1 mg/l and require relatively little time, although the oxygen meter is faster than the titration and in most cases, provides an indication of temperature as well as DO. Winkler titrations require several chemicals and glassware, including a burette if a high degree of precision is desired. In the field an oxygen meter is handier, and once calibrated the device can assist the culturist in numerous determinations with no more trouble than dropping the probe into the water.

In most instances the aquaculturist is interested in determining whether the DO concentration in a particular culture chamber is 1 or 5 mg/l, not whether it is 5.1 mg/l as opposed to 5.2 mg/l. Thus despite the availability of highly sophisticated devices for determining DO to within a small fraction of a milligram per liter, the expense of making such fine measurements is not warranted for aquacultural use. The chemicals and glassware required for several hundred Winkler titrations can be obtained for a few dollars, whereas an oxygen meter costs a few hundred. Batteries and the membranes associated with oxygen meter probes must be replaced at intervals (often many months in the case of batteries, but more frequently for membranes), and chemicals must be replaced as required for Winkler titrations.

Improvements in dissolved oxygen meters over the past few years have made them a good investment, and considering the time that can be saved relative to the Winkler titration, purchase of a good quality meter is recommended. Oxygen meters are available that in addition to measuring temperature along with oxygen, are compensated for salinity and even altitude (Figure 3.9). A salinity-compensated dissolved oxygen meter is desirable for mariculture, although uncompensated meters are generally supplied with conversion tables for saline waters.

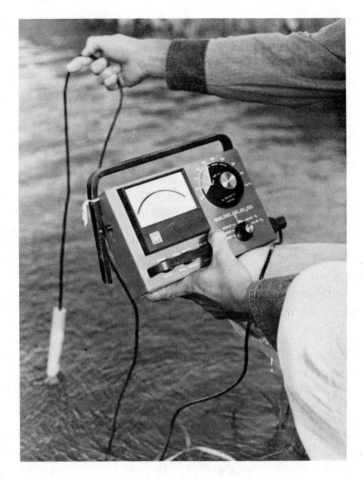

Figure 3.9 Portable dissolved oxygen meters can be utilized to make rapid determinations with adequate precision. This meter can be compensated for altitude, temperature, and salinity.

Oxygen Depletions

If DO is not routinely measured, the culturist may become aware of an oxygen depletion indirectly. Aquacultured animals generally refuse feed under low oxygen stress, and when the level becomes critically low, many species of fishes rise to the water surface and appear to gasp for air. This behavior is not demonstrated by all cultured animals, however. Shrimp often remain on the bottom of the culture chamber when DO is low and may die and decompose there without attracting the attention of the culturist. Fishes, on the other hand, generally float to the surface within a day or two of their death, and mortalities are discovered.

The presence of fish at the surface of a pond should not be routinely attributed to an oxygen depletion, although it is often indicative. Tilapia and goldfish, among other species, are sometimes observed at the surface of the water when DO levels are adequate. The behavior of these animals may be associated with feeding, reproduction, or other activities. Catfish rarely come to the surface except to feed on floating pellets or when DO is critically low.

Throughout most of the year sufficient oxygen is produced by photosynthesis during daylight and absorbed from the air at all times to maintain adequate levels of DO through the night. During spring, when young animals are usually stocked into aquaculture systems, the total biomass in any individual culture chamber is often relatively low and heavy demands are not placed on oxygen. In addition, since the water is fairly cool through much of the spring, the saturation DO level is high relative to the concentration required for the well-being of aquatic animals. In the fall, although the biomass per unit area of water is often very high, declining water temperature again allows more oxygen to become dissolved at saturation than in the summer. Oxygen depletions can occur during the spring and fall but are uncommon. During winter in warmwater fish-growing regions, oxygen depletions are extremely unlikely because the oxygen-carrying capacity of the water is high, metabolism is low, and there are usually only low BODs.

Oxygen depletions are most often associated with the summer season. In summer the metabolic rates of all animals in a culture pond (culture species, benthic invertebrates, zooplankton, and any other animals that are present) are high, large amounts of feed are being added to each pond daily (increasing the BOD level if a significant fraction is not consumed and also contributing to BOD when unabsorbed portions pass through the bodies of the culture animals and enter the water as feces), and photosynthesis is high. In most instances photosynthetic oxygen production is sufficient to offset respiratory demands and there is no net change in oxygen level over 24 hours, or even a gain. The cyclical pattern of DO during a typical summer day was presented in Figure 3.8. If something happens to reduce photosynthetic production of oxygen or increase respiratory demands, there may be a net loss in DO each day until the level gets low enough to stress the aquaculture animals.

Normally DO is at its lowest level of the day just at dawn, since respiratory demands throughout the night have reduced the level, but oxygen production begins again when photosynthesis starts at daylight. The aquaculturist should measure DO in the culture chambers at or close to dawn each day during the summer to ensure that the early morning level is sufficient to maintain the culture animals. Early morning checks on DO are especially important if anything has happened that might have decreased the photosynthesis rate. The most common causes are destruction of vegetation through treatment with herbicides (usually done in the spring to avoid the potential problem of oxygen depletion, as discussed) and extended periods of cloudy weather. Oxygen depletions are also often associated with the collapse of phytoplankton blooms. The presence of overcast skies during daylight limits the amount of light available to culture ponds, hence reducing the rate of photosynthesis and in some cases inducing or hastening the collapse of a phytoplankton bloom. Respiration continues at its normal rate or even at an accelerated rate, with the result that each morning at dawn the DO level in culture ponds may be somewhat lower than it was the preceding day.

If a pond has a high BOD level, is overstocked, receives too much feed, or for some other reason has a higher respiratory demand than its neighbors, an oxygen depletion may occur. It is not unusual for several ponds to experience oxygen depletions on a given morning, and once a depletion has occurred in a pond, the condition may be expected to recur on subsequent mornings. Several days of sunshine often assist in returning the situation to normal; however since oxygen depletions are not always associated with cloudy weather, the culturist must remain vigilant.

Overcoming Oxygen Depletions

When an oxygen depletion is detected, immediate action must be initiated to restore the DO to safe levels. Several means can be used to this end. Often the simplest method of increasing DO is to add copious amounts of well-aerated new water. Spraying water over the surface of ponds in a manner that creates as much turbulence as possible will aid in aeration. At least several hundred liters of water per hectare should be available each minute. Water should be added until the DO level in the affected pond is above 3 mg/l during daylight, when photosynthesis will maintain or increase that concentration.

A second method is to pump water from the affected pond with an electric or gasoline pump and spray it back across the surface in a turbulent manner. No new water is added, but a large amount of oxygen can be carried into the water in a short period. The size of pump required to achieve satisfactory results increases with pond dimensions and the manner in which the water is sprayed through the air and back into the pond. Compressed air, compressed oxygen, air blowers, and air compressors can be utilized to aerate oxygen-deficient ponds, but the volume of gas provided by such devices is often very small relative to the volume of water to be treated. In tanks, raceways, and other small culture chambers those methods of aeration are often as appropriate as any other. Various mechanical agitators are commercially available specifically for fish ponds. Representatives from businesses that market those devices, or the manufacturers, can recommend the size and number appropriate in any particular situation. Design features and operational characteristics of various surface agitators were discussed by Wheaton (1977).

When an oxygen depletion is anticipated, particularly in the case of ponds that have recently experienced the problem, preventive measures can be taken. New water may be added throughout the night (at a rate that will serve to replace the volume of the pond within about 72 hours), a pump may be used to continuously circulate water and spray it over the pond surface, or mechanical aerators may be turned on for several hours daily.

Culture animals should not be fed immediately after an oxygen depletion, since the accompanying stress often causes them to reject feed and adding more organic matter to affected ponds will only increase the BOD. Stress that does not result in death of aquaculture animals may make them susceptible to diseases and parasites. The culturist should observe the animals for at least 2 weeks after an oxygen depletion (or any other type of stress) to ensure that an epizootic does not occur.

Though controversial, the use of potassium permanganate ($KMnO_4$) at a rate of 5 mg/l is felt by some culturists to be effective in oxidizing organic matter in ponds and lowering the BOD, thus reducing the threat of oxygen-depletion recurrence. Some empirical evidence of the effectiveness of this chemical is available (Mathis et al., 1962), but not all aquaculturists agree on its utility in conjunction with oxygen depletions or as a preventive to ward off a potential oxygen depletion. Potassium permanganate has not been cleared for use on food fish (Meyer et al., 1976) but is presently under review for clearance.

Ideally the water utilized in an aquaculture facility is well oxygenated when introduced into the culture chambers. Oxygen problems normally occur as a result of demands that exist in the culture chambers. The aquatic animal producer must be aware of those demands, anticipate changes that may alter the demands, and be prepared to take immediate action when the threat of oxygen depletion exists.

LITERATURE CITED

American Public Health Association. 1975. *Standard methods,* 14th ed. American Public Health Association, Washington, D.C. 1193 p.

Anderson, W. H. 1965. Search for insects in South America that feed on aquatic weeds. *Proc. South. Weed Conf.* **18:** 586–587.

Avault, J. W., Jr. 1965a. Biological weed control with herbivorous fish. *Proc. South. Weed Conf.* **18:** 590–591.

Avault, J. W., Jr. 1965b. Preliminary studies with grass carp for aquatic weed control. *Prog. Fish-Cult.* **27:** 207–209.

Baily, T. A. 1965. Commercial possibilities of dehydrated aquatic plants. *Proc. South. Weed Conf.* **18:** 543–551.

Blackburn, R. D., and L. W. Weldon. 1965. A fresh-water snail as a weed control agent. *Proc. South. Weed Conf.* **18:** 589–591.

Boyd, C. E. 1968a. Evaluation of some common aquatic weeds as possible feedstuffs. *Hyacinth Control J.* **7:** 26–27.

Boyd, C. E. 1968b. Fresh-water plants: A potential source of protein. *Econ. Bot.* **22:** 359–368.

Burrows, R. E. 1964. *Effects of accumulated excretory products on hatchery-reared salmonids.* U.S. Bureau of Sport Fishing and Wildlife Resource Report 66. 12 p.

Chipman, W. A., Jr. 1934. The role of pH in determining the toxicity of ammonium compounds. Ph.D. dissertation, University of Missouri, Columbia. 153 p.

Colle, D. E., J. V. Shireman, and R. W. Rottmann. 1978. Food selection by grass carp fingerlings in a vegetated pond. *Trans. Am. Fish. Soc.* **107:** 149–152.

Culley, D. D., Jr., and E. A. Epps. 1973. Use of duckweed for waste treatment and animal feed. *J. Water Pollut. Control Fed.* **45:** 337–347.

Davis, H. S. 1961. *Culture and diseases of game fishes.* University of California Press, Berkeley. 332 p.

deGruchy, J. H. B. 1938. *A preliminary study of the larger aquatic plants of Oklahoma with special reference to their value in fish culture.* Oklahoma Agricultural Experiment Station, Technical Bulletin 4, Stillwater. 31 p.

Denzer, H. W. 1968. Studies on the physiology of young *Tilapia. FAO Fish. Rep.* **44:** 357–366.

Emerson, K., R. C. Russo, R. E. Lund, and R. V. Thurston. 1975. Aqueous ammonia equilibrium calculations: Effect of pH and temperature. *J. Fish. Res. Bd. Can.* **32:** 2379–2383.

Eyles, D. E., and J. L. Robertson. 1944. *A guide and key to the aquatic plants of the southeastern United States.* U.S. Public Health Service Bulletin 286. 151 p.

Fassett, N. C. 1960. *A manual of aquatic plants.* University of Wisconsin Press, Madison. 405 p.

Fogg, G. E. 1972. *Photosynthesis.* American Elsevier, New York. 116 p.

Fromm, R. O. 1963. Studies on the renal and extra-renal excretion in a freshwater teleost, *Salmo gairdneri. Comp. Biochem. Physiol.* **10:** 121–128.

Gaarder, T., and H. H. Gran. 1927. Investigations of the production of plankton in the Oslo Fjord. *Rapp. Proces.-Verb. Cons. Int. Explor. Mer.* **42:** 1–48.

Hampson, B. L. 1976. Ammonia concentration in relation to ammonia toxicity during a rainbow trout rearing experiment in a closed freshwater-seawater system. *Aquaculture,* **9:** 61–70.

Hart, W. B., P. Doudoroff, and J. Greenbank. 1945. Evaluation of toxicity of industrial wastes, chemicals and other substances to fresh-water fishes. Water Control Laboratory, Atlantic Refining Co., Philadelphia.

Hochachaka, P. W. 1969. Intermediary metabolism in fishes. In W. S. Hoar and D. J. Randall (Eds.), *Fish physiology*, Vol. 1. Academic Press, New York, pp. 351–389.

Inglis, A., and E. L. Davis. 1972. *Effects of water hardness on the toxicity of several organic and inorganic herbicides to fish.* U.S. Bureau of Sport Fisheries and Wildlife, Technical Paper 67. 22 p.

Klussmann, W. G., and F. G. Lowman. 1975. *Common aquatic plants.* Texas Agricultural Extension Service, College Station. 15 p.

Konikoff, M. 1975. Toxicity of nitrite to channel catfish. *Prog. Fish-Cult.* **37:** 96–98.

Lange, S. R. 1965. The control of aquatic plants by commercial harvesting, processing and marketing. *Proc. South. Weed Conf.* **18:** 536–542.

Lawrence, K. J. M. 1968. Aquatic weed control in fish ponds: *Proceedings of the World Symposium on Warm-water Pond Fish Culture. FAO Fish. Rep.* **44:** 76–91.

Lees, H. 1952. The biochemistry of the nitrifying organisms. 1. The ammonia-oxidizing system of *Nitrosomonas. Biochem. J.* **52:** 134–139.

LeRoux, P. J. 1956. Feeding habits of the young of four species of *Tilapia. South Afr. J. Sci.* **53:** 96–100.

Mathis, W. P., L. E. Bardy, and W. J. Gilbreath. 1962. Preliminary report on the use of potassium permanganate to alleviate acute oxygen shortage and counteract hydrogen sulfide gas in fish ponds. *Proc. Southeast. Assoc. Game Fish Comm.* **16:** 357–359.

Mayo, R. D. 1971. Reuse of water in hatcheries to increase capacity. *Am. Fishes Trout News,* January–February, pp. 6ff.

McBay, L. G. 1961. The biology of *Tilapia nilotica* Linnaeus. *Proc. Southeast. Assoc. Game Fish Comm.* **15:** 208–218.

Meade, T. L. 1976. *Closed system salmonid culture in the United States.* Marine Advisory Service, University of Rhode Island, Kingston, Marine Memorandum 40. 16 p.

Meyer, F. P., R. A. Schnick, and K. B. Cumming. 1976. Registration status of fishery chemicals, February, 1976. *Prog. Fish-Cult.* **38:** 3–7.

Mitzner, L. 1978. Evaluation of biological control of nuisance aquatic vegetation by grass carp. *Trans. Am. Fish. Soc.* **107:** 135–145.

Moriarty, D. J. W. 1973. The physiology of digestion of blue-green algae in the cichlid fish, *Tilapia nilotica. J. Zool., Lond.* **171:** 25–39.

Muenscher, W. C. 1944. *Aquatic plants of the United States.* Comstock Publishing Co., Ithaca, N. Y. 374 p.

Odum, H. T., W. McConnell, and W. Abbott. 1958. The chlorophyll "A" of communities. *Publ. Inst. Mar. Sci. Tex.* **5:** 65–69.

Pillay, T. V. R. (Ed.). 1972. *Coastal aquaculture in the Indo-Pacific region.* Fishing News (Books), London. 497 p.

Redner, B. D. 1978. Toxicity and acclimation to ammonia by *Tilapia aurea.* M.S. thesis, Texas A&M University, College Station. 70 p.

Richards, F. A., and T. C. Thompson. 1952. The estimation and characterization of plankton populations by pigment analyses. II. A spectrophotometric method for the estimation of plankton pigments. *J. Mar. Res.* **11:** 156–172.

Robinette, H. R. 1976. Effect of selected sublethal levels of ammonia on the growth of channel catfish (*Ictalurus punctatus*). *Prog. Fish-Cult.* **38:** 26–29.

Ryther, J. H. 1956. The measurement of primary production. *Limnol. Oceanogr.* **1:** 72–84.

Saunders, G. W., F. B. Trama, and R. W. Bachmann. 1962. *Evaluation of a modified C-14 technique for shipboard estimation of photosynthesis in large lakes.* Great Lakes Resources Division Publication 8. University of Michigan, Ann Arbor.

Sguros, P. L., T. Monku, and C. Phillips. 1965. Observations and techniques of the Florida manatee—Reticent but superb weed control agent. *Proc. South. Weed Conf.* **18:** 588.

Smith, G. M. 1950. *Fresh-water algae of the United States.* McGraw-Hill, New York. 719 p.

Smith, H. W. 1929. The excretion of ammonia and urea by the gills of fish. *J. Biol. Chem.* **81:** 727-742.

Spotte, S. H. 1970. *Fish and invertebrate culture.* Wiley-Interscience, New York. 145 p.

Steemann Nielsen, E. 1951. Measurement of the production of organic matter in the sea by means of C-14. *Nature,* **167:** 684-685.

Steemann Nielsen, E. 1952. The use of radioactive carbon (C-14) for measuring organic production in the sea. *J. Cons.* **18:** 117-140.

Steemann Nielsen, E. 1975. *Marine photosynthesis.* Elsevier Oceanography Series, 13. Elsevier Scientific Publications, Amsterdam. 140 p.

Stemler, A., and Govindjee. 1973. Bicarbonate ion as a critical factor in photosynthetic oxygen evolution. *Plant Physiol.* **52:** 119-123.

Stevenson, J. 1965. Observations on grass carp in Arkansas. *Prog. Fish-Cult.* **27:** 203-206.

Stickney, R. R., L. O. Rowland, and J. H. Hesby. 1977. Water quality—*Tilapia aurea* interactions in ponds receiving swine and poultry wastes. *Proc. World Maricult. Soc.* **8:** 55-71.

Swingle, H. S. 1947. *Management of farm fish ponds.* Alabama Agricultural Experiment Station Bulletin 254. 30 p.

Tabata, K. 1962. Toxicity of ammonia to aquatic animals with reference to the effect of pH and carbon dioxide. *Bull. Tokai Reg. Fish. Res. Lab.* **34:** 67-74.

Uchida, R. M., and J. E. King. 1962. Tank culture of *Tilapia. Fish. Bull.* **62:** 21-25.

Weiss, R. F. 1970. The solubility of nitrogen, oxygen and argon in water and seawater. *Deep-Sea Res.* **17:** 721-735.

Weldon, L. W., R. D. Blackburn, and D. S. Harrison. 1969. *Common aquatic weeds.* U.S. Department of Agriculture Handbook 352. 43 p.

Wetzel, R. G. 1975. *Limnology.* Saunders, Philadelphia. 743 p.

Wheaton, E. W. 1977. *Aquaculture engineering.* Wiley-Interscience, New York. 708 p.

White, A., P. Handler, and E. L. Smith. 1964. *Principles of biochemistry.* McGraw-Hill, New York. 1106 p.

Wood, J. D. 1958. Nitrogen excretion in some marine teleosts. *Can. J. Biochem. Physiol.* **36:** 1237-1242.

Wuhrmann, K., and H. Woker. 1948. Experimentelle Untersuchungen über die Ammoniak- und Blausaurevergiftung. *Schweiz. Z. Hydriol.* **11:** 210-244.

Wuhrmann, K., F. Zehender, and H. Woker. 1947. Über die fischereibiologische Bedeutung des Ammonium- und Ammiakgehaltes fliessender Gewässer. *Vierteljahrsschr. Naturforsch. Ges. Zür.,* **92:** 198-204.

Yashouv, A., and J. Chervinsky. 1960. Evaluation of various food items in the diet of *Tilapia nilotica. Bamidgeh,* **12:** 71-78.

Yashouv, A., and J. Chervinsky. 1961. The food of *Tilapia nilotica* in ponds of the fish culture research station at Dor. *Bamidgeh,* **13:** 33-39.

SUGGESTED ADDITIONAL READING

Boyd, C. E. 1976. Nitrogen fertilizer effects on production of *Tilapia* in ponds fertilized with phosphorus and potassium. *Aquaculture,* **7:** 385–390.

Buckley, J. A. 1978. Acute toxicity of un-ionized ammonia to fingerling coho salmon. *Prog. Fish-Cult.* **40:** 30–32.

Hepher, B. 1965. Fertilization of fish ponds. *Bamidgeh,* **17:** 58–59.

Lloyd, R., and D. W. M. Herbert. 1960. The influence of carbon dioxide on the toxicity of un-ionized ammonia to rainbow trout. *Ann. Appl. Biol.* **48:** 399–404.

Merkins, J. C., and K. M. Downing. 1957. The effect of tension of dissolved oxygen on the toxicity of un-ionized ammonia to several species of fish. *Ann. Appl. Biol.* **45:** 521–527.

Murphy, J. P., and R. I. Lipper. 1970. BOD production of channel catfish. *Prog. Fish-Cult.* **32:** 195–198.

Rainwater, F. H., and L. L. Thatcher. 1960. *Methods for the collection and analysis of water samples.* U.S. Geological Survey Water-Supply Paper 1454. 301 p.

Russo, C. E., C. E. Smith, and R. V. Thurston. 1974. Acute toxicity of nitrate to rainbow trout. *J. Fish. Res. Bd. Can.* **31:** 1653–1655.

Ruttner, F. 1953. *Fundamentals of limnology.* University of Toronto Press, Toronto. 295 p.

Schutte, K. H., and J. F. Elsworth. 1952. The significance of large pH fluctuations observed in some South African vleis. *J. Ecol.* **42:** 148–150.

Smith, C. E., and W. G. Williams. 1974. Nitrite toxicity in rainbow trout and chinook salmon. *Trans. Am. Fish. Soc.* **103:** 389–390.

Strickland, J. D. H., and T. R. Parsons. 1968. *A practical handbook of seawater analysis.* Fisheries Research Board of Canada, Bulletin 167, Ottawa. 311 p.

Warren, C. E. 1971. *Biology and water pollution control.* Saunders, Philadelphia. 434 p.

Worsham, R. L. 1975. Nitrogen and phosphorus levels in water associated with a channel catfish (*Ictalurus punctatus*) feeding operation. *Trans. Am. Fish. Soc.* **104:** 811–815.

CHAPTER 4

Conservative Parameters of Water Quality and Physical Aspects of the Aquatic Environment

GENERAL CONSIDERATIONS

Conservative water quality parameters are those that are not affected in any significant way by the activities of organisms. Such factors as temperature, salinity, alkalinity, and hardness are modified in some cases by physical processes, but are not perturbed to any extent by biological activity except at a microcosmic level in some cases. For example, metabolic activity in fishes releases some heat energy into the environment, and very careful measurement may detect minute local changes in temperature. However the oceans, lakes, rivers, and culture chambers around the world are not thermally altered as a result of the dissipation of such heat energy.

Physical aspects of the culture environment that may have to be considered, depending on the type of water system and the species involved, include density of stocking, need for separation of individuals into separate compartments, and substrate requirements. The most important physical parameter for aquatic animals is, of course, temperature. It is technologically possible to control temperature in any type of water system utilized by aquaculturists, but in most cases attempts at alteration of ambient temperature will lead to financial disaster.

Although temperature is often allowed to fluctuate seasonally, many other physical and conservative chemical parameters of water quality can be controlled within limits. Adjustments in hardness, alkalinity, and total dissolved solids can be accomplished relatively easily and inexpensively in some instances. This chapter outlines the most important parameters in the subject areas mentioned and provides the aquaculturist with the background and techniques for implementation of sound management.

TEMPERATURE

Thermal Requirements

Selection of an aquaculture species must be based on knowledge of the temperature requirements of the species. Survival cannot be the only criterion. Channel catfish, for example, will live in both temperate and tropical climates. The species occurs naturally as far north as the Great Lakes region of the United States (Eddy, 1957). This is not to imply that aquaculture of *Ictalurus punctatus* is practical throughout its range. Channel catfish grow most rapidly within a temperature range of 26 to 30 C (Kilambi *et al.*, 1970; Andrews and Stickney, 1972) and can easily be grown from egg to market size in 18 months in the southern United States, whereas several years may be required to produce a marketable (approximately 0.5 kg) fish at the northern edge of the range.

Aquaculture organisms are usually separated into three groups on the basis of optimum or required temperature range. These are coldwater, warmwater, and midrange. Coldwater species have temperature optima at or below about 15 C, warmwater species at or above 25 C, and midrange species between the others. Table 4.1 indicates the temperature requirements for survival or at least for optimum growth of several marine and freshwater species of present or potential aquaculture importance. In some instances (e.g., the American oyster, *Crassostrea virginica*), culture is possible in either warm or relatively cool environments, but the time required to produce a marketable animal, as was seen with channel catfish, varies considerably between the two areas.

Regions that contain coldwater, warmwater, and midrange species are separated not only by latitude, but also by altitude and water source. The low latitudes and low altitudes of the southern United States make the freshwater environments there suitable for warmwater species, whereas coldwater forms are found in mountain streams and high latitude waters. Midrange species are found primarily in northern latitudes, mostly in lakes. In the marine environment the west coast of the United States is fairly cold because of the influence of the California current, which flows in a southerly direction down the western margin of the continent. The Gulf of Mexico is characterized by warm temperatures through much of the year and supports typically warmwater aquaculture species. On the east coast of the United States, warmwater culture can generally be practiced as far north as Cape Hatteras, North Carolina. South of Cape Hatteras, the Gulf Stream has a tempering effect on water temperature, whereas north of Cape Hatteras there is a distinct change in the fauna, with coldwater forms becoming dominant and warmwater ones occurring only during the summer, after

Table 4.1 Classification of Present and Potential Aquaculture Species as a Function of Temperature Requirement for Optimum Growth[a]

Temperature Requirement	Common and Scientific Names
Cold water	Abalone (*Haliotis rufescens*)
	European oyster (*Ostrea edulis*)
	Common scallop (*Pecten irradians*)
	Common crab (*Cancer magister*)
	American lobster (*Homarus americanus*)
	Rainbow trout (*Salmo gairdneri*)
	Brown trout (*Salmo trutta*)
	Brook trout (*Salvelinus fontinalis*)
Midrange	Yellow perch (*Perca flavescens*)
Warm water	American oyster (*Crassostrea virginica*)
	Scallop (*Aequipecten irradians*)
	Blue crab (*Callinectes sapidus*)
	Spiny lobster (*Panulirus argus*)
	Quahog (*Mercenaria mercenaria*)
	White shrimp (*Penaeus setiferus*)
	Pink shrimp (*Penaeus duorarum*)
	Brown shrimp (*Penaeus aztecus*)
	Freshwater shrimp (*Macrobrachium rosenbergii*)
	Red drum (*Sciaenops ocellata*)
	Black drum (*Pogonias chromis*)
	Summer flounder (*Paralichthys dentatus*)
	Southern flounder (*Paralichthys lethostigma*)
	Florida pompano (*Trachinotus carolinus*)
	Dolphin (*Coryphaena hippurus*)
	Channel catfish (*Ictalurus punctatus*)
	Blue catfish (*Ictalurus furcatus*)
	White catfish (*Ictalurus catus*)
	Bigmouth buffalo fish (*Ictiobus cyprinellus*)
	Smallmouth buffalo fish (*Ictiobus bubalus*)
	Grass carp (*Ctenopharyngodon idella*)
	Common carp (*Cyprinus carpio*)
	Blue tilapia (*Tilapia aurea*)
	Green sea turtle (*Chelonia mydas mydas*)

[a] All species are present in the United States; though some are exotics.

which they tend to migrate south into warmer water for the remainder of the year.

Virtually all aquaculture candidates are poikilothermal; thus their metabolic rates are determined by ambient water temperature. For reasons that are largely unknown but undoubtedly entail differences in enzyme systems among the various species, temperature tolerances and optima vary greatly. As a general rule, a coldwater or midrange species may attain a maximum size and weight as large or larger than that of a warmwater species, but the growth rate of the latter will be more rapid because of its higher metabolic rate.

Seasonal Patterns In Water Temperature

Waters of nearly ideal temperature for warmwater species do occur. They are most widespread, of course, in tropical climates. In the United States optimum conditions can be found throughout the year only in Hawaii, although southern Florida and southern Texas have significantly longer growing seasons than other parts of the South. Geothermal springs, warmwater wells (usually rather deep), and power plant cooling waters provide water of suitable temperature over the entire year for a few aquaculturists, but the majority of persons engaged in culture in the United States must accept fairly significant seasonal fluctuations. Site selection should be based on the availability of water at or near the optimum temperature for the species to be reared over as long a growing season as possible.

Aquaculture animals, especially warmwater species, should reach market size from the egg in a single growing season, or at most, two. Channel catfish production in most of the South is based on the rearing of fingerlings to 10 or 15 cm during the first growing season, overwintering those fish in ponds at a low or zero feeding rate (Chapter 5), and rearing them to market size during the second growing season. Some of my experience in Texas indicates that during certain years it may be possible to produce market sized fish from the egg in a single growing season. Further research in catfish nutrition and genetics, coupled with an aggressive management program, may lead to routine single-season channel catfish production in certain parts of the United States.

Deep lakes and the open ocean provide a relatively wide range of temperatures from surface to bottom, especially in temperate and tropical environments where stratification of such water bodies is likely to occur (Ruttner, 1953; Sverdrup et al., 1942). Shallow ponds and estuaries, the water bodies most commonly selected by aquaculturists, tend to cool more rapidly in the fall and warm more quickly in the spring than do large, deep water bodies. Such seasonal temperature changes generally affect the whole water column (although a thermocline is sometimes evident in aquaculture ponds). Ponds and estuaries tend to become colder in the winter and warmer in the summer than large water bodies at the same latitude, and for many aquaculture species the optimum temperature for growth may occur for only a short period. Depending on the source of water and the location of the facility (indoors as opposed to outside), semiclosed and open culture systems may or may not be less subject to water temperature fluctuations than ponds. In most cases seasonal water temperature fluctuations affect all culture systems similarly, except for indoor closed recirculating systems, where new water of a temperature very different from that being maintained is added only infrequently and in small amounts.

For most species the maximum temperature reached in the aquaculture facility during the year is not the same as the optimum water temperature for the culture organisms, although in general water slightly warmer than optimum provides for better growth and food conversion than that which is too cold. Metabolic rate slows as the temperature cools below optimum for a particular species and increases as the temperature rises above the optimum level. Maximum growth and optimum food conversion efficiency may occur at the same environmental temperature, but this does not always happen, and the culturist should attempt to operate closed culture systems at a compromise temperature that allows for both rapid growth and efficient food conversion. Above the optimum temperature the metabolic rate continues to increase and energy begins to be diverted from growth to maintenance of the higher metabolic rate. Finally, as temperature approaches the thermal death point, metabolism often slows, and feeding activity declines and finally ceases.

Shallow aquaculture systems tend to change in temperature more rapidly than larger, deeper water bodies; compared with terrestrial conditions, however, water temperature changes relatively slowly because of its high heat capacity. Even in the fall and winter, the characteristics of water allow it to heat and cool not only slowly, but fairly evenly.

Thermal Shock and Tempering

Temperature shock, the exposure of animals to rapid changes in temperature, does not occur in a well-filled aquaculture chamber unless a copious amount of new water at a vastly different temperature from ambient is added. However the need to handle and transport aquaculture animals during the various seasons of the year and to place them from one body of water into another can expose them to rapid and extreme temperature changes that may lead to stress and even mortality.

When aquatic animals are transported from one facility to another, the water temperature at the point of origin is often somewhat different from that in which the organisms are to be stocked. In addition, some change in water temperature can be expected while the animals are in transit. In some instances transportation water is cooled with ice to lower the metabolic rate of the animals and ensure against overheating of the water during transit.

When aquatic organisms are received at a facility, and before they are stocked, the temperature in the transportation tank (Chapter 8) and that in the receiving water should be compared. If they differ by more than about 2 C, the animals should be tempered from one temperature to the other. For large numbers of organisms or small numbers of large individuals, pond water can be added slowly to the water in the transportation tank until the two temperatures become equalized.

Alternatively, if fry, larval, or small juvenile animals are involved, the organisms can be put into plastic bags, garbage cans, or other suitable containers that will conduct heat, before being placed in the receiving pond. The water in the bags should come to equilibrium fairly rapidly. This technique should not be used if a drastic difference in temperature (e.g., 10 C or more) is found in the two types of water. In general, it is probably not a good idea to temper aquaculture animals at a rate faster than about 5 C per hour, although the actual tolerance for rapid temperature change varies with species.

In some instances culture animals are unavoidably subjected to thermal shock, as for example, when a pond is drained rapidly and the organisms are exposed to small volumes of water on a hot day. The water remaining in a nearly empty pond warms rapidly and may reach lethal temperature in a relatively short time. Furthermore, high temperature reduces the oxygen-carrying capacity of the water. That, coupled with the higher metabolic rate of the aquaculture animals, often leads to death because of low DO. Solving this problem is usually accomplished by the addition of large volumes of new, cool water. The drain may be allowed to remain open so that the volume of water in the pond does not change, but the addition of new water produces cooling and an increase in DO as compared with the initial conditions.

If rapid changes in temperature cannot be avoided, it is usually better to change temperature in the direction of the thermal optimum for the species than to change it rapidly away from the thermal optimum. In the case of fish exposed to very warm water, the addition of cooler water usually is beneficial, since the fish were being exposed to temperatures in excess of their thermal optimum. This type of situation commonly occurs when fish are exposed to the thermal effluents of power plants during periods when ambient water temperature is high, or as previously discussed, when ponds are drained during warm periods of the year.

Aquaculture animals can generally be handled without damage or severe stress in cool weather, but great care should be taken when handling them in the summer. Handling often leads to injury, and the higher metabolic rate of aquatic animals in the summer may make them more active when caught, leading to an increase in the incidence of self-inflicted injury. Also bacterial activity is often higher in warm water than in cold; thus a wound is more likely to become infected during the summer than in the winter. Animals that have been exposed to water temperatures above their thermal optimum may be more subject to parasite and bacterial outbreaks than those that have not, especially when exposure to unusually high temperatures is coupled with handling.

When it becomes necessary to handle animals during hot weather, they should be caught early in the morning after the water has reached its coolest temperature for the day. Care should be taken to ensure that a high level of DO is maintained at all times. The animals should be handled gently and returned to the water as quickly as possible. If the animals are to be weighed, the operation should be accomplished in tared, water-filled containers; to avoid damage, animals should not be packed into the containers during the weighing process.

Overwintering Tropical Species

Two exotic organisms that are gaining popularity among researchers and commercial aquaculturists in the United States, although they cannot be overwintered outdoors in most regions of the country, are *Macrobrachium rosenbergii* and *Tilapia* species. *M. rosenbergii* cannot tolerate temperatures much below 20 C and are produced in ponds only when the temperature ranges from about 22 to 32 C (Bardach *et al.*, 1972). *Tilapia* die when the water temperature falls below 10 to 12 C (Chimits, 1957; McBay, 1961; Avault and Shell, 1968). Intolerance to low temperature can be considered a disadvantage in commercial aquaculture. However it is likely that this drawback can be overcome with only minor difficulty, since when maintained indoors, fairly low numbers of these organisms will provide sufficient offspring the following year to stock a large facility, and the young animals can be reared to market size in a single growing season.

The expense of constructing overwintering facilities is often not great if the fecundity of the species to be maintained is high, the animals are able to accept a degree of crowding, and good survival can be obtained given the available water quality. An open culture system would offer the best possible conditions for overwintering aquaculture animals in relatively small numbers, but a dependable supply of sufficient water volume at the proper temperature is necessary. Since such water is rarely available, most overwintering must be accomplished in heated buildings that feature closed recirculating water systems. Both *M. rosenbergii* and *Tilapia* species are readily adaptable to such systems.

Overwintered animals are often maintained at a low or maintenance feeding level; that is, they are fed enough to keep them alive with loss of neither weight nor growth. Therefore the amount of waste material that accumulates in the system is relatively low, and the demands on water quality are not as great as in a production closed recirculating water system. Temperature can be kept well below optimum for the animals as long as there is no danger of death or greatly reduced resistance to disease.

Small overwintering systems can be constructed quickly with readily available materials. More elaborate permanent systems should be designed for commercial operators, but the principles are the same. We have utilized 120 l plastic garbage cans as biofilters and galvanized stock watering tanks or fiberglass tanks as culture chambers for overwintering *Tilapia aurea*, *T. mossambica*, and *T. nilotica* in a building warmed with natural gas space heaters (Figures 4.1 and 4.2). Ambient air temperature was maintained in

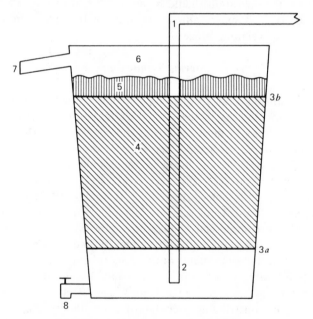

Figure 4.1 Schematic representation of a 120 l plastic garbage container converted into a biofilter. Water flows through PVC pipe (1) from the pump into the settling area (2). The water then flows through a partition into which numerous small holes have been drilled (3a) and enters a chamber filled with filter medium (4), in this case, pieces of scrap PVC. A second partition (3b) separates the filter medium from a layer of oyster shell (5) used to buffer the water. The water finally flows into the uppermost chamber (6) and overflows back through the exit pipe (7) into the culture tank. The filter is fitted with a drain (8) through which settled solid material can be removed and through which the filter can be emptied.

Figure 4.2 Indoor overwintering facility (dimensions variable). The culture tank is fitted with a submersible pump (1), which pushes water into two biofiltration units (2). Water leaves the biofilters (see Figure 4.1) and flows by gravity through PVC pipes (3) back into the culture tank.

excess of 15 C at all times. No water heating was utilized, only heat transfer from the air to the water within the system. Water in the system can be circulated with a small submersible pump of the type utilized in basement sumps. The overall cost of such a system capable of holding up to a few hundred *Tilapia* and somewhat fewer *Macrobrachium rosenbergii* should be no more than $100, plus the cost of the culture chamber. Supplemental aeration should be provided in case of pump failure.

SALINITY

Salinity has been defined as the total amount of solid material in grams contained in one kilogram of seawater when all the carbonate has been converted to oxide, the bromine and iodine replaced by chlorine, and all organic matter completely oxidized (Sverdrup *et al.*, 1942). To determine salinity, the following empirical relationship has been developed:

$$\text{salinity} = 0.03 + 1.805 \times \text{chlorinity} \qquad (15)$$

Chlorinity is defined as the total amount of chlorine, bromine, and iodine in grams, contained in one kilogram of seawater, assuming that the bromine and iodine have been replaced by chlorine. To make the determination, chlorinity is titrated using silver nitrate (resulting in the precipitation of silver chloride). The titration is standardized with seawater of known chlorinity (standard seawater).

Salinity Measurement and Natural Levels

Modern techniques for salinity measurement include use of conductivity, density, and the refractive index of seawater. Of these, the least expensive method involves the measurement of density with a hydrometer. Density can be rapidly converted to salinity by employing a conversion table.

Refractometry is the simplest and most rapid means of obtaining the salinity of a water sample, although refractometers are somewhat expensive when compared to hydrometers. A typical refractometer requires only a drop of water for use in the test, whereas most other techniques call for much greater volumes of water. Once the drop has been placed on the refractometer (which may look something like a pocket telescope), the determination can be made in seconds. Some refractometers are available with salinity scales for direct reading. Others require conversion to salinity from density or some other unit of measurement. Again, appropriate conversion tables are available.

Conductivity is used to measure salinity with high precision. Most devices are expensive and require a fairly large sample. They are useful more often to the researcher than to the aquaculturist. As was the case with DO, the aquaculturist is interested in learning the approximate salinity of the water so that this variable can be maintained within a given range. There is often no great need for a high degree of accuracy.

Salinity is reported in parts per thousand ($^o/_{oo}$, ppt, or "per mille"). The refractometer is generally accurate to within $\pm 0.5^o/_{oo}$, whereas other methods determine salinity with higher precision. The salinity of fresh water is usually less than $0.5^o/_{oo}$ (Table 4.2), and water begins to taste salty at a

Table 4.2   Classification of Aquatic Environments on the Basis of Salinity Ranges (After Hedgpeth, 1957)[a]

Classification of Environment	Salinity Range ($^o/_{oo}$)
Freshwater	Less than 0.5
Oligohaline	0.5–3.0
Mesohaline	3.0–16.5
Polyhaline	16.5–30.0
Marine	Greater than 30.0

[a] Waters in excess of normal seawater salinity (e.g., greater than about 40$^o/_{oo}$) are called brines or hypersaline waters.

salinity of about 2$^o/_{oo}$. The salinity of seawater varies depending on distance from shore, depth, and other factors. The open ocean averages about 35$^o/_{oo}$ salinity (Pearse and Gunter, 1957), although there is some variability. Salinity is not generally a consideration in freshwater aquaculture but is of great importance in mariculture.

Salinity and Its Control in Mariculture Systems

Most mariculture is practiced in oligohaline and mesohaline estuarine waters. An estuary has been defined as a semienclosed coastal water body with unrestricted access to the open ocean, where salt water is diluted to some extent by freshwater runoff (Pritchard, 1967). This definition applies most precisely to positive estuaries (those in which freshwater inputs exceed evaporation), although other types do exist: negative estuaries where evaporation exceeds freshwater input, and neutral estuaries where evaporation is equal to freshwater input. The Laguna Madre of Texas is one of the best examples of a negative estuary in the United States. This body of water is often hypersaline, but it may have potential for use in mariculture in the future. The salinity of the Laguna Madre is often in excess of 40°/oo, and readings of 50 to 80°/oo are not uncommon (Pearse and Gunter, 1957). In most cases neutral estuaries are transient, since a long-term balance of evaporation and freshwater input is uncommon in natural systems.

In some cases estuarine salinities are relatively constant with time, but more often, diurnal, semidiurnal, seasonal, and random fluctuations occur because of tides, seasonal upstream runoff patterns, and local rainfall or drought, leading to high rates of dilution and concentration, respectively. Estuarine organisms, including many mariculture species, are often euryhaline, although optimum rates of growth and survival as well as optimum food conversion efficiency may occur within a limited range of salinity for any given species. The salinity ranges to be utilized in mariculture vary considerably depending on the species being reared.

Although it is probable that estuarine species have specific salinities to which they are best adapted, most animals are exposed daily to salinity fluctuations for the reasons just mentioned. Under most aquaculture strategies no attempt is made to maintain a constant salinity in culture chambers, although this can be achieved in closed recirculating systems if it is desirable. Most aquaculture systems are allowed to remain at the salinity of the incoming water. If that salinity is too high, it is often possible to dilute the saline water in a culture chamber with fresh water from a well or other source; such a supply is not always available however. If the salinity of the incoming water is too low, it can be increased with the use of artificial sea salts (available commercially from several sources), although this can be expensive, especially if a significant change in salinity is desired. Increasing salinity in a closed recirculating system at the time of filling would be economically more feasible than the use of artificial sea salts in open intensive, or extensive culture systems.

A common problem, especially in closed recirculating water systems and in ponds, is the tendency of the system to constantly increase in salinity as a result of evaporation. When new water is added, the total amount of salt is increased and continues to increase each time water is added, as long as that water is salty. The only way to maintain salinity in a system that is becoming saltier because of evaporative losses of water is to add fresh water when replacing the loss. Some increase in salinity with time can occur in semiclosed water systems, especially if the turnover rate is slow, but closed recirculating systems and ponds are most likely to demonstrate problems in this regard.

In general, saltwater wells tend to produce water of more stable salinity than do surface saltwater sources. If the culture facility is located in an area that receives little surface runoff and/or has a source of deep water from which to draw (e.g., along the coastlines of tropical islands in some cases), very little fluctuation in salinity may be observed over extended periods.

Osmoregulation in Aquatic Animals

The blood of vertebrates often contains salts at concentrations quite different from the surrounding medium. For instance, the blood of freshwater vertebrates is hypertonic to the water, whereas that of marine vertebrates is hypotonic. The marine elasmobranchs (sharks, skates, and rays) are the only vertebrates other than hagfishes that have blood isotonic or nearly isotonic with the surrounding water. The salt content of elasmobranch blood is not much different from that of bony fishes; however sharks, skates, and rays often have high levels of blood urea, which increases the osmotic pressure within the body to that of seawater (Hickman and Trump, 1969).

To establish a constant internal environment with respect to the blood, a fish must osmoregulate; that is, it must be able to maintain its body fluids at the same salt concentration at all times. Osmoregulation is a physiological function that requires a significant amount of energy. If euryhaline fish can be maintained in culture at salinities similar to the ionic strength of their blood, more energy may be utilized for growth, and less for osmoregulation, than if the external salt concentration is much higher or lower than that of the blood. Despite mitigating factors that rule out that approach for some species, it is a reasonable first step when determining the proper salinity for aquaculture. This rule of thumb is not valid with elasmobranches for example, since these animals respond to different salinities by altering the level of urea in the blood to compensate for the new salinity. Different amounts of energy are required to maintain blood urea levels at different salinities, and an optimum salinity for any particular species of elasmobranch probably exists. However it may be best to rear such animals at relatively low salinities, since high levels of urea associated with high salinity water may lead to off-flavors in the flesh of these organisms. At present there is little interest in the aquaculture of elasmobranchs, but certain species are consumed throughout the world and may be adaptable to aquaculture in the future.

Crustacean blood, and that of other invertebrates, is isotonic with sea-water, or very nearly isotonic. Osmoregulation is required, however, since the ionic composition of the blood (as is also the case in vertebrates) is distinct in some ways from that of seawater (Lockwood, 1967). Thus invertebrate aquaculture organisms may also have a relatively narrow range of preferred salinity, and the optimum should be determined for each species.

Osmoregulation is controlled by selective absorption of ions through the gills, and in some cases the selective removal of salts by way of the same organs. In marine teleosts, the kidney is responsible for excretion of magnesium and sulfate. Other ions may appear in the urine, but these are not exclusively handled by the kidneys (Hickman and Trump, 1969).

Oysters (*Crassostrea virginica*) can adjust to salinities as low as $3^o/_{oo}$ and at least as high as $35^o/_{oo}$ (Pearse and Gunter, 1957). In low salinity water the growth and flavor of oysters may be adversely affected. Other marine organisms are able to tolerate fresh water—for example, redfish, mullet, and flounder. Channel catfish, normally grown in fresh water, can tolerate up to $14^o/_{oo}$ salinity and have been reared in brackish water ponds (Perry, 1969; Perry and Avault, 1968, 1969, 1971). The salinity tolerance of channel, blue, and white catfish, as well as hybrids among these species, appears to be about the same (Perry, 1967; Perry and Avault, 1969; Allen and Avault, 1969; Stickney and Simco, 1971).

Species that undergo annual migrations on and offshore may require different salinities at different times of the year. A good example is the southern flounder, *Paralichthys lethostigma*, which appears to be physiologically adapted to different salinities at different ages (Stickney and White, 1973). The aquaculturist who deals with species of this kind must determine whether there is any advantage in periodically altering salinity in the rearing chambers to fit the natural salinity regimes in which the animals would be found, or in maintaining constant salinity throughout the growing season. In many cases if the salinity selected is similar to that at which the animals are found during the greatest portion of the year in nature, growth will be adequate and the cost of operating the culture system may be reduced as compared to the additional expenses incurred by frequent, and possibly large-scale salinity alterations.

Stenohaline marine and freshwater organisms have been included among aquaculture candidates around the world, and of those the most widely cultured species are ones that occur only in fresh water. Stenohaline marine animals generally inhabit high salinity waters where aquaculture is extremely difficult. As discussed in Chapter 2, open sea mariculture is not widely practiced, since fencing off portions of the ocean is difficult and uneconomical. The adaptation of oil rigs and other offshore structures offers some potential for future stenohaline animal mariculture, but the majority of all aquaculture ventures will undoubtedly remain in freshwater and estuarine areas.

LIGHT

Light quality, quantity, and photoperiod are all important to plant growth and may exert considerable influence on that of animals as well. The onset of sexual development in animals is often related to temperature but may also be influenced greatly by photoperiod, especially in climates characterized by little or no seasonal fluctuation in water temperature. As is the case with salinity, the photoperiod requirements for various stages in the life cycle of each aquaculture species should be determined experimentally so that adjustments, if required, can be made at the appropriate time. In outdoor culture systems photoperiod and light intensity adjustments are not normally included in the culture strategy, but these factors may be critical in the control of indoor culture systems.

Some species are active during the daylight, whereas others are nocturnal. For example, the white shrimp, *Penaeus setiferus,* is generally active during the daytime, and the brown shrimp, *P. aztecus*, is active at night. Activities of the culturist, such as feeding, may have to be altered to conform with the normal activity of the animals, or (at least in closed systems) the pattern of light and dark could be altered for the convenience of the culturist. For instance, culture tanks could be kept in the dark during the day and receive artificial light at night. Extreme caution should be exercised when a photoperiod phase change is attempted; to ensure that the performance of the animals does not differ from what it would be under normal photoperiod situations.

Most species respond to biological rhythms that are often influenced by diurnal changes in light quality and quantity. Although the duration of photoperiod may be critical for the developing of gametes prior to spawning, other aspects of physiology may also be affected by light. A complete phase change could adversely affect growth and food conversion efficiency, among other factors. For most species there seems to be little need to make such a shift in photoperiod even for the convenience of the culturist. Alterations in photoperiod to induce spawning are acceptable, as discussed in

Chapter 6. These normally do not involve phase changes but are used only to alter apparent day length by a few hours through the mechanism of turning lights on or off early at one or both ends of the normal period of daylight.

Photoperiod, light intensity, and light quality cannot be conveniently controlled outdoors, as previously indicated. Perhaps the only feasible modification in these parameters is the emplacement of lights near ponds to deter poachers and/or to attract insects that might be fed on by the aquaculture animals. If the primary purpose of such lights is insect attraction, careful studies should be made to determine that substantial improvements in growth and food conversion result when the lights are used. If this is not the case, the expense of buying the lights and paying for electricity to operate them may exceed the value of the benefits gained. The natural feeding behavior and food habits of the culture species will have a bearing on the probable efficacy of supplemental feeding with insects attracted to lights. Lights mounted directly over culture ponds would be unlikely to adversely affect growth and food conversion efficiency.

Light intensity and photoperiod are of extreme importance for the growth of aquatic plants. The effects of phytoplankton blooms in terms of shading out rooted aquatic macrophytes and filamentous algae have been described, and the effects of water turbidity on light penetration are discussed in the next section of this chapter.

Some organisms change color when exposed to different quantities or qualities of light or when placed in turbid water. Channel catfish respond to some extent by becoming darker or lighter in response to light (and temperature), whereas flounder are known for their ability to match backgrounds by changing both color and the mottling pattern on the upper side.

Specific studies into the effects of photoperiod on growth rate in aquaculture animals are somewhat limited. A few have been conducted on channel catfish. Kilambi et al. (1970) detected some response of growth to varied photoperiod in channel catfish fry, but Stickney and Andrews (1971) and Page and Andrews (1975) could not demonstrate growth differences in catfish fingerlings with respect to photoperiod. No significant difference has been reported in catfish growth in response to light intensity (Page and Andrews, 1975), although this does not rule out the possible importance of this and other aspects of light quality and quantity on the growth of other fishes and invertebrate culture animals.

Many have noted that aquatic animals become agitated when suddenly exposed to bright light. When aquatic organisms are cultured indoors they may be exposed frequently to instantaneous and drastic changes in light quantity when artificial lights are turned on during the night, or at any time in windowless buildings. The possible effects of rapid changes in light on survival, growth, food conversion efficiency, reproduction, and resistance to disease have not been adequately researched for any species. Some species appear to adapt well to rapid changes in light quantity, but others may not, even after a long period of conditioning.

In nature, light intensity increases slowly as the sun rises, and decreases slowly as the sun sets. Many aquaculturists believe that windowless indoor culture facilities should be equipped with rheostatically controlled lights that automatically come on and go off at a slow rate at preset times of the day and that efforts should be made to avoid turning on the lights during the normal night cycle (except in emergencies).

SUSPENDED SOLIDS

Suspended solids are bits of particulate matter, larger than 0.45 μ found in the water column. Suspended solids are made up of sediment particles, organic material (detritus composed of plant and animal remains, waste food particles, and fecal material), as well as phytoplankton cells and other living microorganisms.

The higher the concentration of suspended solids in the water, the more turbid the water becomes. If turbidity becomes too high as a result of the presence of suspended inorganic solids, primary productivity may be reduced because of shading. This can be an advantage in reducing the growth of filamentous algae and aquatic macrophytes, but it can be a disadvantage if a phytoplankton bloom is required to provide food for young organisms or for such herbivores as *Tilapia* species and bivalve mollusks (e.g., oysters, clams, and mussels).

Sediment particles, composed largely of silt and clay (known to laymen as mud), may become suspended in the water column as a result of currents or wind mixing, or they may be washed into the water from the land during periods of runoff. Suspended solids in large ponds, bays, rivers, and other areas where aquaculture is practiced may vary considerably from time to time as a function of weather conditions, especially with respect to precipitation in the watershed.

Inorganic particles can have detrimental effects on aquatic organisms (Cairns, 1967). The mechanical action of such particles can lead to clogging of the gills or the irritation of gill filaments and other membranes. If a great deal of suspended particulate matter is introduced into an aquaculture pond (e.g., as a result of heavy rains), subsequent settling of the material may lead to the burial of eggs, larvae, or fry of benthic organisms or of certain species or life stages of the culture organisms themselves. For example, salmonids lay their eggs in nests, called redds, in the gravel of streams, and culturists sometimes simulate natural conditions for the hatching of salmon eggs. The incubating eggs generally lie below the surface layer of gravel and depend on a flow of water through the sediments to provide them with oxygen and to remove metabolites.

Heavy rains and the resultant runoff of large amounts of fine sedimentary material can lead to clogging of gravel in the streambed, resulting in the restriction of flow. High mortalities can occur in the developing eggs. This type of situation is most common after forest fires or certain human activities that destroy the natural vegetation adjacent to streams and expose the soils to erosion. More pertinent to aquaculture would be the silting in of centrarchid, ictalurid, or cichlid nests during egg incubation. Goldfish lay adhesive eggs (mats of Spanish moss or similar substances are often provided by goldfish culturists as spawning substrates) that are also sometimes subject to siltation.

Particulate matter suspended in water provides a vast amount of surface area for the growth of fungi and bacteria (Cairns, 1967) and could increase the potential for disease in aquaculture systems. Suspended particles also absorb and adsorb various chemicals, such as phosphates. Thus fertilization may be less effective in turbid water, not only because of shading, but also because the nutrients may not be free to be incorporated into plant tissues.

Turbid water is less subject to temperature alteration over short time intervals, heats more rapidly, and holds heat better than clear water. Whether these characteristics can extend the growing season significantly in turbid ponds is somewhat doubtful. Aquaculturists generally exercise little control over suspended solids in incoming water, although it sometimes happens that the water source is extremely turbid and action to reduce the level of suspended solids must be taken. The simplest solution is to run the incoming water through a settling reservoir before introducing it into the aquaculture facility. If this is not done, damage to the culture animals may occur in some instances.

In addition, the introduction of high levels of suspended material directly into culture ponds can lead to rapid filling in of those ponds. In such cases it may be necessary to rework the ponds at periods of a few to several years and remove the accumulated sediments. This can be an expensive proposition and should be avoided when possible. If high levels of suspended solids are introduced into intensive culture facilities, plumbing can clog and observation of the animals in the culture tanks becomes difficult. Futhermore, the sediments in the tanks begin to build up rapidly, and a great deal of time may be required for cleaning.

One of the first laws of pond aquaculture is that no two ponds are alike.

Ponds that are identically constructed, receive the same amount of water from the same source in an identical manner, and are stocked at the same time with the same number of fish or invertebrates of the same species, often demonstrate radically different levels of primary and secondary production, as well as water quality. This is especially true with respect to suspended solids levels. One pond of a pair may have extremely clear water while the other appears turbid. This distinction may last for several months, or it may be altered after a short time. The following year both the ponds may be turbid, both clear; or the situation may be as it was the previous year (with either pond being the muddy one). It is almost impossible to predict the situation. An aerial view of nearly any aquaculture pond facility will clearly demonstrate the variability that can occur.

In most cases some turbidity is not objectionable, since extremely high levels of suspended solids are tolerated by most aquaculture animals. However if a phytoplankton bloom is required and the pond has a muddy appearance, it may be necessary to attempt to settle the suspended particulate material. There are two simple ways in which this can be accomplished. The first involves spreading chopped hay over the pond surface. As the hay settles, it will cause the clay particles to flocculate out of suspension. The second method involves the application of about 200 to 900 kg/ha of gypsum (CaSO$_4$). The application may be repeated at 7 to 10 day intervals, as required, until the desired effect is achieved (Lee, 1973).

Some animals induce a certain amount of turbidity into otherwise clear water as a result of their feeding or reproductive habits. Examples of fishes that tend to increase water turbidity are common carp (*Cyprinus carpio*), buffalo fish (*Ictiobus* spp.), and tilapia (*Tilapia* spp.). The levels of suspended solids induced by these types of animal are generally not detrimental to other organisms directly, although they may affect primary productivity through shading.

Oysters (*Crassostrea virginica*) feed most efficiently when the ratio of food to water volume is relatively low (Loosanoff and Tommers, 1948). In turbid water the pumping rate of oysters is greatly reduced, thus growth rates can be affected. Many species of fish can survive extremely high levels of suspended solids for at least short periods. Wallen (1951) examined the effects of turbidity on 16 species of fish, including channel catfish (*Ictalurus punctatus*) and common carp (*Cyprinus carpio*) and observed no adverse effects until turbidity was in excess of 20,000 mg/l. Most species were able to withstand turbidities of 100,000 mg/l for a week or more. Thus aquaculturists do not normally have to concern themselves about direct mortality as a result of turbid water, since most aquacultural waters rarely have suspended solids loads in excess of a few hundred milligrams per liter.

ALKALINITY

The capacity of a natural water system to resist changes in pH can be measured in terms of the amount of bicarbonate and carbonate ions that are available in the system. This measurement is called alkalinity. Chapter 3 dealt with the effects of the carbonate buffer system on pH and the importance of photosynthesis and respiration in the availability of carbonate and bicarbonate.

Alkalinity is measured by titrating water samples with dilute sulfuric acid to end points indicated by phenolphthalein and methyl orange. Bicarbonate alkalinity is derived from the difference between carbonate (phenolphthalein) and total (methyl orange) alkalinity values (APHA, 1975).

In aquaculture systems the alkalinity should generally be between 30 and 200 mg/l in fresh water, although water of higher and lower alkalinity has often been utilized successfully by culturists. Water of very low alkalinity has little capacity to resist pH changes and should be avoided under most circumstances. Low alkalinity water may be suitable for intensive culture because a buffering agent is routinely incorporated into closed culture systems and could be added to open or semiclosed systems if necessary. The soils in ponds exchange ions with the water and can have a fairly profound effect on alkalinity until carbonates are depleted. It is possible to add calcium carbonate to ponds to increase alkalinity (Arce and Boyd, 1975).

When alkalinity is extremely high, carbonates may precipitate on surfaces in the culture system (e.g., on the walls of plumbing and culture tanks). This is especially likely to happen when sufficient levels of calcium or magnesium ions are present in the water to react with the carbonate. The alkalinity of seawater is always fairly high because of the abundance of carbonates in marine sediments and dissolved in the water. However it is generally not so high that precipitation of calcium carbonate occurs in culture chambers.

The pH of water affects the percentage of alkalinity contributed by carbonic acid, bicarbonates, and carbonates. Temperature and salinity also affect these relationships. For example, in seawater of pH 8.0 and 24 C temperature, slightly more than 8% of the alkalinity is in the form of carbonates, whereas in fresh water at the same pH and temperature, less than 0.5% of the alkalinity is represented by carbonate ions (Spotte, 1970). In general, most of the alkalinity of fresh water and low salinity estuarine waters may be attributed to bicarbonates.

HARDNESS

The concentration of divalent cations (primarily calcium and magnesium) present in water determines its hardness, which is expressed in terms of milligrams per liter of calcium carbonate (APHA, 1975). Although high alkalinity and high hardness often occur simultaneously, the two are actually independent, especially in fresh water. Salt water is generally hard, but very soft fresh waters do occur, and these may have very high alkalinities. One aquaculture system I used has a well with approximately 10 mg/l total hardness and about 900 mg/l total alkalinity.

Low hardness can adversely affect some aquaculture animals. For example, when euryhaline fishes such as red drum (*Sciaenops ocellata*) are maintained in very soft fresh water, survival is poor (probably because of the animals' inability to osmoregulate efficiently in a solution almost devoid of divalent cations). For most freshwater species a hardness of somewhere between about 20 and 150 mg/l is recommended. Hardness is not a consideration in mariculture systems.

In ponds, the hardness of the incoming water may be increased by the leaching of calcium and magnesium from the sediments, although this should not be relied on as an adequate source of these elements. If hardness must be increased, lime (CaO), can be added. Slaked lime $[Ca(OH)_2]$ should not be used because it will make the water basic, overcoming the buffer system in many cases. Calcium carbonate can be used but would not be effective in waters of high pH because the limestone would not dissolve to any extent. No attempts should be made to increase hardness after the addition of phosphate fertilizers, since the calcium released will combine with the phosphate as insoluable $Ca(PO_4)_2$, making the fertilizer treatment ineffective. Gypsum can also be used to increase water hardness without affecting pH.

The amounts of chemicals to be added to a particular body of water vary considerably with the initial water quality and desired final hardness. The culturist may begin by adding as much as 1000 kg/ha (200 to 500 kg/ha might be preferred). Hardness should be checked after a few days to determine the effectiveness of the treatment. Thereafter, additional treatments at the same, higher, or lower rates may be utilized until the desired hardness is obtained. Water hardness can be determined with relative ease through titration (APHA, 1975).

Arce and Boyd (1975) examined the effects of adding limestone to soft water ponds in Alabama at a rate of 4300 to 4900 kg/ha, after which the ponds were fertilized at 2 week intervals for 13 applications with 45 kg/ha of 20-20-0 fertilizer. Total hardness (initially less than 10 mg/l) was increased by fourfold, as was total alkalinity (initially less than 15 mg/l). Phytoplankton productivity was increased, as was the production of the herbivorous fish, *Tilapia aurea*, in ponds receiving calcium carbonate treatment.

Lime is sometimes used to sterilize the bottom of dry ponds before filling and stocking. A layer of lime is spread over the pond bottom and worked into the sediments with a disc. This technique is effective in killing undesirable benthic animals and may help in the control of pathogens.

SUBSTRATE REQUIREMENTS

Many aquaculture animals adapt as well to fiberglass, metal, or wooden culture tanks, cages, and concrete raceways as they do to ponds, but some species respond with better growth or even improved product quality when provided with a special type of substrate. Outdoor ponds usually have bottoms composed of sand, shell-hash, mud (clay and silt), or mixtures of these and other natural materials. Species that normally burrow in the sediments during all or part of the day may require such materials for rapid growth and survival.

Certain species require a hiding place during particular stages of the life cycle. For example, among both penaeid and palaemonid shrimps of aquaculture interest (*Penaeus* spp. and *Macrobrachium* spp., respectively), cannibalism is common during molting and thereafter until the new carapace has become hard. If the animals have a place in which to hide during molting, their vulnerability is significantly reduced. Taken a step further, the cannibalism that occurs among blue crabs (*Callinectes sapidus*) and lobsters (*Homarus americanus*) during ecdysis is so severe that individuals must be reared in separate culture chambers or in chambers that are partitioned in a manner that prohibits contact between individuals.

Hiding places for shrimps may be made of a variety of materials that are common around the average aquaculture facility or are readily available on the market. Pieces of PVC pipe cut into short lengths, concrete blocks, and other similar objects have been effectively utilized. The provision of hiding places does not eliminate cannibalism, but the phenomenon can be greatly reduced.

Oysters require firm substrates on which to attach or on which they can be supported. Sediments containing high levels of silt and clay may provide poor support, leading to the destruction of these animals from burial and asphyxiation. If not fatal, such environments may lead to impaired growth due to the clogging of gills or decreased rates of pumping. Oysters and clams can be reared in chambers with solid bottoms, in baskets that contain no sedimentary material, on strings, or on firm substrates in natural or pond environments. None of these methods of culture should damage the animals in any way, although each has its advantages and disadvantages in terms of suitability in a particular culture facility or available culture environment.

Flounders have been shown to develop dark pigmentation on the underside if cultured in fiberglass tanks without some type of loose substrate, such as sand or shell-hash (Stickney and White, 1975). This anomaly is depicted in Figure 4.3. Ambicoloration does not affect the flavor or texture of the flesh, but the consumer would probably reject a fish that had pigment on both sides (flounder are often marketed headed and gutted but with the skin and scales intact).

In the culture of benthic animals such as shrimp and flounder, it may be useful to provide more substrate area than would naturally be available in a tank or pond. This can be accomplished by constructing horizontal or even vertical partitions, or hanging netting vertically in the water. Some species utilize the additional substrate, thus increasing the effective surface area available in the culture facility. This can mean a significant increase in animal density per unit volume of water, as long as water quality is maintained.

The commercial culture of *Macrobrachium rosenbergii* has been accomplished in Hawaii in ponds in which such net partitions are hung. The

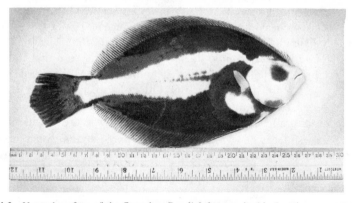

Figure 4.3 Ventral surface of the flounder, *Paralichthys* sp. Ambicoloration can occur when flounders are reared on bare tank bottoms.

vertical partitions are removed during harvesting (Robert W. Brick, personal communication). Flounders have been known to associate with any type of unbroken surface, whether vertical, horizontal, or at an angle. Although netting would not provide a suitable vertical substrate for flounders, it might be useful if placed in the horizontal position off the bottom. Flounders appear to attach to vertical surfaces by forming their bodies into suction cups. They have been observed to hang on the walls of fiberglass tanks for hours without apparent difficulty (Figure 4.4).

Figure 4.4 Flounders (*Paralichthys lethostigma* and *P. dentatus*) spend most of their lives in contact with or in close proximity to substrates. They have been observed to hang motionless for periods of several hours on the sides of tanks like these.

ANIMAL DENSITY

Different species respond in a variety of ways to crowding. Some animals become aggressive at low densities, whereas others may exhibit aggression or cannibalism at high densities. The proper culture density must be determined for each species in terms of numbers of animals per unit area or volume of water. Water quality must, of course, be maintained no matter what density is utilized.

As animal density increases, competition for food may become severe and territorial behavior often breaks down. Channel catfish, trout, and certain other species appear to coexist well under crowded conditions, but some species do not. If highly aggressive or cannibalistic behavior results from crowding (and this may occur even in relatively low density aquaculture), it may be necessary to rear individuals in separate culture chambers. The practice of raising crabs and lobsters individually was outlined earlier. This procedure, though effective, may add greatly to the cost of aquaculture because of the elaborate facilities required to install a separate container for each animal. Usually large culture areas (tanks, ponds, and areas in estuaries or bays) are partitioned off into compartments through which water flow is not impeded and in which individuals can be maintained. Wire or nylon netting may be useful for such partitions, but care must be exercised to ensure that the material selected is not subject to rapid deterioration as a result of corrosion or the attachment of fouling organisms. Blue crabs (*Callinectes sapidus*) have been known to destroy nylon netting within a few hours. The animals may become entangled in the net and will tear large holes in it while trying to escape. If individual compartments are not maintained intact, the animals may escape or eat each other when common walls deteriorate.

Since aquaculture animals housed alone in small containers are not able to move around to any extent, the culturist must feed each individual in the culture system rather than allowing the animals to congregate at feeding times, as is often the practice with gregarious species. Thus more time may be required for feeding and certain other activities associated with culture than would otherwise be necessary. Harvesting of individual culture containers may be easier than the harvesting of some other types of culture system because each individual is confined in a known, small area and can be captured relatively easily.

For tolerant species, culture chamber densities can often be raised to extremely high levels. Reports of more than 10^6 kg/ha have been claimed for flowing water systems. I have observed tanks containing fish at such high densities that the animals were able to maintain their relative positions in the water but could swim hardly at all laterally or vertically because of the proximity of their neighbors. Such conditions in and of themselves can place considerable stress on certain species, although others may be able to adapt unless water quality deteriorates from optimum. Even in open systems with rapid turnover rates, some deterioration of water quality can be expected as the water flows from one end of a raceway to the other or from the inflow point to the center drain of a tank. The result of water curtailment in such systems, even for a few minutes, may be manifested in high levels of mortality.

Ponds can be overstocked to the point that natural replacement of dissolved oxygen becomes insufficient. Supplemental aeration can be supplied in ponds, but the chance of a mechanical failure having a disastrous effect on the culture animals, as well as the cost, must be considered before an elaborate aeration plan is implemented in a pond system.

It is also important to realize that crowded conditions often lead to rapid transmission of diseases. Bacterial infections and parasitic epizootics can occur virtually overnight after stress has occurred in overcrowded environments. Many disease treatments require short baths in chemicals of fairly high concentration or exposure of the animals to lower concentrations for long periods. In running water systems the flow of water must be stopped if bath treatments are to be effective and affordable; yet shutting down the water input can lead to increased stress because of water quality deterioration. Treatment in running water can be accomplished but requires much more chemical and can be prohibitively expensive. Treatment of diseases in closed recirculating systems can lead to decimation of the bacterial flora in the biofilter, with consequent degradation of water quality. Finally, the use of chemicals in ponds may require large amounts of expensive treatment agents. Proper water system management involves the maintenance of high density, low stress conditions, but such a balance is always tenuous and can easily be upset.

It is assumed by many aquarium enthusiasts that the dimensions of the culture chamber control the ultimate size to which the fish in that chamber will grow. For example, many aquarium species that never get heavier than a few grams in aquaria may grow to several kilograms in nature. The difference may not be related to the tank volume as much as to water quality. Even an aquarium fitted with the best subgravel filter may permit an accumulation of metabolites in the water that effectively limits the growth of animals.

If the same aquarium is set up to receive several turnovers a day of new water, or if an external biological filter of suitable capacity is utilized in conjunction with the system, it may be possible to rear an animal to a size that will virtually fill the tank. Thus space is generally not the limiting factor. However in any type of culture system, it generally is possible to reach such a density or biomass of animals that some aspect of water quality becomes limiting to further growth. It is an unusual case indeed, when the walls of the tank become the determining factor, although this is possible.

PESTICIDES, HERBICIDES, AND TRACE METALS

The chronic and acute toxicity of various pesticides, herbicides, and trace metals on aquatic animals have been determined for a variety of species, including several of aquaculture interest. As discussed in Chapter 2, the water utilized for aquaculture should be essentially devoid of herbicides and pesticides, and it should not contain high levels of such trace metals as cadmium, zinc, copper, silver, lead, and mercury. The U.S. Environmental Protection Agency (EPA) has established a maximum allowable concentration of 0.5 mg/l of mercury in the flesh of fish tissues. This value has posed problems for commercial swordfish and tuna interests, but there have been no reports of excessively high levels of mercury in aquaculture animals.

Trace metal contamination is most likely to occur when an aquifer is contaminated by recharge water into which large amounts of metallic substances are introduced (e.g., from an industrial plant or even a municipal waste treatment plant). The analysis of water samples for trace metals by atomic absorption spectrophometry is somewhat expensive (up to several dollars per element); but such an investigation could save the prospective culturist considerable expense (if there is any reason to believe that the proposed water source might be contaminated).

Pesticides such as DDT are known to undergo a phenomenon called biological magnification, in which the pesticide enters the food web at a low trophic level and at a low concentration (e.g., by way of absorption from the water by plants) and increases in concentration as it passes through each successive trophic level. Trace metals sometimes exhibit the same phenomenon, but often they do not. Copper, for example, is found in much higher concentration in crustaceans than in vertebrates because the blood pigment of those invertebrates is based on the copper-containing compound hemocyanin. The copper requirement for vertebrates is not the same as that of the crustaceans because the former have blood pigment containing iron in the hemoglobin molecule. Even so, a fish that feeds primarily on crustaceans should be expected to have a higher level of copper in its body than a species that eats, for example, plants. This does not seem to be the case. Rather than biological magnification, there is often a diminution of copper levels as one passes from the crustaceans to the teleosts (Stickney et al., 1975). The difference between the biological magnification of pesticides and that of trace metals (at least in the case of copper) probably relates to the presence or absence of a biochemical pathway to handle high levels of substances that are not required by the body.

Not all trace metals respond in the same way as copper when consumed in relatively high concentrations by fish. Mercury, lead, arsenic, and others are not required for normal metabolism and are all stored, often until they reach toxic levels if the fish are exposed to unusually high dosages. Cadmium, a metal that is toxic when present in high concentration, also has no metabolic function in animals, but its chemistry is very similar to that of zinc. Cadmium appears to compete with zinc for enzyme active sites, and as the ratio of zinc to cadmium decreases, the cadmium begins to outcompete zinc and eventually denatures the enzymes, leading to a disruption of normal metabolism. Hypertension in man appears to be related to high levels of cadmium in the kidneys (Schroeder, 1965).

Pesticides and herbicides can enter relatively shallow aquifers from surface recharge and could conceivably contaminate deep wells if the water table is able to approach the surface at any point. Most incidents of pesticide and herbicide contamination are likely to occur in conjunction with the use of surface waters in aquaculture. Locating aquaculture facilities in close proximity to cropland should be avoided, if possible; however much of the aquaculture conducted in the United States occurs in areas that are intensively farmed, both because of climate and because many aquaculturists are also involved in terrestrial agriculture. It is not unusual for aquaculturists to produce fish in ponds adjacent to fields planted in corn, rice, soybeans, cotton, or a variety of other crops that require applications of pesticides and/or herbicides. Cotton fields, for example, may be sprayed more than 10 times during a single growing season. If the wind is blowing strongly, or even gently from the wrong direction, the effect on an aquaculture crop can be devastating.

Some agriculturists, especially those interested in rearing channel catfish and crayfish, have practiced crop rotation that includes an aquatic organism. For example, certain farmers in Arkansas rotate rice, soybeans, and catfish. Since most of the herbicides and pesticides in use today are short-lived organophosphates rather than the infamous chlorinated hydrocarbons that dominated the industry a few years ago, there is little chance for residues to carry over from one crop to the next, although some regions may contain soils in which unusually high levels of chlorinated hydrocarbons still exist. Suspected areas should be tested before aquatic animals are introduced. Pesticides and herbicides can be analyzed by gas-liquid chromatography; this is expensive, but again it is better to be safe than sorry if contamination is suspected.

Fish are highly susceptible to most pesticides, but crustaceans are especially vulnerable, being more closely related to the insects the chemicals are designed to eradicate. Herbicides may kill aquaculture animals directly, or they can lead to an increase in BOD by destroying the primary producers in the system. The result may be an oxygen depletion that will also lead to high mortality unless countermeasures are taken (see Chapter 3).

Mercurial fungicides were once widely used in the United States but are no longer available because of the dangers posed to man. Many algicides still employ copper as the active ingredient. Since copper is more toxic to algae and other types of plant than it is to fish, it can be used safely, but caution should be exercised in the calculation of treatment rates. An error could lead to the addition of sufficient copper to create a toxic conditon, resulting in the destruction of the culture species. Such water quality papameters as pH, hardness, alkalinity, and temperature often have a significant effect on the amount of herbicide, pesticide, or other chemical that can be safely utilized in conjunction with an aquaculture operation. Before applying any chemical, the label should be carefully read and the directions precisely followed.

Many of the data reported in the literature with respect to the lethal concentrations of pesticides, herbicides, and trace metals on aquatic organisms are based on acute toxicity tests in which the concentration of chemical that will kill 50% of the population is determined for 24, 48, 72, or 96 hours. Such data are generally more meaningful to the ecologist investigating the effects of intermittent or individual doses of a toxicant flowing down a stream or into an estuary than to an aquaculturist who might be continually pumping minute amounts of a chemical into a culture system.

More valuable to the aquaculturist are chronic bioassays in which the animal is exposed for extended periods to low levels of a toxicant. Death does not necessarily occur under chronic bioassay conditions, but the physiology of the animals may be affected to the extent that growth or other attributes are detrimentally altered. Exposure for long periods to low levels of pesticides, herbicides, or trace metals may result in the presence of these substances in the animal at levels that make the consumption of the culture organism dangerous for human beings. In most instances where careful site selection and water system management are employed, the risks can be minimized.

DISPOSAL OF AQUACULTURAL EFFLUENT

Water must be released from aquaculture systems of all types, at least on occasion. Flow-through systems may utilize large volumes of new water daily, and the effluent from such systems is often discarded rather than being recycled through a holding reservoir. The effluent from open systems is often affected only slightly in terms of water quality as compared with its quality on entering the culture system, especially if the residence time in the system is short. At the other extreme is the effluent from closed recirculating water systems, which may be released infrequently but is usually high in nutrients and has been significantly changed from its original quality. The effluent from ponds and semiclosed intensive culture systems may lie between these extremes. In some cases the water quality of the effluent is actually better than that of the incoming water.

The quality of water that is released from aquaculture systems into public waters is controlled by the EPA and by various state agencies (depending on the state in which the culturist establishes a facility, it may be necessary to obtain permission to operate from one or several entities). Each state is free to impose its own water quality standards for aquaculture, although these may not be more liberal than those of the EPA. If the water from an aquaculture operation is released into a privately owned pond or reservoir, there may be no minimum standards set; however in some states even these waters are under the legal control of one or more state agencies. In many cases aquaculture effluents are disposed of by allowing them to flow into rivers or streams, all of which are subject to control by some state and/or federal agencies.

For several years the EPA has been considering minimum water quality standards for aquaculture effluents. Early indications were that a variety of tests would have to be run on all effluents, possibly including an array of trace metal analyses as well as tests of such parameters as BOD, nutrients, and suspended solids. In certain instances such rigid standards may still be imposed, but by the mid-1970s it appeared that because of the difficulties and expense involved in obtaining much of the information just listed, and primarily because most aquaculture facilities did not appear to be contributing significantly to the degradation of receiving waters, the proposed standards would be limited to the determination of settleable solids.

This determination can easily be made by taking a grab sample of effluent water and introducing it into a device called an Imhoff cone. The water is allowed to stand in the inverted, graduated cone for a specified time, after which the analyst measures the amount of solid material that has accumulated in the bottom of the cone. If the effluent is found to exceed a certain level of settleable solids, action must be taken to reduce the level before the water can be released from the culture system.

Further studies of aquaculture effluents appear to have provided support for the contention that most aquaculture effluents are of similar, or even higher quality than the receiving water body. It now appears that no rigid federal water quality standards for aquaculture effluent will be imposed on freshwater systems, although the situation may be somewhat different in the marine environment. Mariculture firms must obtain permits to construct facilities, and these permits must be reviewed not only by the EPA, but by the U.S. Army Corps of Engineers (concerned with effects of construction on navigation), the National Marine Fisheries Service (concerned with effects of proposed construction on commercial fisheries), and the U.S. Fish and Wildlife Service (primarily concerned with the protection of sport fish and wildlife). An objection by one or more of these agencies may be sufficient to permanently halt the establishment of a mariculture operation or to throw it into lengthy litigation. The developing pattern in Congress not only to recognize aquaculture, but to support it in various ways, implies that certain ambiguities and problems with current laws relating to aquacultural development will eventually be addressed.

State standards, where they exist, must be met in all cases. Aquaculturists should consult the appropriate agency or agencies to determine whether permits for aquaculture are required either for removing water from surface sources or for flowing effluent water into surface water receiving areas.

It is not always necessary to release water from aquaculture operations into other water bodies. One alternative that may be especially appealing if the water is reasonably high in nutrients, is the use of the effluent for irrigation of pastures or crops. The effluent from sewage treatment plants has been effectively utilized in this manner, and this method of disposal should apply equally well to aquaculture effluents, although the availability of the water may be intermittent in many cases. One solution is to provide an additional storage reservoir in which the effluent can be placed until it is required for irrigation. This solution to the effluent problem should be especially attractive to culturists who are also farmers or ranchers.

If an aquaculture facility produces an effluent that fails to meet state standards, it must be treated before disposal into public waters. In many cases a period in settling basins will be sufficient to return the water to acceptable quality. More elaborate treatment may be prohibitively expensive. Prospective aquaculturists should pay careful attention to potential problems of effluent disposal.

LITERATURE CITED

Allen, K. O., and J. W. Avault, Jr. 1969. Effects of salinity on growth and survival of channel catfish, *Ictalurus punctatus*, *Proc. Southeast. Assoc. Game Fish Comm.* **23:** 319-331.

American Public Health Association. 1975. *Standard methods*, 14th ed. American Public Health Association, Washington, D.C. 1193 p.

Andrews, J. W., and R. R. Stickney. 1972. Interactions of feeding rates and environmental temperature on growth, food conversion, and body composition of channel catfish. *Trans: Am. Fish. Soc.* **101:** 94-99.

Arce, R. G., and C. E. Boyd. 1975. Effects of agricultural limestone on water chemistry, phytoplankton productivity, and fish production in soft water ponds. *Trans. Am. Fish. Soc.* **104:** 308-312.

Avault, J. W., Jr., and E. W. Shell. 1968. Preliminary studies with the hybrid tilapia *Tilapia nilotica* × *Tilapia mossambica*. *FAO Fish. Rep.* **44:** 237-242.

Bardach, J. E., J. H. Ryther, and W. O. McLarney. 1972. *Aquaculture*. Wiley-Interscience, New York. 868 p.

Cairns, J., Jr. 1967. Suspended solids standards for the protection of aquatic organisms. *Purdue Univ. Eng. Bull.* **129:** 16-27.

Chimits, P. 1957. The tilapias and their culture. A second review and bibliography. *FAO Fish. Bull.* **10:** 1-24.

Eddy, S. 1957. *The freshwater fishes.* Wm. C. Brown, Dubuque, Iowa. 253 p.

Hedgpeth, J. W. 1957. Classification of marine environments. In J. W. Hedgpeth (Ed.). *Treatise on marine ecology and paleoecology,* Vol. 1. Memoir 69, Geological Society of America, New York, pp. 17-27.

Hickman, C. P., Jr., and B. F. Trump. 1969. The kidney. In W. S. Hoar, and D. J. Randall (Eds.), *Fish physiology,* Vol. 1, Academic Press, New York, pp. 91-239.

Kilambi, R. W., J. Noble, and C. E. Hoffman. 1970. Influence of temperature and photoperiod on growth, food consumption and food conversion efficiency of channel catfish. *Proc. Southeast. Assoc. Game Fish Comm.* **24:** 519-531.

Lee, J. S. 1973. *Commercial catfish farming.* Interstate Printing & Publishing, Danville, Ill. 263 p.

Lockwood, A. P. M. 1967. *Aspects of the physiology of crustacea.* Freeman, San Francisco. 328 p.

Loosanoff, V. L., and F. D. Tommers. 1948. Effect of suspended silt and other substances on the rate of feeding of oysters. *Science,* **107:** 69-70.

McBay, L. G. 1961. The biology of *Tilapia nilotica* Linnaeus. *Proc. Southeast Assoc. Game Fish Comm.* **15:** 208-218.

Page, J. W., and J. W. Andrews. 1975. Effects of light intensity and photoperiod on growth of normally pigmented and albino channel catfish. *Prog. Fish-Cult.* **37:** 121-125.

Pearse, A. S., and G. Gunter. 1957. Salinity. In J. W. Hedgpeth (Ed.). *Treatise on marine ecology and paleoecology,* Vol. 1. Memoir 69, Geological Society of America, New York, pp. 129-157.

Perry, W. G., Jr. 1967. Distribution and relative abundance of blue catfish, *Ictalurus furcatus,* and channel catfish, *Ictalurus punctatus,* with relation to salinity. *Proc. Southeast. Assoc. Game Fish Comm.* **21:** 436-444.

Perry, W. G., Jr. 1969. Food habits of blue and channel catfish collected from a brackish-water habitat. *Prog. Fish-Cult.* **31:** 47-50.

Perry, W. G., Jr., and J. W. Avault, Jr. 1968. Preliminary experiments on the culture of blue, channel, and white catfish in brackish water ponds. *Proc. Southeast. Assoc. Game Fish Comm.* **22:** 397-406.

Perry, W. G., Jr., and J. W. Avault, Jr. 1969. Culture of blue, channel and white catfish in brackish water ponds. *Proc. Southeast. Assoc. Game Fish Comm.* **23:** 592-605.

Perry, W. G., Jr., and J. W. Avault, Jr. 1971. Polyculture studies with blue, white and channel catfish in brackish water ponds. *Proc. Southeast. Assoc. Game Fish Comm.* **25:** 466-479.

Pritchard, D. W. 1967. What is an estuary: Physical viewpoint. In G. H. Lauff, (Ed.), *Estuaries.* American Association for the Advancement of Science, Washington, D.C., pp. 3-5.

Ruttner, F. 1953. *Fundamentals of limnology.* University of Toronto Press, Toronto. 295 p.

Schroeder, H. A. 1965. Cadmium as a factor in hypertension. *J. Chron. Dis.* **18:** 647-656.

Spotte, S. H. 1970. *Fish and invertebrate culture.* Wiley, New York, 145 p.

Stickney, R. R., and J. W. Andrews. 1971. The influence of photoperiod on growth and food conversion of channel catfish. *Prog. Fish-Cult.* **33:** 204-205.

Stickney, R. R., and B. A. Simco. 1971. Salinity tolerance of catfish hybrids. *Trans. Am. Fish. Soc.* **100:** 790-792.

Stickney, R. R., and D. B. White. 1973. Effects of salinity on growth of *Paralichthys lethostigma* postlarvae reared under aquaculture conditions. *Proc. Southeast. Assoc. Game Fish Comm.* **27:** 532-540.

Stickney, R. R., and D. B. White. 1975. Ambicoloration in tank cultured flounder, *Paralichthys dentatus.* Trans. Am. Fish. Soc. **104:** 158-160.

Stickney, R. R., H. L. Windom, D. B. White, and F. E. Taylor. 1975. Heavy-metal concentrations in selected Georgia estuarine organisms with comparative food-habit data. In F. G. Howell, J. B. Gentry, and M. H. Smith (Eds.), *Mineral cycling in southeastern ecosystems.* U.S. Energy Research and Development Administration, CONF-740513, Springfield, Va., pp. 257-267.

Sverdrup, H. N., M. W. Johnson, and R. H. Fleming. 1942. *The oceans.* Prentice-Hall, Englewood Cliffs, N.J. 1087 p.

Wallen, I. E. 1951. The direct effect of turbidity on fishes. *Bull. Okla. Agr. Mech. Coll.* **48:** 1-27.

CHAPTER 5

Feeds, Nutrition, and Growth

TROPHIC LEVEL AND FEEDING REQUIREMENTS

Once an aquaculture species reaches the juvenile stage, the major expense involved in culture is often associated with the need to provide a suitable diet to the growing organisms. This may also be true of larvae and fry, but in species that will accept prepared feeds early in the life cycle, little expense may be incurred until the organisms attain larger size. Most aquaculture animals, upon reaching the juvenile stage (and in many cases before), are omnivores or carnivores. The exceptions reside largely in the phylum Mollusca (clams, scallops, mussels, and oysters), although a few fishes must be included (e.g., tilapia and grass carp). Virtually all the marine fishes that have been examined as potential aquaculture candidates in the United States are carnivores. The mullet, *Mugil cephalus,* thought by many to be a herbivore, actually is omnivorous in its food habits (Odum, 1966, 1968). Other potential mariculture fishes of a more carnivorous nature are the red drum, the black drum, and flounders of the genus *Paralichthys.*

The cost of feed for animals of different trophic levels depends on the species and the location of the culture facilities. In theory marine mollusks can be fed at no cost if they are stocked into a natural environment that features a high level of primary productivity to provide the algal cells those organisms feed on. If it were necessary to furnish laboratory-grown phytoplankton as food throughout the culture lives of the animals, the cost would undoubtedly become prohibitive. Natural phytoplankton populations can, of course, be induced by adding fertilizer to the water. This may be inorganic fertilizer (Chapter 3) or it may be organic, usually in the form of animal wastes. Organic fertilization has served as a phytoplankton food base for a variety of aquaculture organisms, primarily in foreign lands but also in the United States (Hickling, 1962; Swingle *et al.,* 1965; Schroeder, 1974; Boyd, 1976; Stickney *et al.,* 1977a, 1977b; Stickney and Hesby, 1978).

Omnivores and carnivores can be expected to obtain at least some food from the natural environment when reared under extensive conditions, but when raised intensively, the demand for food far exceeds the ability of the natural environment to provide it at a rapid enough rate. Fertilization will lead to high levels of secondary productivity, but in virtually every case where a carnivorous or omnivorous species is under high density culture, some type of supplementary food must be supplied. This can be in the form of natural feeds (meat scraps, trash fish, etc.) or it may be a prepared diet, comprised of various grains, processed animal meals, and possibly supplemental vitamins and minerals.

With the exception of herbivores, all species being reared commercially or under study as potential aquaculture candidates are fed prepared diets. The formulation and the feeding of such diets are discussed later in this chapter. In some cases it is necessary to train an animal to accept prepared feeds, but this has been accomplished without too much difficulty for most species. For example, White and Stickney (1973) attempted to convert flounders (*Paralichthys dentatus* and *P. lethostigma*) from natural foods to prepared diets in the laboratory. The process was accomplished by first feeding postlarval flounders brine shrimp (*Artemia salina*), then introducing in steps: frozen penaeid shrimp cut into fine pieces, freeze-dried penaeid shrimp pieces, and finally a commercial trout feed. A simpler way of converting from living to prepared diets would be to introduce one fish that is already established on prepared diets into each flounder culture chamber. The feeding activity of the trained fish would probably initiate a similar response in the naive animals, especially since White and Stickney (1973) have observed that when an individual flounder in a particular tank began to consume pelleted feed, others soon followed that example.

Some aquaculture species—for example, channel catfish, tilapia, and rainbow trout—immediately accept prepared diets after yolk-sac absorption and the initiation of feeding. Many other species, both vertebrate and invertebrate, and notably the majority of mariculture species, require living food for a certain period after hatching. The primary reason for the living food requirement appears to be related to the extremely small size of marine species at birth and the amount of larval development that must take place after the organisms hatch. Organisms with very small eggs (consequently small amounts of yolk) often undergo a great deal of larval development after hatching and usually require very small living food organisms, whereas animals with large eggs (containing large amounts of yolk) produce newly hatched offspring that resemble the adult and often can be introduced to prepared diets immediately with a good deal of success. Fecundity as it relates to egg size in aquatic animals is discussed further in Chapter 6.

NATURAL FOODS FOR LARVAE AND FRY

Because a larval invertebrate or fry fish refuses to accept a prepared diet at one point is no reason to believe that the provision of natural food organisms will always be a condition for successful aquaculture of that species. The texture, odor, color, and means of presentation of a diet may all contribute to acceptance or rejection; yet aquaculturists have little or no information about many of these factors. As research and food technology advance, prepared diets suitable for many species now thought to require living food may be developed.

The initial food source of many larval organisms is phytoplankton. This requirement is linked to the small size of the culture animals at hatching. Algae culture is a well-developed field, the details of which are beyond the scope of this book. In short, phytoplankton mass cultures for feeding larval aquaculture organisms require a certain amount of space, a good deal of skill and knowledge, and considerable time. If large amounts of algae are required, this cost may become a significant portion of the total operating budget.

Phytoplankton Rearing and Algae Culturing

Phytoplankton rearing under bacteria-free conditions is not feasible in aquaculture facilities where large volumes of algae must be available each day during the larval growing period. Algae are often cultured in glass carboys or open fiberglass or plastic (nontoxic) tanks. A suitable nutrient medium (which may vary considerably depending on the species of algae to be cultured) is added to filtered water in each culture tank; then the tanks are inoculated with the desired species of algae. Different species may be required for different aquaculture organisms, or a variety of species may be suitable for a particular culture animal. The cultures are exposed to bright artificial light and are usually aerated to keep the water circulating. This circulation exposes all the cells in each culture chamber to the light as those cells pass across the surface. Under the proper conditions, the cells will rapidly multiply until the concentration reaches 10^6 cells per liter or higher. Common practice is to initiate new cultures at staggered times so that food will be available at the required cell density on a daily basis. It is also good practice to have at least two culture tanks available for each species, in case a bloom fails to develop in one of them. The time required from inoculation to harvest of algae cultures does vary but tends to be around 72 to 96 hours under optimum conditions. More rapid turnover can be obtained if fairly low cell densities are acceptable.

In continuous algae culturing the culture chambers are always being augmented with new growth medium, and a centrifuge removes algal cells at a steady rate. Such systems will undoubtedly find wide acceptance among aquaculturists in the future, but the techniques involved are somewhat more advanced than standard algal culture methods. Continuous algal culture systems must be carefully monitored to ensure that proper nutrient levels are maintained, and the culturist must recognize the importance of not harvesting the cells either too rapidly or too slowly.

For most nonmolluskan species of aquaculture organisms that are initially fed phytoplankton (alone or in combination with animal food), the feeding of algae can be discontinued after some period of time. Thereafter, such organisms as marine shrimp (*Penaeus* spp.) can be fed animal matter exclusively. This occurs at the postlarval stage in penaeids (Hanson and Goodwin, 1977). Marine fishes may also be offered phytoplankton during the larval stage, although these organisms may act in the removal of fish metabolites exclusively, or in addition to their role as food (Arnold *et al.*, 1977).

Zooplankton Rearing

A large variety of zooplanktonic species of sizes suitable for fish and invertebrate larvae can be found in nature. Some culturists have attempted to supply their demands by capturing wild zooplankton. The disadvantages of that technique include patchy distribution, variation in abundance temporally, the presence of forms of improper size, and the presence of undesirable species that may prey on or compete with the aquaculture species. Thus most aquaculturists who utilize zooplanktonic feed maintain unispecific cultures. The most commonly utilized zooplankters are a rotifer, *Brachionus* sp., and the brine shrimp, *Artemia salina*.

Both *Brachionus* sp. and *Artemia salina* are easily maintained in the laboratory, although the methods employed for culture are somewhat different. *Brachionus* sp. is smaller than *A. salina,* thus the former is utilized at an earlier stage than (and sometimes in conjunction with) brine shrimp. *Brachionus* sp. are reared in static tanks and maintained on algal cultures (e.g., *Tetraselmis* sp.). McGeachin (1977) described the technique utilized to culture *B. plicatilis* for feeding larval spotted sea trout (*Cynoscion nebulosus*). The rotifers were allowed to reproduce and maintain their own populations in 1.8 m diameter fiberglass tanks with water 30 cm deep. Salinity was maintained at approximately 28⁰/₀₀ and temperature about 26 C. The algae, *Tetraselmis* sp., were cultured in separate tanks under similar conditions, although the algal culture tanks received the light from four fluorescent light bulbs over each tank, 60 cm above the water. Algal cell concentration in the *B. plicatilis* tanks was maintained by adding the plants at densities of 5000 to 20,000 cells/ml. To maintain logarithmic growth in the rotifers, their numbers were reduced by harvesting 25 to 75% of the population daily and maintaining the level in the tanks at below 200 rotifers/ml.

Artemia salina have been the most popular zooplankter for feeding larval aquatic organisms. Brine shrimp cysts can be purchased in bulk and hatched as required in the laboratory. Generally, newly hatched *A. salina* nauplii are utilized as food, although brine shrimp can be grown to adulthood if suitable algae are provided. When *Cynoscion nebulosus* were fed either newly hatched *A. salina* nauplii or nauplii that had been provided with algae for one or more days, there appeared to be little difference in the nutritional value of the brine shrimp to the larval fish (McGeachin, *et al.,* 1977).

Artemia salina can be obtained from two traditional sources and are also available from newly discovered locations, although the cysts from the latter may be somewhat less abundant. The traditional locations for brine shrimp cysts are Great Salt Lake in Utah and San Francisco Bay, California. Since aquaculturists as well as aquarium hobbyists around the world rely on brine shrimp cysts for feeding larval animals, the demand has often exceeded the supply in recent years and alternative foods have been sought. Brine shrimp cysts are expensive, when they can be found. To date, the techniques for producing large numbers of cysts from self-sustaining populations of *A. salina* in the laboratory have not been perfected, so cysts must be purchased from suppliers. The brine shrimp from wild populations in Great Salt Lake, San Francisco Bay, and other areas are collected along the beaches, where they occur in tremendous concentrations at certain times.

Brine shrimp rearing is relatively simple and requires no elaborate equipment. If the animals are fed within 24 hours of hatching, there is no need to maintain an algal culture to feed them, thus the expense of the cysts is somewhat offset by savings in time, equipment, and energy. Several thousand *A. salina* cysts will fit easily in a teaspoon, and literally millions can be hatched in a few liters of water. The cysts are placed in seawater (usually within the salinity range 35–40⁰/₀₀) at room temperature (about 25 C). The seawater can come from a natural source (in which case it should be filtered to remove foreign organisms), or it can be prepared from artificial sea salts, which are readily available from commercial sources. The cysts are liberally aerated to keep them in suspension during hatching (otherwise they would all float at the surface of the water) and to provide oxygen to the nauplii after they hatch. Hatching usually occurs within about 24 to 36 hours, although in any group some cysts will fail to hatch at all.

To separate the hatched nauplii from dead cysts and shells, it is convenient to attract the positively phototaxic animals toward a light. Shells float at the surface following hatching, while the unhatched cysts often sink after aeration is discontinued; therefore a light placed near the side of the culture vessel will attract the nauplii away from dead or unhatched cysts and shells. Once the nauplii have been concentrated they can be siphoned into another container or fine meshed net and then presented to the larval aquaculture animals. Figure 5.1 gives a simple, effective design for a brine shrimp cyst hatching chamber, which simplifies the harvest of nauplii.

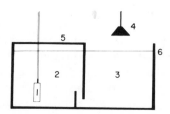

Figure 5.1 Schematic representation of an *Artemia salina* hatching chamber. Brine shrimp cysts are aerated with an airstone (1) in the chamber on the left (2). Following hatching, the shells and unhatched cysts remain in the hatching chamber while the swimming nauplii are attracted to the chamber on the right (3) by a light suspended over the water (4). A cover (5) over the hatching chamber prevents light from entering that side. The water level is indicated (6). All dimensions are variable, but the largest dimension would rarely exceed 1 m.

If *A. salina* nauplii are fed to marine organisms, they can be directly transferred from the hatching container to the larval rearing tanks. In the case of freshwater organisms, however, it is generally considered wise to wash the nauplii with fresh water before introducing them to the larval aquaculture animals. This is easily accomplished by pouring the brine shrimp nauplii into a fine mesh dip net, then flushing them with fresh water. The brine shrimp nauplii survive long enough to be preyed on by the aquaculture species, and the technique avoids the introduction of salt water into larval rearing containers.

PREPARED DIETS

Prepared diets, also referred to as artificial diets, are either complete (containing all the required nutrients) or supplemental (meant to provide additional protein, carbohydrate, and fat to animals receiving some natural food, but not fortified with vitamins and minerals). They are of two general types: moist and dry. The most common moist diet in use today is the Oregon Moist Pellet (OMP) (Hublou, 1963), which is used extensively by salmon culturists, particularly in the Pacific Northwest, and has been adopted in some federal and other hatcheries to feed warmwater and midrange sport fishes that will not readily accept a dry, hard food pellet. Thoroughly ground offal (the heads and viscera of processed fish and other animals) has been utilized as fish feed. The latter type of diet requires freezing to prevent spoilage. The OMP, which until recently also required freezing for storage, will presently be available in a shelf-stable form that can be stored at room temperature for extended periods. With the availability of the OMP in the new form, that diet can be shipped throughout the United States and abroad safely and less expensively than was the case when freezing was required. From a nutritional standpoint, moist diets are not necessarily more desirable than dry ones, provided the animal being fed will accept either or both.

Prepared diets usually come in the form of pellets, crumbles, or flakes. The size of the particle used depends on the species being fed, its feeding habits, and its size. Dry diets, as well as moist ones, are subject to vitamin decomposition and fat rancidity on long-term storage unless sufficiently protected by antioxidizing agents and proper storage. Later sections of this chapter cover the vitamins that require protection and discuss fat rancidity.

Three basic types of prepared diet have been utilized by aquaculturists: practical, semipurified, and purified. A practical diet is one that is formulated from natural ingredients such as cereal grains, oilseed meals, fish meals, and meat by-products. Supplemental vitamins and minerals are usually added to fortify the natural ingredients. Semipurified and purified diets are used in nutritional research and are significantly more expensive than practical diets.

A semipurified diet may contain some natural ingredients in a relatively pure form. For example, casein (milk protein) is a nearly pure protein source with a good amino acid balance. A typical semipurified diet might be formulated with casein as the protein. Corn oil might be used as the lipid source in this semipurified diet because that fat is nearly 100% lipid, having little in the way of contaminants. The carbohydrate in a semipurified diet might consist of corn starch. This diet would then be supplemented with sufficient levels of vitamins and minerals to ensure against the development of nutritional deficiency symptoms that might occur if these components were deleted. Each primary ingredient in this type of diet is relatively pure compared to such practical ingredients as fish meal, corn meal, soybean meal, and fish oil, which not only contain varying levels of proteins, lipids, and carbohydrates, but are also often high in fiber or other indigestible substances.

A purified diet contains only ingredients of precisely known composition, and for this reason such diets are very expensive and only rarely utilized, even in aquaculture research. A purified diet is formulated from a group of individual synthetic amino acids, fatty acids, simple sugars, or other carbohydrates of precise composition, and vitamin and mineral mixtures made from individual chemicals of high grade.

In nutritional research a series of semipurified, or less often, purified diets may be prepared to test one or more dietary components in relation to the growth and food conversion efficiency they elicit. For example, a corn starch, corn oil, and casein diet may be altered to replace the casein by individual amino acids. Then one or more of the amino acids would be deleted from various diets, to test the requirements for each. In this way the essential components of the diet can be elaborated. Later a series of diets might be formulated with casein once again, but the percentage of total protein could be varied to determine the quantitative needs of the animals. Various modifications in this basic scheme can be made depending on how specific the research becomes.

ENERGY AND GROWTH

Metabolism is the result of all the chemical and energy transformations that occur within a living organism. All phases of metabolism require energy, which animals obtain from food. Energy is expended for the maintenance of life and for growth and reproduction. Metabolism includes the storage of energy (anabolism) as fat, protein, and carbohydrate, and the transformation of those storage products into free energy (catabolism).

Food energy is measured in calories. In nutrition, reference to the calorie generally implies the large or kilocalorie (kcal), which is equivalent to the amount of heat required to raise the temperature of 1 kg of water 1 C, as distinguished from the small or gram-calorie $(g \cdot cal)$, which is the amount of heat required to raise the temperature of 1 g of water 1 C.

The immediate source of free energy in biochemical systems is adenosine triphosphate (ATP). During anabolism, adenosine diphosphate (ADP) traps energy in phosphate bonds, forming ATP. Conversely, during catabolic reactions ATP is converted back to ADP and free energy is released. Since both ADP and ATP are present in the tissues of aquatic organisms and appear to function in the same manner as in mammals, it is assumed that the metabolic pathways in the poikilotherms and homoiotherms are similar.

The amount of energy expended for growth compared with that required for maintenance is dependent on the species of animal, its age, the environmental conditions, dietary composition, reproductive state, and other factors (all of which influence basal metabolic rate). Maintenance energy is a combination of basal metabolism and specific dynamic action (SDA). SDA is the amount of heat produced in addition to that of basal metabolism as a result of food ingestion (White et al., 1964). SDA is considerably higher for protein than for carbohydrate and lipid.

Since metabolic rate in poikilothermic animals is highly dependent on environmental temperature, energy requirements can be expected to vary both seasonally and diurnally as fluctuations occur in water temperature. Small animals generally have higher metabolic rates than larger ones, and the growth rate of young animals is more rapid than that of older individuals. Based on the rate of oxygen consumption, the metabolic rate of channel catfish increases shortly after feeding, indicating increased energy utilization as a function of digestion, absorption, and assimilation (Andrews and Matsuda, 1975).

Classical fishery biology examines the increase in size of aquatic organisms in terms of age and growth studies that employ length-weight relationships. Samples from natural populations are obtained, and the individuals in the sample are measured and weighed. If isometric growth is assumed (all parts of the animal enlarge at similar rates), the pattern of growth can be described by the relationship

$$W = aL^b \tag{16}$$

where W is the weight, L the length, and a and b are constants determined from the empirical data (Everhart et al., 1975). Figure 5.2 plots a typical growth curve of this type for a wild population of white shrimp, Penaeus setiferus, from a Georgia estuary. Such growth curves may be generated from data for a single age class, but more commonly they are obtained from organisms of several ages.

One of the goals of aquaculture is to produce a crop as rapidly as possible. Thus it is necessary to accelerate growth through water quality management, dietary manipulation, and in other ways. Length-weight relationships are not much used by aquaculturists because such relationships provide information on what has occurred in the past, whereas the culturist is generally interested in predicting growth. The most commonly used short-term predictive growth equation is in the form:

$$W_t = W_0 e^{gt} \tag{17}$$

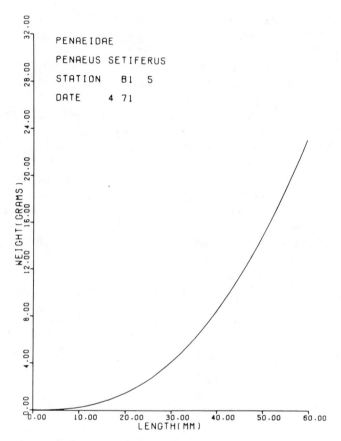

Figure 5.2 Computer-generated plot of the growth of *Penaeus setiferus* as derived from field measurements and equation 16.

where W_t is the weight at time t, W_0 is the initial weight, e is the natural logarithm, and g is the growth coefficient. The growth coefficient is determined as follows:

$$g = \ln \frac{W_t}{W_0} \tag{18}$$

where ln is the natural logarithm and W_t and W_0 are as before (Everhart *et al.*, 1975). This model generally holds fairly well for a growing season but cannot be extended to cover more than one year. A modification of this formula can be used to calculate the feeding rate required to maintain a desired growth rate of aquaculture animals, as discussed later in this chapter.

Optimum harvest size for many aquaculture animals occurs at or near the size where growth rate begins to decrease significantly. For example, channel catfish can reach a maximum weight in excess of 20 kg, but the most rapid growth occurs between hatching and 500 g. The time required to double the weight of a 500 g channel catfish may be as long as that needed to bring the fish from the egg to the first 500 g. This growth pattern is generally followed by aquatic animals that live for several years, although the actual size at which growth rate decreases varies from species to species. Reduction in the rate of growth is often accompanied by a decrease in food conversion efficiency; therefore the costs of doubling the weight of a marketable catfish, trout, tilapia, shrimp, or other species can become high. Not only must feed costs be considered, but also the costs of water, energy, and labor, and the risks associated with attempting to maintain animals beyond the minimum time possible to get them to marketable size.

Optimum harvest size is not based only on a decline in growth rate. Marketing has been set up for catfish of about 0.5 kg in most parts of the United States. This is not only a convenient size for the seller and consumer, but also represents the average size of the fish at the end of the second growing season when they are reared under typical culture strategies. Market sizes of shrimp, tilapia, and other species are also determined, at least in part, by the size the animals have reached at the end of the growing season. This is especially critical to species that cannot survive the winter in temperate areas and must be marketed before they die from exposure to cold.

Phillips (1972) discussed some of the factors relating to energy requirements in fishes. In general, the requirements for energy are somewhat greater for carnivores than herbivores because more energy is needed to eliminate high levels of nitrogenous wastes that build up when animal protein is digested than those resulting when dietary protein comes primarily from plants. Similarly, animals that consume diets high in protein use higher levels of energy than do animals reared on low protein diets. Diets with high mineral levels also tend to increase metabolic rates because of increased demands on osmoregulation to rid the body of excess salts.

Energy requirements are increased during periods of gametogenesis and spawning. The energy drain due to reproduction may be so great that the adults die after spawning (e.g., various species in the family Salmonidae), or the quality of the adult as a human food product may be poor (oysters and various fishes fall into this category). Most aquaculture species are harvested before reaching adulthood, since so much energy is diverted from growth to gamete production that food conversion efficiency suffers. Since a few brood animals will satisfy the needs of most culturists, there is no requirement for large numbers of adult animals in the typical aquaculture operation except when sexual maturity occurs before harvest size is attained. Tilapia and a few other species exemplify this concept. Chapter 6 outlines the problems that arise with respect to tilapia reproduction prior to harvest, and mentions some solutions.

Physical and chemical aspects of water quality other than temperature may have impacts on energy utilization. Low dissolved oxygen (DO) leads to an increased respiration rate, at least to a point, although when the DO level becomes extremely low some animals compensate by reducing their metabolic rate, including their rates of respiration and food consumption. Soft water may lead to increased metabolic rates because of the scarcity of divalent cations. Increased energy expenditure is required to move those ions into the body against the concentration gradient. High ammonia levels increase the metabolic rate, as does exposure to various types of organic pollutants. Under severe stress, metabolism may decrease as a means of conserving energy, and this often occurs in the case of low DO. Energy requirements are increased in flowing water because the animals must expend energy in station-keeping. Exercise resulting from orientation with currents may be important to the improvement of product quality, however. Aquaculturists who rear fish in cages in the static waters of lakes and reservoirs have reported a loss of muscle tone in the fish, leading to a product that was somewhat less desirable than that of fish reared in ponds or tanks where more exercise was possible. Fish reared in cages in flowing water may also be of better quality because of the need to station-keep.

The physiological fuel value (PFV) or metabolizable energy (ME) value of a food is the amount of energy that can be extracted during digestion and absorption. Because of physiological differences both within and among animals, the PFV of the same food item varies from species to species. Furthermore, in the case of poikilothermic animals, environmental conditions may greatly affect PFV. In the past, diets for many warmwater aquaculture animals have been formulated utilizing ME values derived from livestock and poultry feed tables on the assumption that the energy extractable from feed ingredients by cattle, hogs, and poultry would be similar to that extractable by fish and aquatic invertebrates. Some data are presently available on the ME level of certain feedstuffs to channel catfish, although little information is available for other warmwater aquaculture species. In some cases the ME values for catfish are similar to those of terrestrial animals, but in others they are quite dissimilar (Taggart, 1974; Cruz, 1975; Lovell, 1977a).

The energy-containing components of foodstuffs are protein, lipid, and carbohydrate. The gross energy contained in these food groups is 5.65, 9.40, and 4.15 kcal/g, respectively (White *et al.*, 1964). Utilizing pure sources of the energy-bearing foodstuffs to feed a variety of terrestrial animals, it has been determined that the average PFV for both protein and carbohydrate is 4 kcal/g, and that of lipid is 9 kcal/g. These levels are virtually never obtained from practical feed ingredients, since most are not pure protein, lipid, or carbohydrate, and if they were, it is unlikely that they would be 100% digestible. If the PFV of a diet is based on the 4, 4, 9 values for energy contained in protein, carbohydrate, and lipid, there may be a considerable overestimate of the available energy in that diet. Formulation of an experimental diet is considered at the end of this chapter, but as an example, it has been determined that although the PFV of starch in livestock feeds is nearly 4 kcal/g, channel catfish are able to extract only about 2.5 kcal/g from the same feedstuff (Wilson, 1977b).

Most of the energy used by fish in nature comes from protein. Protein is required for growth, but the energy for maintenance metabolism may also be derived from lipids and carbohydrates. To a limited extent, lipids and carbohydrates can be used to spare protein (i.e., to release it from immediate use in meeting the energy requirements of an animal, allowing it to be laid down for growth). The concept of protein sparing by carbohydrate in salmonid diets has been discussed by Halver (1972). Stickney (1977) and Wilson (1977a) have treated this subject with respect to channel catfish in terms of both lipids and carbohydrates. When lipids and carbohydrates are provided in excess, nutritional diseases may develop. For example, salmonids may store high levels of glycogen in the liver and exhibit symptoms of diabetes when fed too much carbohydrate (Halver, 1972), whereas high levels of dietary lipid can lead to fatty livers in a variety of animals. In channel catfish high levels of dietary lipid may give rise to the presence of excessive concentrations of visceral fat, which is lost during processing and decreases the dress-out percentage of the crop. When diets containing the same amount of lipid are fed to channel catfish at different environmental temperatures, a trend has been noted toward increased lipid concentration in the body with increasing temperature (Andrews and Stickney, 1972).

Practical animal diets are composed of minerals, vitamins, water, and fiber in addition to the energy-bearing components. Vitamins and minerals are important to proper nutrition, but the requirements for many of them are known only poorly for many aquaculture species. The same is true for amino acids, fatty acids, sugars, and starches. In terms of aquaculture species, more is known about the salmonids than nearly any other group, although a considerable body of nutritional information has been obtained on channel catfish and a few other species. Because very little is known about the actual nutritional requirements of many aquaculture species, most of the examples and information presented below come from the channel catfish literature. For other warmwater fishes and invertebrates, the data are speculative, based on salmonid or terrestrial animal requirements.

VITAMINS

Vitamins are organic compounds that are required by at least some species in small quantities for normal growth and health (White et al., 1964). Vitamins are considered to be catalytic, since they take part in biochemical reactions but are not contained in the end products of those reactions. Most vitamins serve as coenzymes in biological systems. Animals are capable of producing their own enzymes biochemically but are unable to synthesize the vitamin coenzymes (or cofactors, which are mineral components of enzyme systems). All the known vitamins have been chemically identified and can be synthetically produced in the laboratory (Figure 5.3).

Vitamins were originally assigned letters of the alphabet, and these are still in common usage, especially with respect to the fat-soluble vitamins. Since the structures of the vitamins, especially the water-soluble vitamins, have been determined, the nomenclature has changed largely to one based on chemical characteristics. The fat-soluble vitamins, however, are still commonly referred to by the letters A, D, E, and K. The water-soluble vitamins were originally called the B-complex and vitamin C. Most B-complex vitamins are now commonly known by other names, except for vitamin B_{12} (cyanocobalamin), which has retained its original numerical designation in common usage. Table 5.1 presents the common names, sources, and symptoms associated with both the fat- and water-soluble vitamins. Hypovitaminosis (symptoms associated with vitamin deficiency) may occur with vitamins of both types; hypervitaminosis (symptoms associated with vitamin excess) occurs only with respect to the fat-soluble vitamins. If excessive amounts of water-soluble vitamins are ingested, they can be eliminated from the body by the kidneys because they are carried in solution in the blood. The fat-soluble vitamins are stored in association with lipids and are not excreted when present in excess.

The potency of vitamins can be expressed in four ways: (1) international units (IU), in which vitamin activity is compared with an international standard controlled by the Expert Committee on Biological Standardization of the World Health Organization, (2) United States Pharmacopoeia (USP) units, which use standards maintained in the United States for comparison

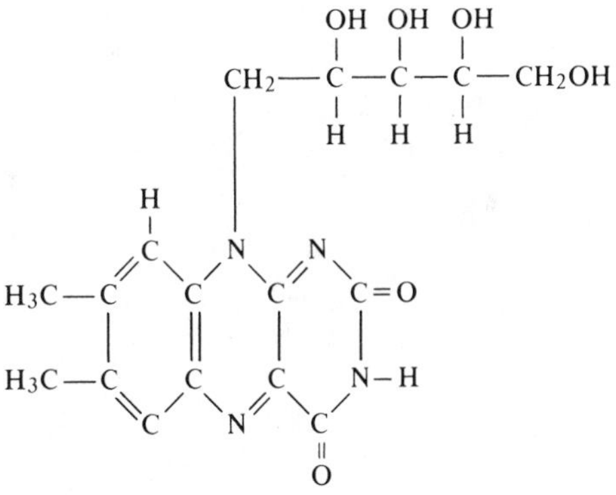

Figure 5.3 Structural formulas of some of the vitamins that have been found to be important in animal nutrition. Adapted from White *et al.* (1964) and used with permission.

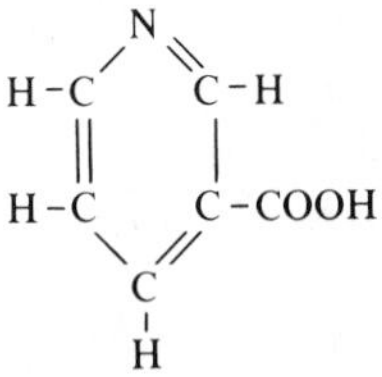

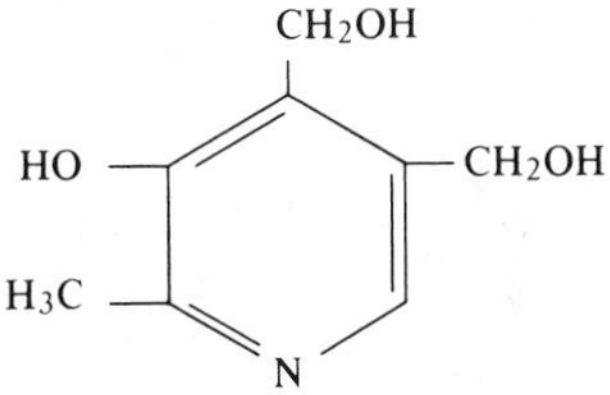

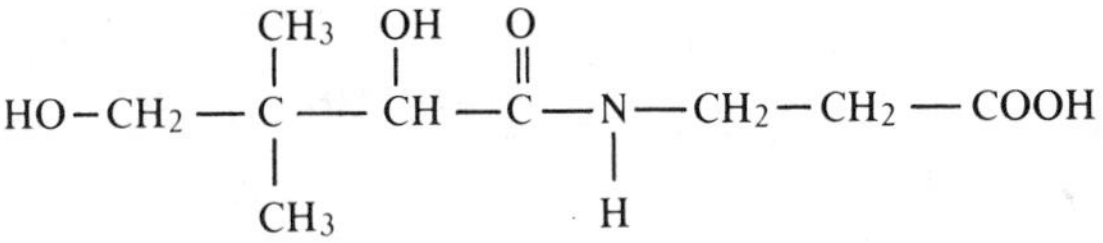

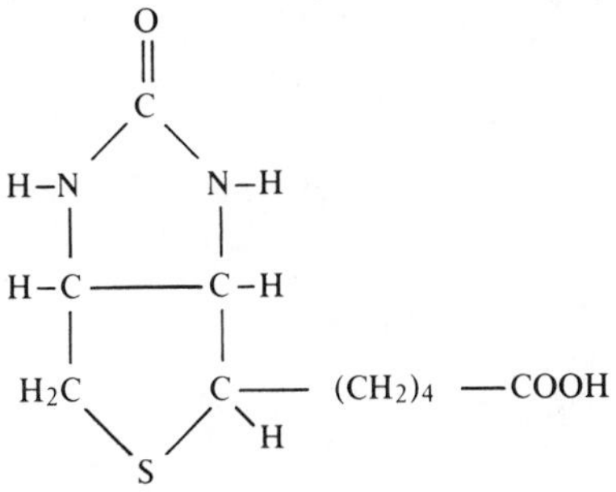

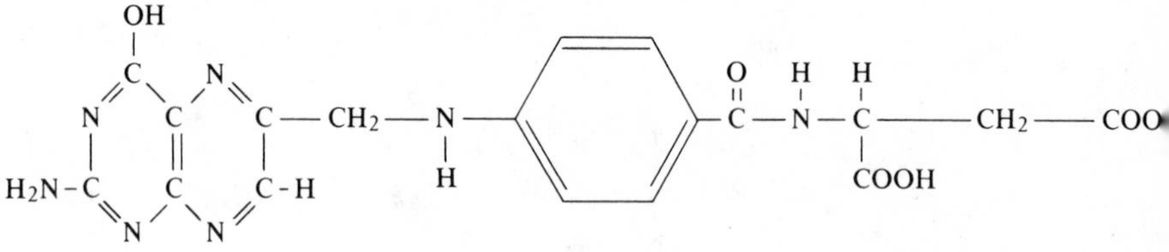

Folic acid

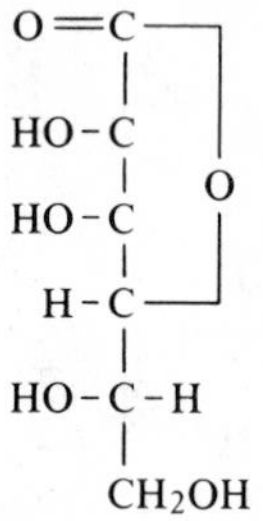

L-Ascorbic acid

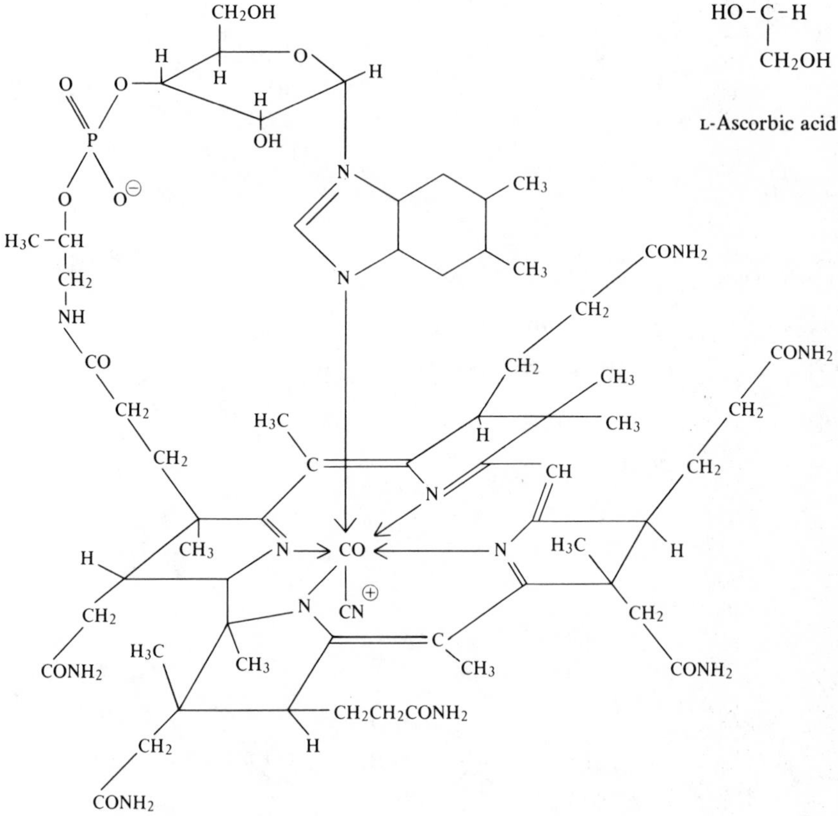

Cyanocobalamin

Figure 5.3 (continued)

176

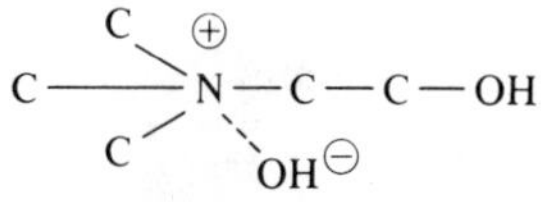

Choline

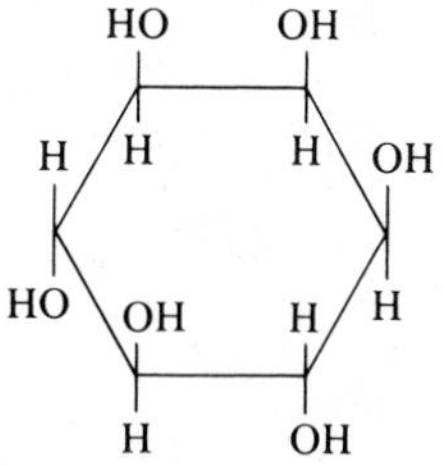

Inositol

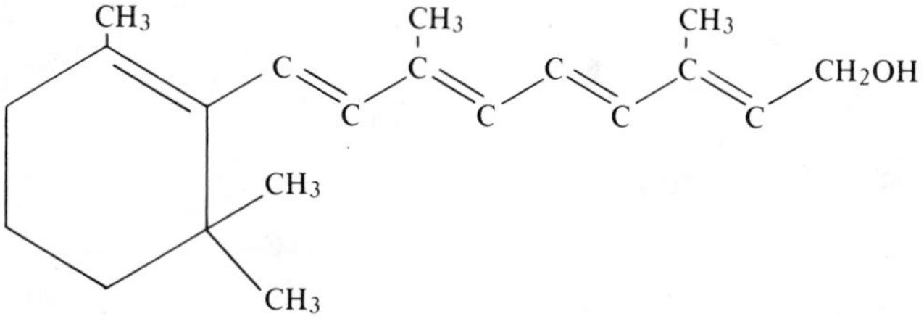

Vitamin A₁

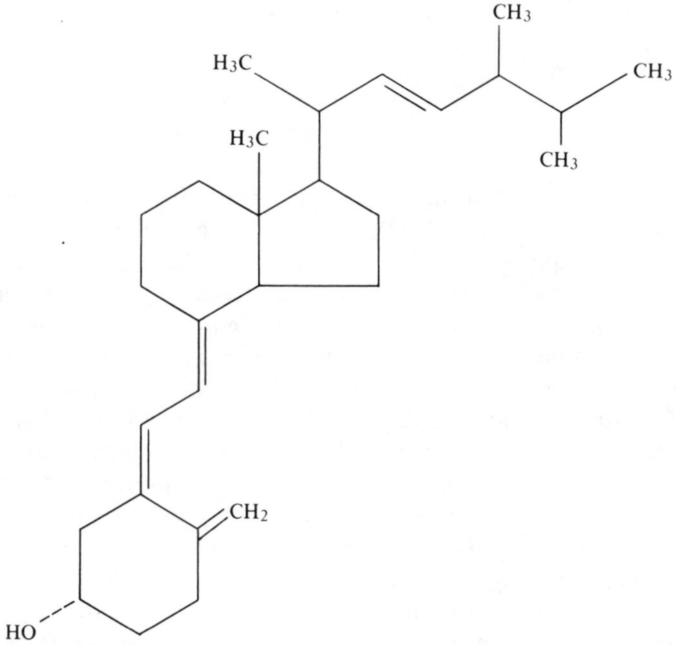

Vitamin D₂

Figure 5.3 (continued)

177

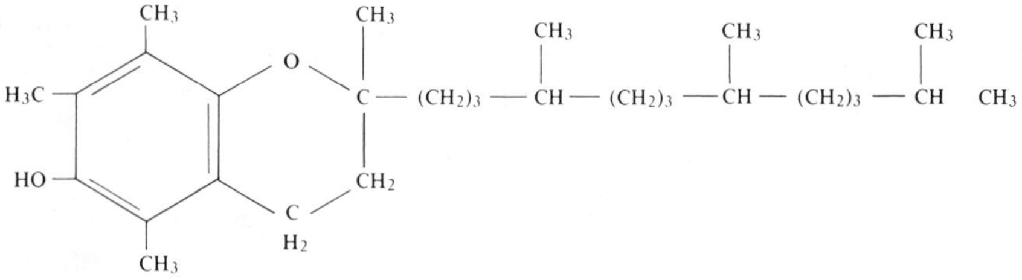

α-Tocopherol

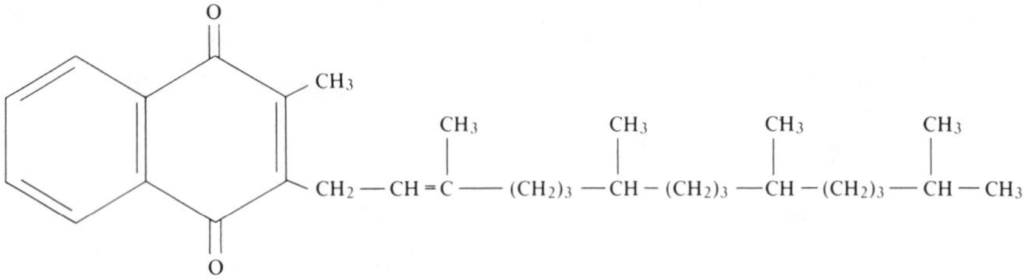

Vitamin K₁

Figure 5.3   (continued)

with vitamin activity (in most cases IU and USP units are identical), (3) international chick units (ICU), in which vitamin activity is measured in terms of the response elicited in chickens, and (4) as a weight measurement (e.g., activity may be presented in terms of milligrams per kilogram of feed). The latter method of measuring vitamin activity is presently being utilized to a large extent in the formulation of animal feeds.

Many vitamins are heat, moisture, and air labile, and although vitamins are often stabilized against deterioration, diets should be stored in a cool, dry place to prevent degradation. Ascorbic acid is especially susceptible to improper storage. Long-term storage of aquacultural feeds should be avoided. In most cases only sufficient diet is purchased to accommodate the needs over a few months. In no case should feed be stored for longer than a single growing season unless that feed is stored frozen. Since feed manufacturers try to make no more ration than they anticipate selling in a given growing season, it is often a good idea to purchase sufficient feed early and not take a chance on its being available late in the year.

Antioxidants are generally added to diets to retard the rate of vitamin degradation and prevent oxidation of lipids, which can lead to rancidity

Table 5.1 **Sources and Symptoms Associated with Fat- and Water-Soluble Vitamins in Fish**[a]

Vitamin	Sources	Symptoms of Hypervitaminosis or Hypovitaminosis
Fat-soluble vitamins		
A	Fish liver oils, fish meals containing fish oil residues	*Hypovitaminosis:* poor growth, poor vision, night blindness, hemorrhagic areas at base of fins, abnormal bone formation *Hypervitaminosis:* enlargement of liver and spleen, abnormal growth and bone formation, epithelial keritonization
D	Fish oils (some synthesis may occur in certain fishes when skin is exposed to sunlight)	*Hypovitaminosis:* not described in fish *Hypervitaminosis:* impaired growth, lethargy, dark coloration
E (α-Tocopherol)	Wheat germ, soybeans, and corn	*Hypovitaminosis:* exophthalmia, reduced growth, anemia *Hypervitaminosis:* poor growth, toxic liver reaction, death
K (Menadione)	Green, leafy vegetables; soybeans, animal livers	*Hypovitaminosis:* prolonged blood clotting time, anemia, hemorrhagic gills and eyes *Hypervitaminosis:* no information available
Water-soluble vitamins		
Thiamine	Cereal bran, beans, peas, yeast, fresh organ meats	*Hypovitaminosis:* poor appetite, loss of equilibrium, lethargy, poor growth
Riboflavin	Milk, liver, yeast, cereal grains, fresh meats	*Hypovitaminosis:* opaque eye lens, hemorrhagic eyes, poor appetite, poor growth
Pyridoxine	Cereal brans, yeast, egg yolk, liver	*Hypovitaminosis:* reduced weight gain, nervous disorders
Pantothenic acid	Cereal bran, yeast, organ meats, fish flesh	*Hypovitaminosis:* clubbed gill filaments, eroded tissues, lethargy, loss of appetite, reduced weight gain

Table 5.1 (*Continued*)

Vitamin	Sources	Symptoms of Hypervitaminosis or Hypovitaminosis
Niacin	Yeast, legumes, organ meats	*Hypovitaminosis:* loss of appetite, tetany, poor coordination, lethargy, poor growth
Folic acid	Yeast, green vegetables, organ meats	*Hypovitaminosis:* lethargy, poor growth
Biotin	Liver, kidney, egg yolk, yeast, milk products	*Hypovitaminosis:* loss of appetite, skin and colon lesions in trout; no significant effects on catfish
Cyanocobalamin (vitamin B_{12})	Meats and meat by-products, fish meal	*Hypovitaminosis:* reduced weight gain, poor appetite
Ascorbic acid (vitamin C)	Citrus fruit, beef liver and kidney, fresh fish tissues	*Hypovitaminosis:* lordosis and scoliosis, dislocated vertebrae, reduced growth
Inositol	Present in all plant and animal tissue	*Hypovitaminosis:* poor growth and skin lesions in trout; no significant effects on catfish
Choline	Wheat germ, soybean and other vegetable meals, organ meats	*Hypovitaminosis:* poor growth and food conversion, hemorrhagic kidney and enlarged liver

[a] Hypovitaminoses and hypervitaminoses are those observed in fish, especially salmonids (Halver, 1972) and channel catfish (Dupree, 1977). Different responses may occur in other species of fishes and in invertebrates.

and the formation of toxic peroxides in the feed. Lecithin (phosphatidyl choline), ethoxyquin, BHT, BHA, and vitamin E (α-tocopherol) are among the substances utilized as antioxidizing agents in diets (vitamin E also acts as an antioxidant in the bodies of the animals). Overfortification of feeds with vitamin E, which is also required for proper nutrition, may aid in protecting the feed but is expensive and can lead to hypervitaminosis if present in too high a concentration. Experiments on the benefits of over-fortification of channel catfish diets with Vitamin E (O'Keefe, 1976) have revealed that storage life of catfish filets was improved when 80 to 120 g/ton of *dl*-α-tocopherol was added to the feed. The vitamin E in this case inhibited oxidative rancidity.

Vitamin requirements vary from species to species. The requirements have been fairly well established for salmonids and catfish (Dupree, 1966, 1977; Halver, 1972; NAS, 1973, 1977) but have not been determined for many other aquaculture species. If natural food organisms are available to culture animals (as is often the case in extensive culture), prepared feed may require no vitamin supplementation, whereas when natural food is not available, or when the density of culture animals is so great that natural food items contribute insignificantly to the dietary intake of any individual, complete fortification of prepared diets is essential.

Species for which no nutrient requirement information is available are being reared successfully in aquaculture, largely under research conditions. In most cases a good quality commercial fish or invertebrate diet will sustain the lives of animals that have not previously been in culture. Once growth and survival have been assured by the use of such commercial diets, experimental modifications of that ration can be prepared in the form of purified or semipurified diets to study the actual nutritional requirements of the animal. In the long run such experiments may lead to a less expensive diet (or perhaps one that is more costly) that will provide optimum growth.

MINERALS

Minerals are required by all animals for various life processes, including the formation of skeletal tissue, respiration, digestion, and osmoregulation. Marine animals live in a medium containing many minerals in concentrations at or above those required, whereas freshwater animals live in a dilute medium and often must receive minerals in the diet. Although marine invertebrates are generally isotonic with the external medium (Lockwood, 1967), fishes are not. Since the ionic strength of the internal fluids of the bodies of freshwater organisms is higher than that of the water, there is a tendency for water to flow into the tissues from the surrounding medium. Thus to maintain the internal medium, freshwater forms must continuously rid their bodies of excess water. The opposite is true among marine fishes, in which intracellular fluids are dilute compared with the external medium and water tends to leave the cells against the concentration gradient as the organism attempts to equalize the ionic strength of the two solutions. Osmoregulation in many fishes seems to be related to calcium concentrations, which affect the permeability of membranes (Podoliak and Holden, 1965, 1966).

Freshwater fish generally do not drink, since they are continuously absorbing excess quantities of water. They do produce large amounts of dilute urine to maintain the proper internal salt concentration. Marine fish, on the other hand, drink copious amounts of water and excrete very small quantities of concentrated urine as a means of ridding their bodies of excess salt. The gills are also extremely important in maintenance of salt balance.

Kidney function varies not only with salinity but also with the physiological state of the animal. Many species are unable to tolerate large changes in salinity because they are not physiologically adapted to such stresses. Others may be able to adapt quite readily to a very rapid change from fresh water to full strength seawater.

Southern flounder, *Paralichthys lethostigma,* show a seasonal pattern in kidney function mirroring that which might be expected from knowledge of their normal migration pattern, although this species is very euryhaline at all times. *P. lethostigma* spawn offshore along the east coast of the United States in the fall. The postlarvae enter the estuaries in the winter, and the juveniles sometimes move up into fresh water during the spring and summer. In the fall adult flounders move back offshore to spawn. The kidney function of this species has been shown to accommodate this seasonal pattern of diverse salinity exposure even when the fish are maintained in constant salinity throughout the year (Hickman, 1968). Therefore less stress, as reflected by lower metabolic rates and more rapid growth, may be achieved by altering the salinity of the environment seasonally to reflect the salinity that exists in areas where wild fish would naturally occur. The economic benefits of this type of environmental manipulation must be determined, especially since it is known that the fish grow reasonably well at constant salinity (Stickney and White, 1973).

Minerals are elements that are not produced or destroyed as a result of their functions in life processes. Because of the abundance of many of them in most waters, it may be unnecessary to supplement aquacultural diets with minerals. Exceptions may occur in the case of certain minerals that are required in relatively high concentration, such as calcium and sodium, especially in soft fresh waters. Phosphorus may also require supplementation in the diets of aquatic organisms because of its generally rapid removal from dissolution in the water (as described below).

Seven minerals, called the major minerals, are required by animals for proper nutrition and contribute up to 80% of the inorganic components of animal dry weight. These are calcium, phosphorus, sulfur, sodium, chlorine, potassium, and magnesium. Additional required elements, which must be present in only small amounts and are known as trace minerals are iron, copper, iodine, manganese, cobalt, zinc, molybdenum, selenium, and fluorine. The general requirements for fishes have been reviewed by Phillips (1969), but specific mineral requirements vary among species.

Calcium and phosphorus are necessary in large amounts for skeletal formation and various metabolic functions. Freshwater fish such as channel catfish appear to be able to extract calcium from water unless the alkalinity is extremely low (Lovell, 1977b). Because of the high mobility of phosphorus, natural waters are often low in this mineral, and culture animals must obtain it from their food. If little or no natural food is available in a culture pond or tank, it may be necessary to supplement aquaculture diets with salts such as dicalcium phosphate. Significant amounts of calcium and phosphorus are available in such natural feed ingredients as fish meal and animal by-product meal.

Most practical diets contain suitable levels of trace elements to fulfill the requirements of aquaculture animals. Freshwater aquaculture diets are generally supplemented only with dicalcium phosphate, and in some cases sodium chloride. Lovell (1977b) recommended that a mineral premix containing a variety of other minerals be added to catfish diets if less than 15% of the ration is made up of animal products. Such recommendations have not been made for most other warmwater culture animals because there has been so little information available on them.

PROTEINS

Proteins are the structural components of the body. Although proteins may be metabolized for energy, one goal in aquaculture is to utilize as much of the dietary protein as possible for growth, allowing the carbohydrates and lipids to provide metabolic energy. Several types of protein have been identified. Those that make up connective tissues and tendons in animals are called fibrous proteins. Conjugated proteins have another food group attached to them. For example, mucoproteins and glycoproteins are attached to carbohydrates, and lipoproteins are attached to lipids. The enzyme systems present in living organisms also contain proteins. Enzymes act as catalysts in biochemical reactions.

Proteins are composed of long chains of amino acids. Each amino acid has the general formula:

$$R-\overset{\overset{\displaystyle H}{|}}{\underset{\underset{\displaystyle NH_2}{|}}{C}}-COOH \tag{19}$$

where R is an organic radical or hydrogen (in the case of glycine).

In addition to carbon, hydrogen, oxygen, and nitrogen, an amino acid may contain sulfur (methionine, cystine, and cysteine). A particular protein molecule may contain thousands of amino acid residues, but in most cases only about 20 different amino acids are involved (Figure 5.4).

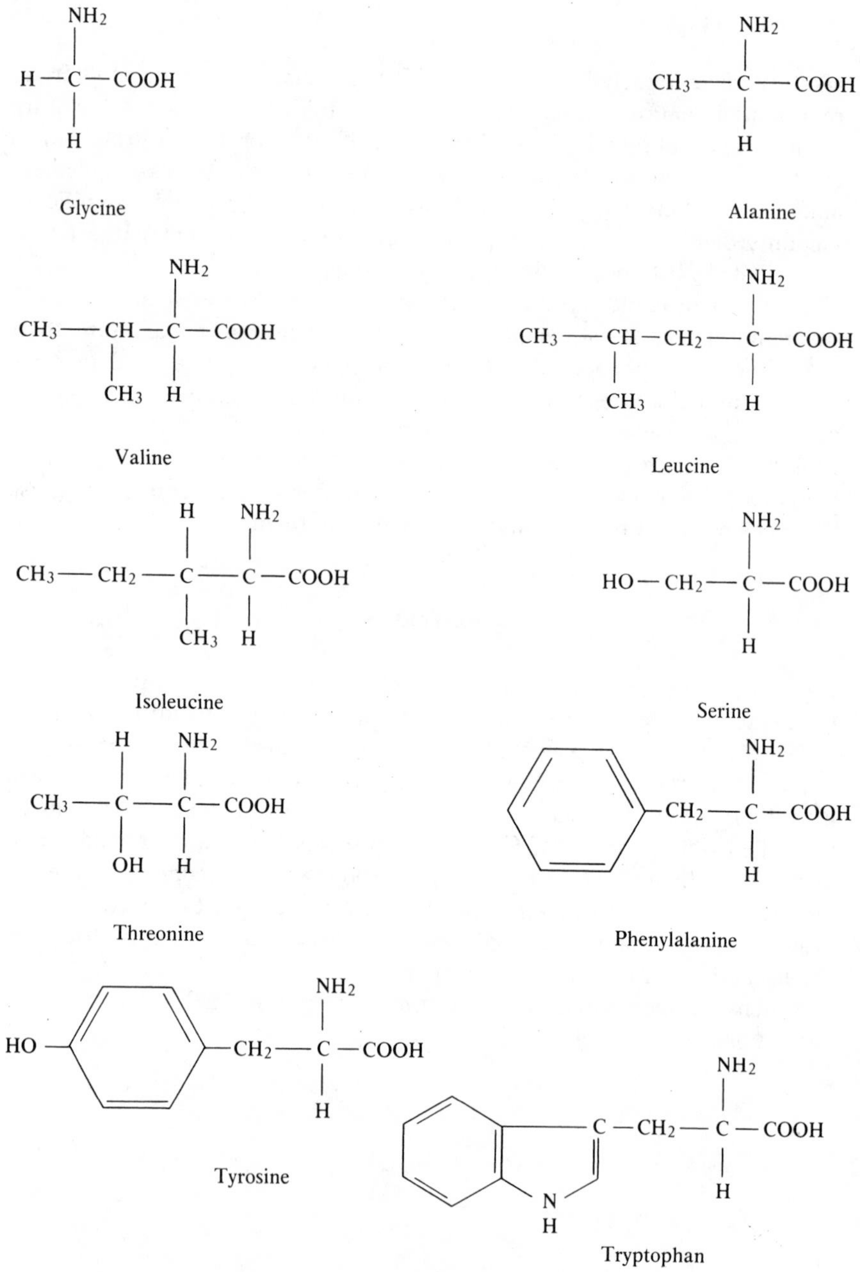

Figure 5.4 Structural formulas of representative amino acids. Adapted from White *et al.* (1964) and used with permission.

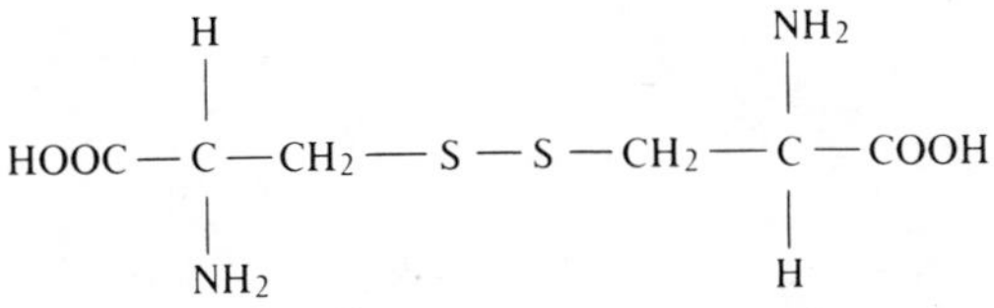

$$\text{HOOC} - \underset{\underset{\text{NH}_2}{|}}{\overset{\overset{\text{H}}{|}}{\text{C}}} - \text{CH}_2 - \text{S} - \text{S} - \text{CH}_2 - \underset{\underset{\text{H}}{|}}{\overset{\overset{\text{NH}_2}{|}}{\text{C}}} - \text{COOH}$$

Cystine

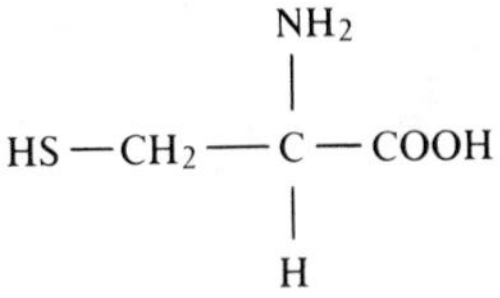

$$\text{HS} - \text{CH}_2 - \underset{\underset{\text{H}}{|}}{\overset{\overset{\text{NH}_2}{|}}{\text{C}}} - \text{COOH}$$

Cysteine

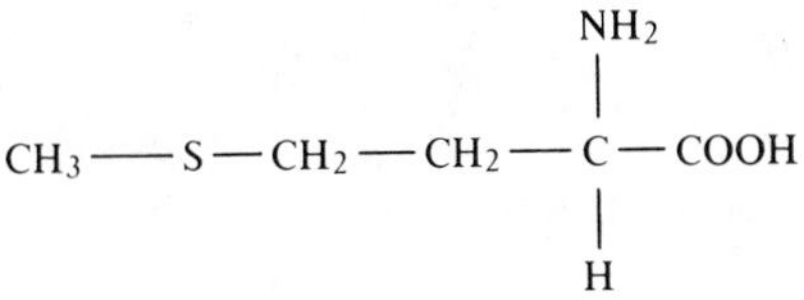

$$\text{CH}_3 - \text{S} - \text{CH}_2 - \text{CH}_2 - \underset{\underset{\text{H}}{|}}{\overset{\overset{\text{NH}_2}{|}}{\text{C}}} - \text{COOH}$$

Methionine

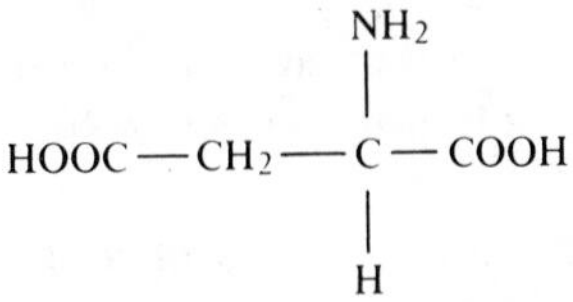

$$\text{HOOC} - \text{CH}_2 - \underset{\underset{\text{H}}{|}}{\overset{\overset{\text{NH}_2}{|}}{\text{C}}} - \text{COOH}$$

Aspartic acid

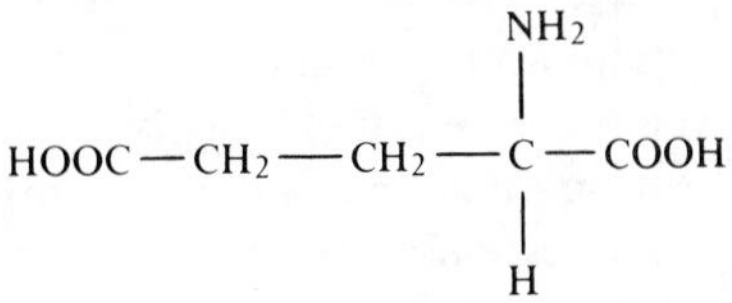

$$\text{HOOC} - \text{CH}_2 - \text{CH}_2 - \underset{\underset{\text{H}}{|}}{\overset{\overset{\text{NH}_2}{|}}{\text{C}}} - \text{COOH}$$

Glutamic acid

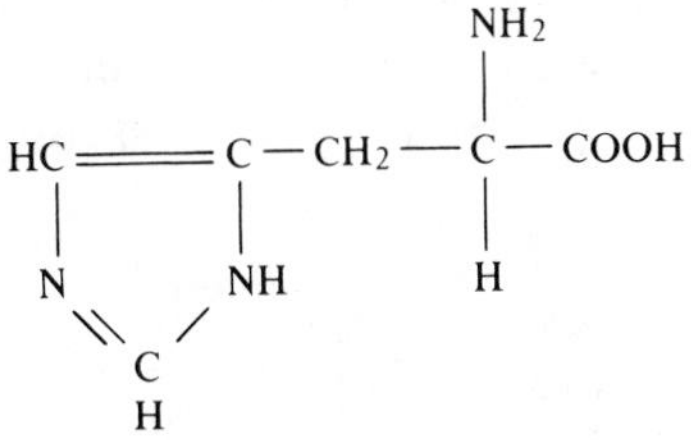

$$\text{HC} = \text{C} - \text{CH}_2 - \underset{\underset{\text{H}}{|}}{\overset{\overset{\text{NH}_2}{|}}{\text{C}}} - \text{COOH}$$

Histidine

$$\text{H}_2\text{N} - \underset{\underset{\text{NH}}{\|}}{\text{C}} - \text{NH} - \text{CH}_2 - \text{CH}_2 - \text{CH}_2 - \underset{\underset{\text{H}}{|}}{\overset{\overset{\text{NH}_2}{|}}{\text{C}}} - \text{COOH}$$

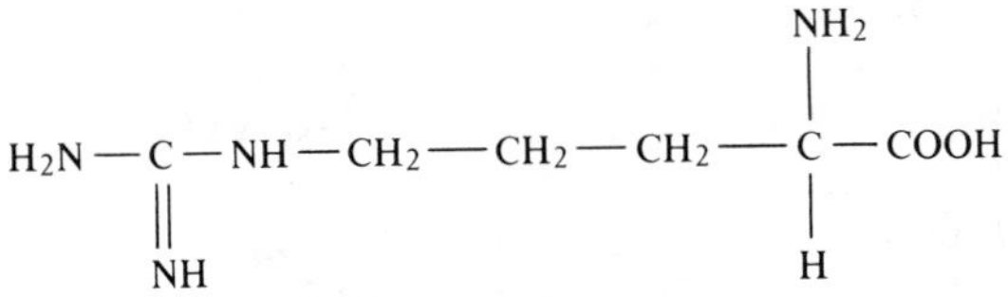

Arginine

Figure 5.4 (continued)

185

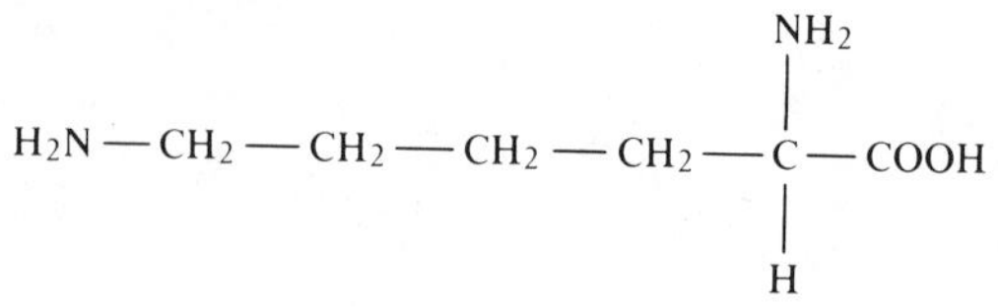

$$H_2N - CH_2 - CH_2 - CH_2 - CH_2 - \overset{\displaystyle \overset{NH_2}{|}}{\underset{\displaystyle \underset{H}{|}}{C}} - COOH$$

Lysine

Figure 5.4 (continued)

   Animals are able to synthesize some of their amino acids (nonessential amino acids) from carbohydrates and lipids, and other nitrogen compounds, or from other nonessential amino acids. Most species of animals require a dietary source of at least one sulfur-bearing amino acid, although when the required one is available in sufficient quantity, other sulfur-bearing amino acids can be synthesized from it.

   The essential amino acid requirements for most invertebrates of aquaculture importance remain to be established; however where information is available it appears that the essential amino acid requirements of the animals are similar to those of vertebrates, and indeed the requirements may be nearly universal in the animal kingdom (Harrison, 1975). Salmonid fishes and channel catfish and the rat share the same 10 essential amino acids (Table 5.2), and this list presumably applies to most other fish species.

Table 5.2   Qualitative Amino Acid Requirements for Trout and Channel Catfish (NAS, 1973)

Essential Amino Acids	Nonessential Amino Acids
Arginine	Alanine
Histidine	Aspartic acid
Isoleucine	Cystine
Leucine	Glutamic acid
Lysine	Glycine
Methionine	Proline
Phenylalanine	Serine
Threonine	Tyrosine
Tryptophan	
Valine	

Diets should always be formulated to include the essential amino acids in the proper ratios when such ratios are known. The diet for a particular species is usually chosen to fulfill the minimum daily nutritional requirements of the animal and to lead to a reasonable rate of growth. Since protein is the most expensive component of the diet, it is important not to supply quantities above those needed for optimum performance. In most cases determination of the protein quantity and quality requirements is the first step in outlining the nutritional requirements of an animal, since proper protein levels are critical for rapid growth and may reduce the cost of feed.

Fish meal generally promotes good growth of aquaculture animals when it is supplied as the primary protein source. It may be generally assumed that the amino acid balance in fish meal approximates the required ratios for cultured fish, and more than likely, for invertebrates also. When plant proteins are substituted for fish meal or are used to augment fish meal as a protein source in aquaculture diets, growth may be retarded, even though the percentage of total protein is not altered. This type of response has been attributed to low levels of methionine and lysine in some vegetable meals, although low mineral levels may also be a factor (Ketola, 1975). Whether the supplementation of diets deficient in amino acids with synthetic amino acids results in a return to optimum growth rate depends on the ability of the culture animal to absorb and metabolize the supplemental amino acids. The means by which dietary proteins are processed by the feed manufacturer prior to their incorporation into the diets may also affect the performance of animals offered such feed ingredients as soybean and other vegetable meals.

The utilization of synthetic free amino acids in channel catfish diets to supplement the amino acid composition of all-plant protein feeds has met with little success (Andrews, 1977). However when a diet containing only free amino acids is utilized, absorption appears to be acceptable (Dupree and Halver, 1970), although the growth generally is not as good as that demonstrated by fish on practical diets. The problem with supplementation of practical diets with purified amino acids may be related to the immediate availability for absorption as compared to practical protein sources. Absorption sites may become saturated, whereupon much of the amino acid component of purified diets would pass through the intestine and be evacuated from the animal. Amino acids present in practical feed ingredients are released much more slowly through enzymatic activity; thus only a small amount of any particular amino acid is available for absorption at one time. Other explanations for the observed response may be equally plausible.

For many years, while South American anchoveta meal was inexpensive, there was little interest in utilizing plant proteins in most aquaculture diets. With the dramatic increases in price that resulted from the failure of that fishery during two consecutive years in the early 1970s, other sources of protein were sought. Because all other animal proteins are as expensive, or more costly than South American fish meal, plant proteins, especially soybean meal, received a great deal of attention. Andrews and Page (1974) suggested that some type of growth factor was present in fish meal and absent from soybean meal, and replacement of fish meal with soybean meal in practical diets resulted in poorer growth of catfish. In addition, soybean meal contains a trypsin inhibitor, although that substance may be inactivated by heating the soybean meal during processing or pelleting procedures. Soybean meal protein diets have elicited poor growth not only from channel catfish (Andrews and Page, 1974), but also from plaice, *Pleuronectes platessa* (Cowey et al., 1971). The digestibility of soybean meal as it comes from the processing plant following normal treatment has been shown to be lower in rainbow trout, *Salmo gairdneri*, than that of animal proteins (Kitamikado et al., 1964). Even if all the problems with soybean meal cannot be worked out, in a few years it may be more economical to utilize this plant protein and accept a reduced rate of aquatic animal growth than to feed fish meal at high levels (Lovell et al., 1974). However the prices of soybean meal and fish meal have undergone rapid and sometimes extreme price fluctuations in recent years. In some instances, even when fish meal is in short supply, the difference in price between these items is not great. Thus it appears that diets containing at least some fish meal will continue to dominate the market.

The use of soybean meal and other vegetable proteins in the feeds of herbivorous aquatic animals may present fewer problems than have been encountered in the feeding of carnivores. Davis and Stickney (1978) fed *Tilapia aurea* diets containing various levels of fish meal, soybean meal, and combinations of the two proteins at various levels and demonstrated that growth was not significantly affected by protein quality as long as the proper quantity was administered.

The quantity of protein required by aquaculture animals is highly variable, but of primary importance to nutrition. The optimum protein level for a species changes as a function of age, condition, and reproductive state, and with variations in the environment. Young, rapidly growing animals need more protein than do older ones, and adult breeding animals may require higher levels during gamete formation than during periods of reproductive inactivity. Channel catfish fingerlings are generally fed diets containing between 25 and 35% protein, whereas the fry may receive as much as 50% protein (although the actual requirements for catfish fry have not been determined with any precision). Most trout diets contain approximately 40% protein, although young fish are fed diets with protein levels of 50% or more. Increased growth has been demonstrated in plaice fed casein when the dietary protein levels reached as high as 70% (Cowey *et al.*, 1970). Later experiments with rations containing animal protein sources other than casein resulted in optimum growth of plaice at protein levels of 52% of the diet (Cowey *et al.*, 1972).

Because of the high price of all types of conventional protein sources during certain years, and because protein does represent the most expensive ingredient in the diet, other, more exotic proteins have been evaluated for their use in aquaculture. Many of them are presently waste products left over from the processing of more conventional crops. Among those that have been used experimentally for feeding aquaculture animals are mixtures of blood meal and cattle rumen contents (Reece *et al.*, 1975), dried poultry waste (Stickney *et al.*, 1977a, egg-waste protein (Davis *et al.*, 1976), paper processing sludge (Orme and Lemm, 1973), brewers single-cell protein (Windell *et al.*, 1974), and coffee pulp (Bayne *et al.*, 1976). Such items may be available only locally and difficult to store, but they offer potential in some instances.

LIPIDS

Lipids are defined as the portion of animal or plant tissue that can be extracted in nonpolar solvents—for example, ether, chloroform, and benzene (White *et al.*, 1964). Lipids occur as fatty acids, triglycerides (neutral fats), phospholipids, glycolipids, aliphatic alcohols and waxes, terpenes, and steroids. For many aquaculture animals the main source of dietary lipids in prepared diets is triglycerides. Triglycerides are formed when three free fatty acids are combined with glycerol:

$$
\begin{array}{ccc}
R-COO-CH & R-COOH & CH_2OH \\
R'-COO-CH_2 + 3H_2O \rightarrow & R'-COOH + & HCOH \\
R''-COO-CH & R''-COOH & CH_2OH \\
\text{Triglyceride} & \text{Free fatty} & \text{Glycerol} \\
 & \text{acids} &
\end{array}
$$

During digestion, certain enzymes cleave the fatty acids from the glycerol molecule through hydrolysis. Following absorption into the blood, the free fatty acids may once again be converted to triglycerides. The three fatty acids associated with each glycerol molecule may be the same or different (as indicated by R, R′, and R″). At least in the case of channel catfish, there seems to be no advantage in feeding free fatty acids as opposed to triglycerides (Stickney and Andrews, 1972), since the two forms result in similar growth rates when present in diets that are otherwise identical.

Saturated fatty acids are those that contain no double bonds, whereas unsaturated fatty acids may contain one (monounsaturated) or more (poly-unsaturated) double bonds. All fatty acids are composed of carbon, hydrogen, and oxygen and are generally acyclic, unbranched molecules containing an even number of carbon atoms. Table 5.3 lists examples of various fatty acids common in animal tissues.

Table 5.3   Common Names, Structural Formulas, and Shorthand Notations for Representative Fatty Acids (Adapted from White *et al.*, 1964)

Common Name	Structural Formula	Shorthand Notation
Caproic acid	$CH_3(CH_2)_4COOH$	$6:0$
Caprylic acid	$CH_3(CH_2)_6COOH$	$8:0$
Capric acid	$CH_3(CH_2)_8COOH$	$10:0$
Lauric acid	$CH_3(CH_2)_{10}COOH$	$12:0$
Myristic acid	$CH_3(CH_2)_{12}COOH$	$14:0$
Palmitic acid	$CH_3(CH_2)_{14}COOH$	$16:0$
Oleic acid	$CH_3(CH_2)_7CH{=}CH(CH_2)_7COOH$	$18:1\omega9$
Stearic acid	$CH_3(CH_2)_{16}COOH$	$18:0$
Linoleic acid	$CH_3(CH_2)_4CH{=}CHCH_2CH{=}CH(CH_2)_7COOH$	$18:2\omega6$
Linolenic acid	$CH_3CH_2CH{=}CHCH_2CH{=}CHCH_2CH{=}CH(CH_2)_7COOH$	$18:3\omega3$

The polyunsaturated fatty acids (PUFA) have been divided into three families named after the shortest chain length fatty acid that is representative of each: oleic, linoleic, and linolenic acids. To simplify the classification of fatty acids and to aid in the recognition of members of the three families of PUFA, the omega (ω) system of nomenclature is commonly utilized. Members of the oleic acid family are called $\omega9$ fatty acids, and those in the linoleic and linolenic fatty acid families are referred to as $\omega6$ and $\omega3$ fatty acids, respectively. In the omega system linoleic acid is abbreviated $18:2\omega6$ (Table 5.3), where 18 is the number of carbon atoms in the fatty acid molecule, 2 is the number of double bonds, and $\omega6$ is the location of the first double bond relative to the methyl (CH_3) end of the molecule.

Aquaculture animals appear to be able to interconvert fatty acids within families but are unable to convert from one family to another (Owen *et al.*,

1975). For example, $18:3\omega3$ may be converted to $20:5\omega3$ and $22:6\omega3$ even if the latter two acids are not provided in the diet, but $18:3\omega3$ cannot give rise to fatty acids in the $\omega6$ or $\omega9$ families.

The melting point of lipids is proportional to their degree of unsaturation, except for the very low molecular weight fatty acids, which are all liquids at room temperature. Beef tallow, which is high in oleic acid and saturated fatty acids, is a solid at room temperature, whereas corn oil (high in linoleic acid) is a liquid under the same conditions. Table 5.4 gives the fatty acid compositions of some typical dietary lipids. Fats that become solids at relatively low environmental temperatures might be thought of as poor lipid sources for aquaculture diets, especially during cold weather, since the body temperature of the animals is essentially the same as that of the environment. However beef tallow has been shown to elicit growth rates equivalent to those achieved with fish oil and better than those obtained with safflower oil (high in linoleic acid) at environmental temperatures as low as 20 C (Stickney and Andrews, 1971).

Most commercial aquaculture diets do not exceed about 8% lipid, largely because of manufacturing problems encountered when high levels of lipid supplementation are attempted. Dupree (1968) fed channel catfish semipurified diets containing lipids at 8 and 16% of the ration and reported no significant difference in growth. Differences probably are related to total dietary energy and percentage of protein. Channel catfish have demonstrated excellent growth rates on diets containing 10% lipid (Stickney and Andrews, 1971, 1972). Other species of aquaculture animals may require higher or lower levels of lipid, but for most the optimum has not been determined.

Nutritional diseases, such as fatty liver, may occur in response to excessive dietary lipid levels. In addition, the amount of visceral fat deposited in an animal may be a function of dietary lipid levels. Such fat is discarded during processing and leads to a lower dress-out percentage for the crop than might otherwise be realized.

Trout appear to have little or no requirement for $\omega6$ fatty acids, but do require $\omega3$ fatty acids in the diet (Lee et al., 1967; Yu and Sinnhuber, 1975). The essential fatty acid requirements of rainbow trout have been outlined (Castell et al., 1972a, 1972b); little is known about the requirements of other aquaculture species, however, although it is likely that the requirements of most fishes will be similar (Cowey and Sargent, 1977).

The differences between the probable fatty acid requirements of fish and higher vertebrates become readily apparent when the ratio of $\omega6$ to $\omega3$ fatty acids is examined for various types of organism. In general, fish show $\omega6:\omega3$ ratios of about 0.1. Similarly, marine birds and seals, which consume large quantities of fish, show ratios of 0.6 and 0.8, respectively,

Table 5.4 Fatty Acid Composition of Several Lipids that Have Been Incorporated into Aquaculture Diets (From Stickney, 1971)

Dietary Lipid	Fatty Acid	Percentage in Diet
Coconut oil	8:0	3.1
	10:0	4.6
	12:0	27.1
	14:0	22.3
	16:0	17.9
	18:0	6.1
	18:1ω9	14.6
	18:2ω6	3.8
	Others	0.4
Beef tallow	14:0	2.9
	16:0	18.8
	16:1ω7	5.1
	18:0	10.9
	18:1ω9	57.5
	18:2ω6	2.1
	Others	2.5
Safflower oil	16:0	5.7
	16:1ω7	1.9
	18:0	1.9
	18:1ω9	12.4
	18:2ω6	72.5
	18:3ω3	2.9
	Others	2.7
Fish oil	14:0	5.1
	16:0	17.0
	16:1ω7	9.4
	18:0	3.2
	18:1ω9	16.8
	18:2ω6	2.5
	18:3ω3	3.1
	20:3ω9	0.2
	20:3ω6	0.8
	20:4ω3	2.0
	20:5ω3	17.2
	22:5ω3	2.9
	22:6ω3	13.2

reflecting the high levels of ω3 fatty acids available in their diets (Richardson *et al.*, 1962). The same study reported that the ratio in chicken livers is 9, which may reflect the fish meal in poultry diets, whereas the ratios in rat and beef livers are 200 and 500, respectively.

Because of the high energy level present in lipids and their relatively low cost compared with protein, it is important that the lipid requirements of aquaculture animals be established, along with the degree to which lipids can be utilized to spare protein. Certain lipids may lead to rapid growth but could result in off-flavors in the final product if they were the only fats present. Lipids such as beef tallow can lead to a product that is high in saturated fat, for example. It would be possible to design a feeding regime under which diets containing high levels of saturated fats or fish oil would be provided during most of the growing season (which could lead to a highly saturated product or one that had an undesirable, fishy flavor). At some time prior to harvest, the diet could be changed to one containing corn oil or some other lipid high in linoleic acid, leading to an alteration in the body lipid composition of the fish and producing a more acceptable flavor.

In addition to more rapid growth on diets containing beef tallow or fish oil as the primary lipid (Stickney, 1971; Stickney and Andrews, 1971, 1972), a reduction in the cost of the diets could be achieved in some cases (Stickney, 1977). Recent information on practical diets supplemented with beef tallow and fish oil (with the basal lipid also being beef tallow) indicate that for channel catfish fry there is sufficient residual linoleic acid in such ingredients as corn meal to offset the undesirable influence on body composition of high levels of ω9 or ω3 fatty acids (Yingst, 1978).

CARBOHYDRATES

Chemically speaking, carbohydrates are the simplest of the energy-containing food groups. Carbohydrates contain the elements carbon, hydrogen, and oxygen and are represented by sugars and starches in the plant and animal kingdoms. The types of carbohydrates are simple sugars (monosaccharides) such as glucose, fructose, and galactose; compound sugars (disaccharides) formed by the chemical union of two simple sugars, such as sucrose (glucose + fructose), maltose (glucose + glucose), and lactose (glucose + galactose); and complex sugars (polysaccharides), which contain the high molecular weight carbohydrates such as starches and celluloses. The polysaccharides are formed by the combination of large numbers of simple sugars.

Carbohydrates provide a significant amount of energy in mammals but may be less important in at least some aquaculture animals, partly because the carbohydrate present in many aquaculture diets is largely in the form of high molecular weight polysaccharides, which may be poorly digested (NAS, 1973). Although no specific recommendations for carbohydrate levels have been put forth for most species, it is known that excessive carbohydrate can lead to dangerous levels of liver glycogen in trout (NAS, 1973) and that this condition may be fatal. Since relatively high levels of carbohydrate in channel catfish diets do not lead to glycogen problems (Wilson, 1977a), catfish diets are often higher in carbohydrate than are diets formulated for the salmonids, accounting to some extent for the lower cost of the former feed type. Trout appear to absorb simple sugars better than starches (NAS, 1973), whereas catfish seem to perform well on diets containing high (up to 40% of the diet) levels of starch.

Channel catfish have been shown to possess intestinal microflora containing cellulase activity (Stickney and Shumway, 1974). Cellulase is an enzyme that breaks cellulose into simpler compounds. Vertebrates do not seem to be able to synthesize the enzyme, but they may harbor microorganisms that do produce cellulase. The degree to which the presence of cellulase may augment the efficiency of utilization of cellulose in the diet appears to be extremely limited. Leary and Lovell (1975) could find no identifiable value when they added fiber to channel catfish diets.

The presence of cellulase activity in the intestinal tract of fishes appears to have little relationship to the feeding habits of a particular species of fish (Stickney and Shumway, 1974; Stickney, 1975). For example, *Tilapia aurea*, a rather strict herbivore, does not show cellulase activity. The presence of cellulase in carnivores is probably related to the occurrence of the enzyme in the food organisms being consumed by the fish, although this relationship may be somewhat obscured in aquaculture. Stickney and Shumway (1974) could not demonstrate cellulase activity in the feed of *Ictalurus punctatus* maintained for several months in an open water system supplied from a well, and they were able to destroy the activity in catfish that were given antibiotics. Thus the microorganisms containing the enzyme apparently were maintained in the digestive tract of the fish long after consumption (the fish had been moved into the facility from outdoor culture ponds as fingerlings).

Carbohydrate, like lipid, may spare protein in aquaculture diets. It is usually necessary to determine the specific type and amount of carbohydrate that should be included in the diet of a particular species; however a reasonable approach would be to provide carbohydrate of a quantity and quality normally consumed by the animal (implying a fairly low level for carnivores). Effective delimitation of the actual requirements for any nutrient will be available only after intensive research, which must be accomplished species by species.

PHYSICAL AND CHEMICAL PROPERTIES OF DIETS

Before discussing the formulation and production of prepared diets, we consider the physical and chemical properties of animal diets that are related to flavor, odor, texture, water stability, and color. Many species will swallow nearly any type of suitable-sized feed particle once they have learned to feed on prepared diets. In such cases odor, flavor, and texture may have little significance. However other species refuse diets of one type while readily accepting the same formulation if one or more characteristics of that diet are changed with respect to the factors mentioned above.

Very little research has been conducted on the effects of flavor and odors in aquaculture diets. A great deal more study should be undertaken, especially with species that seem to be reluctant to accept prepared diets, and those for which certain life stages now must be fed natural living foods. Texture may also be an important aspect of a diet. Some species appear to avoid hard pellets, although once that type of diet has been recognized as food, it may become acceptable. Species that tear their food apart may have to be provided with feed of a texture different from that given to fish that swallow the food whole or nibble off small pieces. Obtaining the proper texture while retaining water stability may pose an additional problem.

The relationship between color and the acceptability of a feed has not been studied in any detail, although it is known that various species of aquatic animals are able to see color. Practical diets are usually brownish; purified and semipurified diets are often white. The aspect of feed attractiveness should be studied, especially with respect to larval animals, which might take a prepared diet if properly presented. For example, larval marine fishes and other types of aquaculture animals often refuse feed that is not moving. The feed may not have to be living at all if it can be kept in motion to simulate a live food organism. This might be done by preparing neutral density feed particles, which would remain suspended in the water column and would move about with the circulation pattern in the culture chamber. Aeration devices of various kinds would aid in suspending such particles.

The feed manufacturer can make diets with a virtually infinite variety of flavors and odors. These may be similar to flavors and odors enjoyed by man, or they may be quite different. Examination of the food habits of the species to be fed may provide some indication of the type of flavors and odors that might be attractive to the culture animal. A fish that is not attracted to a typical food pellet might respond more positively if that pellet had the odor of shrimp or some other preferred food item.

BINDERS

Many practical diets hold together well when exposed to the pelleting process, and thereafter to the water in a culture chamber. Others do not hold up well under either condition. If the diet is difficult to pellet or is unstable in water, it is often necessary to add a binding agent to the formulation. Binders are required for most purified and semipurified diets; in pelleted practical diets the starch serves as the binding agent.

Carboxymethyl cellulose (CMC), algin (an extract from kelp), gelatin, and a variety of other binding agents have been successfully utilized in experimental diets. The binder is often an indigestible substance that imparts no metabolizable energy to the feed.

The actual amount of binding agent required in a diet depends on the composition of the diet and the type of binder used. In most cases 1 or 2% of the diet is composed of binder; however added water stability can be achieved by increasing the level. To determine the proper percentage to add to a particular diet formulation, it is necessary to prepare several small batches of the experimental diet with different percentages of the binding agent selected. Then each mixture is pelleted separately. Binder in excess of that required to achieve the desired pellet texture and water stability should not be utilized.

TOXICANTS

Various toxic substances may occur in both natural and prepared diets. A variety of trace metals can lead to toxicity as discussed in Chapter 4, although in most cases the levels in aquaculture diets should be within acceptable limits unless a contaminated batch of feed ingredients is obtained.

Aflatoxin, a toxic metabolite of a mutant blue-green mold *Aspergillus flavus*, sometimes occurs as a contaminant of oilseed meals. This substance is known to cause cancer of the liver (hepatocarcinoma or hepatoma) in trout. Hepatoma and a variety of other mold-induced toxicity problems of fish have been reviewed by Friedman and Shibko (1972). Mold growth may be induced by improper feed storage. Moldy feed should not be fed to aquaculture animals.

Toxic algae, primarily blue-green species, have been reported in fish ponds (Gorham, 1964) under normal circumstances although most species of blue-green and other types of algae that occur in aquaculture chambers and other surface waters are not associated with toxicity problems. The promotion of algal blooms through fertilization normally does not lead to blue-green dominance except when high levels of organic fertilizers are used; then the species that blooms is generally not toxic. Such species as *Anabaena flos-aquae*, *Aphanizomenon flos-aquae*, and *Microcystis aeruginosa* (reviewed by Gorham, 1964) have been linked with toxicity in a variety of organisms, including the higher vertebrates.

In saline environments dinoflagellates such as *Gymnodinium brevis* and *Gonyaulax tamarensis* are responsible for the so-called red tides, which have contributed directly to the death of millions of fish or have led to the accumulation of toxins in the tissues of oysters and other mollusks, causing paralytic shellfish poisoning in man. Toxins produced by these normally occurring and generally harmless dinoflagellates reach lethal concentrations when the algae bloom (Gorham, 1964).

Incidents of mortality in aquaculture due to toxic algae have not been widely reported, although certain losses of animals for reasons yet unknown may be attributable to this problem. A more common occurrence, especially among channel catfish producers, is the production of fish with off-flavors. Muddy-tasting catfish can result from chemicals produced by certain blue-green algae and a variety of other microorganisms. Some work has gone into determining the cause of off-flavors in a variety of fishery products, not only those raised by aquaculturists. At present there are no management guidelines for the aquaculturist to ensure that a crop will be free of off-flavors. Intensive culture systems seem to be much less likely to develop the flora responsible for off-flavors than do extensive systems.

It is difficult to predict which of several ponds might produce fish with off-flavors during a given year. The simplest test is to cook and eat representatives from each culture chamber. If off-flavors are detected, it may be possible to reduce or eliminate the problem by placing the affected population of fish in tanks through which flowing well water is run. Removed from the influence of the contaminating substance, the culture animals often metabolize the undesirable chemical and recover in flavor.

Generally the presence of microorganisms that cause off-flavors is not detrimental to growth or survival of aquaculture animals, but it certainly affects consumer acceptance. Since many producers must develop local markets for their product and must depend on repeat customers to support their business, it is imperative to offer the highest quality product at all times.

FEED FORMULATION AND PELLETING

Once the desired nutrient content of a diet is delimited, practical feeds may be formulated to provide those nutrients at the proper levels. Usually the nutritionist obtains an indication of the required energy level for a particular species of animal, along with its amino acid, fatty acid, carbohydrate, vitamin, and mineral requirements. Then a selection is made of practical feed ingredients that will, when mixed, give those nutrients at the proper levels.

Many feed manufacturers are utilizing computers to design least-cost formulations. Since protein can be supplied by a variety of grains and animal products, a least-cost formulation may include a high percentage of soybean meal one day, corn meal another, and fish meal the third, depending on the price of each product at the time of purchase. When high levels of grain are used in fish diets, a small amount of fish meal is generally present to provide sufficient levels of certain amino acids in which vegetable protein is deficient. Similarly, other components of the diet may be adjusted on a day-to-day basis to retain a certain set of minimum standards for each batch of feed.

Bags of feed are labeled with a tag that tells the minimum amounts of protein and lipid contained in the diet and the maximum amount of fiber. These labels also list the ingredients of the feed but do not indicate the amount of each, since actual amounts change with each batch of feed produced. Proximate analyses of different batches of feed from the same manufacturer vary to some extent but usually meet the guaranteed analysis stated on the label. The performance of such feeds is not greatly affected by alterations that occur in the formula.

Dry feed rations are available in floating and sinking pellets of various sizes. For fry and larvae very small particles must be used to allow the animals to feed successfully. Small particles are known as granules and crumbles. For very young animals the particles are often less than 0.5 mm in diameter. If they were larger, the fish or invertebrates being fed would be unable to swallow them. The size of feed particles is gradually increased as the animals grow until pellets of several millimeters length and diameter are fed. Usually pellets of a diameter larger than about 5 mm and length exceeding 10 to 15 mm are not utilized, even in conjunction with large brood animals (Figure 5.5).

Figure 5.5 Feed particle size requirements depend on the species under culture and its size. The smallest granules shown are starter feeds for fish fry, whereas the largest may be fed to certain species from juvenile to adult sizes.

Pellets may be round, cylindrical, or a variety of other shapes. One experimental shrimp ration produced by a commercial manufacturer is in the shape of concave discs several millimeters in diameter. Shrimp have also been reared successfully on spaghetti-shaped pellets several centimeters long and on pellets of the standard sizes. Because the rate of consumption of food by shrimp is very low, the pellets must be highly water stable.

Sinking pellets are produced by forcing the ground and mixed dry feed ingredients through a die of selected diameter under pressure in a pellet mill. In some cases steam is applied to assist in the binding process. Extremely high temperatures and pressures are not reached, and there is little or no cooking of the product during the process. As the material leaves the die, a knife blade cuts the pellets into the desired length. The pellets produced generally have a water stability of a few minutes to an hour, although with suitable binders this period can be extended significantly.

In the extrusion process, by which floating pellets are manufactured, the ground and mixed feed ingredients are once again forced through a die, but in this case they are exposed to much higher temperature and pressure than in the pellet mill. As the material exits from the die, the pressure drop causes the starches in the pellets to expand. The resulting feed particle will often float in the water for up to 24 hours.

The processes of pellet manufacture for sinking and floating fish feeds have been described by Robinette (1977). Various steps in the manufacturing process other than the pelleting operation itself will affect pellet quality. The ingredients used, the extent of grinding, the moisture content, and even the individual responsible for operating the equipment can greatly influence the final product. Purified and semipurified diets are especially difficult to pellet by either process. It is often necessary to run fairly large quantities of material through the pelleting process before food particles with acceptable characteristics are obtained. A series of experimental diets may call for only a few kilograms of each ration, and although the different formulations may vary only slightly, each may present different problems to the manufacturer.

Floating rations are somewhat more expensive than sinking diets because the extrusion process is more elaborate and involves the expenditure of considerably more energy than does the pellet mill. The higher cost is the primary disadvantage of floating rations and it can be crucial; when hundreds of tons of feed are involved over the period of a growing season, even a few dollars per ton can make a significant difference in feed cost between sinking and floating rations. Floating feeds are often avoided when demersal animals are reared, since it is difficult to train such animals to approach the surface to eat.

Floating pellets have two important advantages over sinking rations. First, floating feed has, in general, much better water stability (although certain binders can be used to stabilize sinking pellets); thus the diet will not dissolve before the animals have had an opportunity to eat. Culture species that feed slowly must have water-stable pellets. A pellet that dissolves before it can be consumed represents a waste of money and can increase the BOD of the water system, leading to oxygen depletions. In addition, feed particles offer surfaces for the colonization of bacteria and fungi. Disease epizootics could be encouraged in culture systems that have accumulations of waste feed particles.

Aquaculturists who select floating feeds over those that sink can observe the animals eating. This is of primary importance because unless the feeding of animals provided with sinking feeds is accompanied by a great deal of activity, or the feeds are attacked immediately upon distribution in the water, the culturist may not know whether the feed is being accepted. When animals come to the surface to feed, they can be observed directly and the amount of the food consumed can be determined. When animals that are not overfed become sluggish in their feeding behavior, it is often a sign of water quality deterioration or the onset of a disease problem. Recognition of that symptom can allow the culturist to diagnose and treat the specific condition before it leads to heavy mortality. If, on the other hand, the animals are fed sinking pellets and cease to accept them, the culturist may not suspect that this has happened until mortalities begin to appear.

In some cases, of course, the water may be clear enough to permit the culturist to observe feeding activity even when sinking pellets are utilized. Relatively small open and closed recirculating tank culture systems are examples of systems in which sinking pellets may well be appropriate.

Floating rations are usually employed by cage culturists, since it is assumed that if sinking pellets were offered, much of the feed would fall through the bottom or be carried through the sides of the cages. This would not necessarily be true for species that rapidly consume their feed, especially if the feed were offered slowly. Special formulations for cage-reared channel catfish are available from commercial feed manufacturers and are generally more expensive than pond formulations. The additional cost of employing such rations should be considered when the economics of establishing a cage culture facility are evaluated.

Pelleted feeds can be purchased in 22.7 kg (50 lb) bags or in bulk. Bulk feeds are several dollars per ton less expensive than bagged feeds, and the former are certainly desirable if large numbers of animals are being fed. Bulk feeding does require investment in a storage container, and bags must also be stored in a building to prevent them from getting wet.

Various commercial feed companies that market throughout the United States have developed formulations especially for aquaculture animals. At present, the most common diets for aquatic animals are those formulated specifically for trout, channel catfish, and shrimp (both freshwater and marine). Small local feed mills may be willing to formulate and manufacture aquaculture diets if the demand is of sufficient volume to make production economically profitable. It may be feasible for a group of culturists to agree on a feed formulation and approach their local feed mill on a cooperative basis, since the total quantity of feed required may be sufficient to entice the mill to enter the aquaculture feed production business. In some parts of the United States it is difficult to obtain fish or invertebrate feed locally because the area feed dealers are unaware of the demand for such diets. Generally, however, one or more of these dealers represents a company that does manufacture aquaculture diets, and the local dealer can obtain details of ordering from the supplier. The aquaculturist should determine the amount of time required for delivery of feed in the area in question so that orders may be placed far enough in advance to ensure delivery by the time feed is required.

Some types of feed are difficult to obtain other than at specific, often brief periods during the year. For example, very fine feed particles for fry production are usually available only from the major feed companies in the spring, since that is when most warmwater and trout culturists require feeds of that type. Orders often must be placed well in advance of the season so that the companies can determine the total amount of starter feed to prepare. A culturist who is rearing a fall-spawning marine fish along the Gulf of Mexico coast may have to obtain feed from an obscure source, or may be forced to store feed for a fairly long time prior to use. The latter option is generally the less favorable because of storage degradation.

FEEDING STRATEGIES

In general, feeding techniques used for channel catfish culture can be applied to other species of warmwater animals, both vertebrate and invertebrate. Thus this section uses *Ictalurus punctatus* as the primary example to demonstrate feeding practices. For many species special strategies have not been worked out, but techniques unique to each may evolve as more knowledge is obtained regarding its nutritional requirements and feeding behavior.

When channel catfish are hatched indoors the fry will accept prepared diets immediately upon initiation of feeding following yolk-sac absorption. In ponds the general practice is to fertilize prior to spawning and allow the newly hatched fry to graze on plankton. Prepared feed is also provided; however it may not be consumed as readily as natural feed early in the lives of the fish. If laboratory-hatched fish are introduced to prepared diets in a plankton-free environment and are later released into ponds, fertilization of the ponds receiving the fish should still be done because the fish will be able to utilize natural foods as well as prepared diets. In addition, fertilization will reduce the chances for invasion of the ponds by aquatic macrophytes.

Tilapia spp. will accept prepared diets immediately upon initiation of feeding, as will some other species. However many species will not, and as discussed previously, it is often necessary to provide living foods for the young of such animals as *Macrobrachium rosenbergii*, *Penaeus* sp., *Sciaenops ocellata*, and *Paralichthys* spp. In each case conversion to prepared diets must be put off until the animals enter the juvenile stage. In the future it may be possible to design prepared diets that will be accepted by those animals immediately after they begin to eat.

Channel catfish fry, as well as a number of other nonsalmonid species, are often initially fed trout starter rations (finest particles shown in Figure

5.5) that contain about 50% protein in most formulations. The pellet size is increased as the fish grow, and the protein percentage is reduced until, by the time the fish are a few weeks of age, they are placed on the typical catfish ration containing about 30% protein. Fish reared in intensive culture systems can be carried from hatching to market size on prepared diets without difficulty (Stickney et al., 1972).

Complete rations are utilized when catfish are produced under intensive culture conditions such as in tanks, raceways, and ponds that have very high stocking densities. Supplemental feed may be utilized in extensive culture systems such as low density ponds if there is ample natural food available to furnish the vitamin and part of the protein requirements of the animals. Supplemental catfish diets often contain about 25% protein as compared with 30% protein in complete diets.

One of the primary rules in feeding any species of aquatic animal is that the organisms should not be overfed. Some feed may be consumed in excess of that which is required for optimum growth, but food conversion efficiency will suffer, and food that is passed through the digestive tract unabsorbed, plus that which is not ingested, will add to the BOD of the water system.

Channel catfish and other aquatic species can exist for long periods without feed (except during the first few weeks of life, when starvation may occur within a very short period because growth is extremely rapid, metabolic rate is high, and energy reserves are small), and they can be maintained in excellent condition without allowing growth for extended periods if fed at a maintenance level. Thus except during the first few weeks of life, it is always better to underfeed than to overfeed. If fish do not consume all the feed offered within 15 to 30 minutes of presentation, the feeding rate may have to be reduced. In addition, the culturist should attempt to determine whether poor water quality or disease is responsible for the animals' refusal to accept feed. Some organisms, such as shrimp, feed very slowly, and no time limit for complete ingestion of a meal can be imposed.

Channel catfish fry weighing less than 1.5 g should be fed about 8 times daily (every 3 hours) if they are maintained in an intensive culture system that does not provide a source of natural food. The rate of feeding can be decreased to 4 times daily after the fish exceed 1.5 g (Murai and Andrews, 1976). Rapidly growing fry may require as much as 10% of their body weight daily at a weight of 0.25 g, but this is reduced to about 5% of body weight daily when the fish reach 4 g (Murai and Andrews, 1976). Some culturists feed fry at up to 50% of body weight daily to ensure that each fish is exposed to a sufficient quantity for its needs before the feed settles to the bottom or leaves the system. Overfeeding at this time is not unusually expensive, since the total quantity utilized to maintain even several hundred thousand fry is relatively small. To limit bacterial production and fouling of the water, however, extra precautions should be taken to clean excess feed and waste products from fry tanks every day.

During the summer fingerling channel catfish in ponds should be fed in the morning after the dissolved oxygen has reached an acceptably high level, and in the afternoon when the water is not at its temperature maximum for the day. Studies have shown that catfish grow best when fed to satiation twice daily (Andrews and Page, 1975), and that there is no advantage in terms of growth rate or food conversion efficiency in feeding only once daily or more frequently than twice daily.

The percentage of body weight to feed daily varies with water temperature because the metabolic rate of the fish is under environmental temperature control. In general, if the temperature of the water is above 32 C (90 F), catfish should be fed no more than 1% daily. Between 21 and 32 C (70 to 90 F), catfish are generally fed 3% of body weight daily. This amount is reduced to 2% between 16 and 21 C (60 to 70 F), and to 1% between 7 and 16 C (45 to 60 F). Below 7 C feed may not be accepted (Bardach et al., 1972).

Although many catfish culturists do not feed at all during the winter, some growth can be expected if the fish are fed on warm days or on alternate days in the southern United States (Lovell and Sirikul, 1974). The expense of low quantity winter feeding is not great, and such feeding will prevent the loss of weight that often occurs when no feed is offered over the winter months. During warm winters increased growth may be significant, and the time required for overwintered fish to reach market size can be considerably reduced compared with that of unfed fish. A feeding rate of 1% of body weight daily on days fed is probably sufficient during the winter.

Aquaculture animals should be fed at about the same time of day. Many species soon become accustomed to a particular feeding schedule, and often to the person feeding them. A certain amount of shyness may be exhibited by such species as channel catfish when a stranger accompanies the person who normally does the feeding, or if a stranger appears alone to feed. This shyness rapidly disappears, and under most circumstances the fish begin normal feeding behavior within a few minutes. Even large groups of people generally do not keep fish from feeding once the animals have become accustomed to the situation.

Many species anticipate feeding time to the point of waiting in the particular part of the pond in which they know feed will be provided. *Tilapia aurea* that are given prepared feed will respond to the vibrations made by a person walking toward the pond. Many culturists believe that the animals are able to recognize them as individuals when the same person is present on a routine basis.

Aquaculture animals, like people, prefer to eat seven days a week. Many culturists do not feed on Sundays, or in some cases on any day during the weekends or holiday periods. A few claim that the animals require at least one day a week without feed to allow their digestive systems to rest. This belief has little validity and probably reflects only the laziness of the culturist.

MANUAL AND AUTOMATIC FEEDING

For small ponds the amount of feed required daily can be easily distributed from the pond bank by hand (Figure 5.6). In large ponds some type of feed blower, auger, or other device that can be towed behind a tractor or truck, carried in a boat, or placed on an airplane may be necessary. Alternatively, various types of automatic or self-feeding device have been designed, primarily for channel catfish; but they may also be useful in the culture of other species.

Automatic feeders dispense a given amount of feed at preset intervals during each 24 hour period. A timing device, usually electrically activated, is an essential component of such feeders. They are probably most often

Figure 5.6 Hand feeding channel catfish from the pondbank.

utilized in conjunction with fry rearing because they can be set to deliver feed in small quantities several times daily, allowing the culturist to feed without being physically present. For larger animals self-feeding devices are often utilized.

A self-feeder is activated by the culture animals and requires no electrical circuitry. A foot plate in the water, when bumped by a hungry fish, causes the feed hopper to release a few pellets into the pond. As long as fish continue to bump into the foot plate, feed will be made available (assuming that the hopper is not emptied during feeding activity). In theory, the fish will not consume more than they require for good growth and food conversion.

Extensive comparisons between the efficiency of hand feeding as opposed to self-feeding have not been made, but some culturists feel that more feed is used when self-feeders are employed than when premeasured amounts of feed are offered by hand. Other culturists have the opposite view. Although it cannot be stated with certainty that one method is more feed-efficient than the other, the savings in labor realized by utilizing self-feeders can be significant, assuming that each feeder does not have to be filled daily. Offsetting this advantage is the high cost of the self-feeding devices, and large ponds may require several.

The major disadvantage associated with automatic or self-feeding devices is the inability of the culturist to observe the fish feeding. When the fish are fed by hand, the culturist can be assured that they are feeding actively and that they are consuming the feed offered (especially when a floating ration is utilized). If self-feeders have the capacity to hold feed for several days and are not checked daily, the fish could go off feed and the culturist would not know it until the time came to refill the feeders, and by then serious problems could exist.

CALCULATION OF FEEDING RATES

The daily percentage of feed to offer an aquaculture animal must be determined independently for each species. As previously discussed, channel catfish are generally fed at a rate of 3% of body weight daily under optimum temperature conditions. Sedentary animals may grow rapidly at a lower feeding rate, whereas very active species may require higher rates of feeding. High feeding rates may also be needed for animals that are not efficient in locating prepared feeds or for species that tend to waste significant amounts of feed while eating.

The optimum rate of feeding can be determined by examination of the food conversion ratio (FCR). FCR is the dry weight of feed offered, divided by the wet weight gain of the animals. FCR becomes lower as the efficiency of feed utilization increases. By rearing a given species under controlled conditions and offering various percentages of feed to different experimental groups, a variety of FCR values can be obtained. At feeding rates above and below the optimum for the conditions that existed during the experiment, the FCR will increase, with the lowest FCR value occurring at the feeding level where food conversion efficiency was highest (Figure 5.7).

Daily feeding rates can be calculated based on a percentage of the biomass to be fed daily, or the animals can be fed *ad libitum*; that is, feed can be offered until the animals become satiated. If sinking pellets are utilized, or if the culture animals consume feed so slowly that it is impractical for the culturist to observe the feeding activity until complete, *ad libitum* feeding is usually avoided.

For purposes of calculation, let us assume that an aquaculturist wishes to feed a group of animals at 3% of biomass daily. Since the animals increase in biomass daily, the feeding rate must be adjusted at some interval of time to account for growth. For example, if fish were stocked at 5000/ha at an average weight of 10 g, there would initially be 50,000 g or 50 kg of fish in the pond. The first day of feeding would require 0.03 × 50 = 1.5 kg of feed with a feeding rate of 3%. The next day the fish will have grown, presumably, and 1.5 kg of feed would be slightly less than

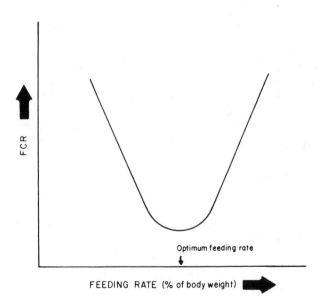

Figure 5.7 Theoretical plot of the change in FCR with changing feeding rate percentage under set experimental conditions.

3% of the total weight of fish in the pond. If it was determined that the FCR on the first day of feeding was 2.0, the population would have increased by 0.75 kg (1.5 kg/2.0 = 0.75 kg); thus the total weight of fish on the second day would be 50 kg + 0.75 kg = 50.75 kg. To feed at 3% on the second day, 50.75 × 0.03 = 1.52 kg of feed would have to be provided. The disparity between the weight of 3% of biomass on the first day of feeding and that on each subsequent day will increase as the time interval between feeding level adjustments increases.

The most precise way of adjusting feeding rates would be to catch all the animals in each pond and weigh them daily. This of course is impractical. Moreover many species will not eat for a day or more following the stress of handling. Tilapia commonly feed shortly after being handled, but that species is an exception. Even in the case of tilapia, the amount of work involved in catching and weighing all the animals at intervals of a week or two would be inordinate for most culturists. In addition, handling leads to injury and promotes susceptibility to diseases and should be avoided to the extent possible. Aquatic animals cultured in experimental tanks are often weighed at intervals of one to several weeks; but since they are generally present in relatively low numbers, capture, counting, and weighing do not impose a great deal of difficulty. But even in tanks where it would be theoretically possible to capture and weigh animals at frequent intervals, feeding rates are generally adjusted either by calculation or by subsampling the population at intervals of two weeks to a month (Figure 5.8).

In most cases where *ad libitum* feeding is not practiced, the subsampling feed adjustment technique is probably superior to other methods in warmwater aquaculture. Trout culturists can utilize tables that specify how much feed to offer per thousand fish each day of the growing season, since growth rates are well established and conditions in trout rearing facilities (water temperature and flow rates in particular) are often constant over long periods. Insufficient information exists to allow most warmwater culturists to follow a set schedule of feeding over a growing season, however, so relatively frequent adjustments must be based on actual or assumed growth rates and FCRs. When the biomass in a pond or large culture tank is estimated by means of subsampling, it may be possible to base feeding rate adjustments on a relatively small sample. The ponds are often seined and a few percent of the fish captured (subsamples of less than 1% of the population should be considered too small). The animals collected are counted and weighed, and their average weight is multiplied by the number of fish stocked into the pond (less any known mortalities), to obtain an estimate of standing crop biomass. Feeding rates are adjusted accordingly.

Estimates obtained in this manner are not completely accurate, but they

Figure 5.8 A subsample of the animals from a pond (in this case channel catfish, *Ictalurus punctatus*) can be captured at intervals to provide an indication of how well the population is growing. The subsample is utilized to calculate a new feeding rate.

are precise enough for purposes of feeding. By comparing the figures from one subsampling period to another, the FCR over each interval can be determined. If the animals are subsampled at intervals of a month, it may be wise to feed at 1% or so above the optimum level, since though the fish might be overfed early in the new feeding period, they will be somewhat underfed at the end.

If the FCR for a species is assumed to be known and constant, it should be possible to calculate the change in feeding rate on a daily basis by assuming a specific growth rate. Two such calculation techniques are presented; however it should be remembered that FCRs are not constant but vary significantly in response to any environmental change that may occur in the culture system, and for a variety of other reasons.

Method 1. $W_t = W_0 + (W_0 \times \dfrac{F}{C})$, where   (20)

$$W_t = \text{weight of animals on day 1}$$
$$W_0 = \text{weight of animals on day 0}$$
$$F = \text{feeding rate percentage}$$
$$C = \text{food conversion ratio}$$

Example.   Calculate the feeding rate at W_t if the initial weight of the animals (W_0) is 1000 kg, food conversion ratio is 1.3, and feeding rate is 4% daily.

$$W_t = 1000 \text{ kg} + \left(1000 \times \frac{0.04}{1.3}\right)$$

$$W_t = 1000 + \frac{40}{1.3}$$

$$W_t = 1000 + 30.8$$

$$W_t = \underline{1030.8 \text{ kg}}$$

The feeding rate at W_0 was $1000 \times 0.04 = 40$ kg/day. The feeding rate at W_t is: $1030.8 \times 0.04 = \underline{41.23 \text{ kg}}$. Thus there is a 1.23 kg increase in the amount of feed required after the first day of feeding.

Method 1 provides a mathematical expression for the example presented on page 207–208; the accompanying example differs only with respect to the values for initial weight, feeding rate, and food conversion ratio. Method 1 assumes that the growth of aquatic animals is linear, although in fact it is generally logarithmic, following a sigmoid growth curve. Swingle (1967) adapted Beer's law to the growth of fish under aquaculture conditions and developed the second method of determining feeding rates.

Method 2. $W_t = W_0 e^{kt}$, where   (21)

$$W_t = \text{weight at time } t$$
$$W_0 = \text{weight at time } 0$$
$$e = \text{natural logarithm}$$
$$k = \text{ratio of feed percentage to food conversion}$$
$$t = \text{time in days of adjustment period}$$

Example.   Calculate the feeding rate at W_t utilizing the same data provided in the previous example ($W_0 = 1000$ kg, FCR $= 1.3$, feeding percentage $= 4\%$, $t = 1$ day).

$$W_t = 1000\ e^{(0.04/1.3)(1)}$$
$$W_t = 1000\ e^{0.0308}$$
$$\ln W_t = \ln 1000 + \ln e^{0.0308}$$
$$\ln W_t = 6.9078 + 0.0308$$
$$\ln W_t = 6.9386$$
$$\underline{W_t = 1031.3\ \text{kg}}$$

Calculate the weight of feed required at W_t.
$$1031.3 \times 0.04 = \underline{41.25\ \text{kg}}$$

Thus the two methods of calculation compare favorably during the first day insofar as the approximation of biomass and feeding rate are concerned. Table 5.5 shows variation in these parameters over a 30 day period utilizing the initial data in the examples. One advantage of method 2 is that any number of days can be plugged into the time parameter, whereas method 1 requires a daily calculation based on information generated the preceding day (for purposes of Table 5.5, all figures were generated from data of the previous day).

Since both methods of calculation assume a constant FCR, neither can be relied on for an extended time. The best practice is to obtain subsamples from each culture unit periodically to update estimates of both biomass and FCR. These can then be put into either formula to form the basis of a new set of daily adjustment calculations. Because animals being reared in different tanks, raceways, or ponds will experience somewhat different environmental conditions, it is generally necessary to subsample each culture chamber and to maintain separate data for purposes of feeding.

FORMULATION OF AN EXPERIMENTAL DIET

To establish the nutritional requirements of any aquaculture species, a series of experiments must be run in which dietary composition is varied in a number of ways. This may require several years and a large number of diets, and for most species a great deal of research remains to be conducted before all the nutritional requirements become well established. An additional problem associated with poikilothermic animals is related to the great variety of environmental parameters that may affect the experimental results, which means that the nutritional requirements of a species may vary as a function of culture conditions. The response of aquaculture

Table 5.5 Comparison of Total Biomass and Feeding Level Values for 30 Days Utilizing Methods 1 and 2: Initial Weight of Population, 1000 kg; FCR, 1.3; Feeding Rate, Adjusted Daily, Based on 4% of Biomass Daily

	Method 1 $W_t = W_0 + (W_0 \times F/C)$		Method 2 $(W_t = W_0 e^{kt})$	
Day	Biomass (kg)	Feeding Rate (kg)	Biomass (kg)	Feeding Rate (kg)
0	1000.0	40.00	1000.0	40.00
1	1030.8	41.23	1031.3	41.25
2	1062.5	42.50	1063.6	42.54
3	1095.2	43.81	1096.9	43.87
4	1128.9	45.16	1131.2	45.25
5	1163.6	46.55	1166.6	46.66
6	1199.4	47.98	1203.1	48.12
7	1236.3	49.45	1240.7	49.63
8	1274.3	50.97	1279.5	51.18
9	1313.5	52.54	1319.5	52.78
10	1353.9	54.16	1360.8	54.43
11	1395.6	55.82	1403.4	56.13
12	1438.5	57.54	1447.3	57.89
13	1482.8	59.31	1492.6	59.70
14	1528.4	61.14	1539.3	61.57
15	1575.4	63.02	1587.4	63.50
16	1623.9	64.95	1637.1	65.48
17	1673.9	66.95	1688.3	67.53
18	1725.4	69.02	1741.1	69.64
19	1778.5	71.14	1795.6	71.82
20	1833.2	73.33	1851.8	74.07
21	1889.6	75.58	1909.7	76.39
22	1947.7	77.91	1969.4	78.78
23	2007.6	80.31	2031.0	81.24
24	2069.4	82.78	2094.5	83.78
25	2133.1	85.32	2160.0	86.40
26	2198.7	87.95	2227.6	89.10
27	2266.3	90.65	2297.3	91.89
28	2336.0	93.44	2369.2	94.77
29	2407.9	96.32	2443.3	97.73
30	2482.0	99.28	2519.7	100.79

animals to nutritional variables is generally measured in terms of growth and food conversion efficiency under carefully controlled environmental conditions.

As a first step in determining the nutrient needs of an aquaculture organism, it is often possible to study the quality and quantity of protein that result in the most rapid growth rate. This can be done by preparing a series of diets containing different commercial protein sources at various percentages or, if more basic information is being sought, a series of diets can be prepared utilizing various combinations of synthetic amino acids. In either case, to help standardize comparisons among diets, all should be isocaloric (contain the same number of calories). Since diets containing different protein quantities will have different amounts of protein energy, the total calorie level of the diet is adjusted with lipids or carbohydrates to make it isocaloric with all other diets prepared.

The calculation of energy levels in aquaculture diets cannot be done with precision unless the digestibility of each ingredient is known. If the gross energy values for feedstuffs are used (4, 4, and 9 kcal/g for protein, carbohydrate, and lipid, respectively), the amounts of energy put into the diets will be quite similar and this can be checked by bomb calorimetry. Since the various ingredients in a diet have different digestibilities, however, there is generally no way to determine whether the metabolizable energy in a series of diets will be equivalent. Metabolizable energy values are available for poultry, swine, and other types of livestock and have been determined for trout, and in part, for catfish, based on the digestibility of various feedstuffs (Wilson, 1977b). The process of determining digestibility involves feeding specially prepared diets and collecting fecal material and other waste products. Such collections are often difficult to obtain from aquatic animals, and few digestibility studies have been undertaken. Metabolizable energy values are determined by multiplying the percentage of ingredient digestibility by the initial gross energy content of the ingredient of interest.

Because metabolizable energy values are largely unavailable for aquaculture animals, nutritionists often utilize the values available from poultry and swine when calculating the available energy in diets. Others utilize the 4, 4, 9 kcal/g values for gross energy and do not attempt to estimate the fraction of energy that is metabolizable. Feed tables are available from numerous sources including feed manufacturers, chemical companies, and feeding manuals.

To demonstrate the techniques involved in preparing isocaloric diets to test the effects of protein source and quantity on the growth of aquaculture animals, let us formulate a group of experimental diets. The information required to formulate these theoretical diets can be obtained from Table

5.6, a partial theoretical feed table. The values for lipid and protein in Table 5.6 are representative of those that might be found on a typical commercial feed table, except that the table is greatly abbreviated both in terms of numbers of feedstuffs considered and amount of information given on each. A commercial feed table might include information on selected amino acids, minerals, fiber, and vitamins in addition to lipid, protein, and carbohydrate content in each ingredient.

Carbohydrate percentages in diets are not determined directly. Therefore it is difficult to assign a percentage for carbohydrate to a feed ingredient unless a complete proximate analysis has been run. A proximate analysis will provide information on the amount of protein, lipid, ash, fiber, and moisture in a feed ingredient (or sample of tissue). Digestible forms of carbohydrate (those other than fiber) are assumed to make up the difference between the actual analysis and 100% of the sample weight. All this information is not always available on a specific batch of feed ingredients, and manufacturers must rely on feed tables similar to Table 5.6.

Table 5.6   Theoretical Metabolizable Energy (ME) Values and Lipid and Protein
Percentages for Selected Feed Ingredients Similar to
Those in Commercial Feed Tables

Ingredient	ME (kcal/kg)	Lipid (%)	Protein (%)
Alfalfa meal	1400	3	17
Cellulose	0	0	0
Anchovy meal	3000	1	65
Casein	4100	1	85
Corn meal	3400	4	9
Corn oil	8900	100	0
Corn starch	3700	1	0
Cottonseed meal	2100	5	41
Fish oil	8200	100	0
Glucose	3400	0	0
Herring meal	3200	1	72
Lard	8600	100	0
Menhaden meal	3000	1	60
Minerals	0	0	0
Soybean meal	2200	6	44
Sunflower meal	1800	43	14
Wheat shorts	2600	5	17
Vitamins	0	0	0

The percentages of various components and the metabolizable energy values of dietary ingredients vary from one batch of feed to another. For example, not all soybeans are exactly the same chemically; thus some variability from values in the table should be expected. The values in any feed table represent averages based on several chemical determinations. Tables may be altered slightly from year to year as new information becomes available.

Assuming that virtually nothing is known about the nutritional requirements of the animal to be tested except that it is a carnivore, the decision has been made to determine first the percentage of protein required for rapid growth. Assuming that preliminary tests on related species have indicated that the requirement for protein may lie between 20 and 50%, a series of diets with varying protein levels within that range will be formulated. Since it is desirable to make the diets isocaloric, the highest protein diet is formulated first, and those with lower protein percentages are made isocaloric with the 50% protein diet.

Since the animal being tested is a carnivore, menhaden meal has been chosen for the protein source. To make sure that the energy levels in each diet are sufficiently high, all diets are supplemented with 5% corn oil. A standard vitamin and mineral premix is added to each diet (2.5% vitamins, and 2.5% minerals), and 1% agar is added to each diet as a binder. Corn starch is used as the carbohydrate and is varied in the diets to adjust the calorie levels. Cellulose serves as a filler, since it is indigestible. One kilogram of each diet will be formulated.

To determine the grams per kilogram of menhaden meal to use in preparing the 50% protein diet, the following calculation is made: [0.50 (protein percentage) $\times$ 1000 (g/kg)]/0.60 (percentage protein in menhaden meal) = 500/0.60 = 833.3 g of menhaden meal. The number of calories contained in the menhaden component of the diet (using 3000 kcal/kg from Table 5.6) is: 0.833 $\times$ 3000 kcal/kg = 2499 kcal/kg. If 5% corn oil is added, the diet will contain 833.3 g of menhaden meal + 50 g of corn oil = 883.3 g, with an energy level of 2499 + (0.05 $\times$ 8900 kcal/kg = 445) = 2944 kcal/kg. The vitamins, minerals, and agar will add an additional 6% or 60 g of weight, making the total 943.3 g. The latter ingredients have no metabolizable energy. To fill the remainder, we add 1000 − 943.3 = 56.7 g of cellulose. The final kilogram of diet contains 1000 g of ingredients having an energy value of 2944 kcal/kg (Tables 5.7 and 5.8).

If each diet successively lower in protein has 10% less protein, the next diet in the series will contain 40% protein: 0.40 $\times$ 1000/0.6 = 666.7 g/kg of menhaden meal. Adding the same amount of vitamins, minerals, agar, and lipid used in the 50% protein diet, we see that the 40% protein

**Table 5.7   Contribution of Each Feed Ingredient (g/kg) Contained in Isocaloric
Diets Formulated at Four Protein Levels**

	Dietary Protein Percentage			
Ingredient	50	40	30	20
Agar	10	10	10	10
Cellulose	56.7	88.7	120	151.3
Corn oil	50	50	50	50
Corn starch	0	134.6	270	405.4
Menhaden meal	833.3	666.7	500	333.3
Minerals	25	25	25	25
Vitamins	25	25	25	25
Total	1000	1000	1000	1000

**Table 5.8   Caloric Content of Each Feed Ingredient (kcal/kg)
Contained in Diets of Varying Protein Percentage**

	Dietary Protein Percentage			
Ingredient	50	40	30	20
Agar	0	0	0	0
Cellulose	0	0	0	0
Corn oil	445	445	445	445
Corn starch	—	498	999	1500
Menhaden meal	2499	2001	1500	999
Minerals	0	0	0	0
Vitamins	0	0	0	0
Total	2944	2944	2944	2944

diet will contain 776.7 g, with a total energy value of $445 + (0.667 \times 3000) = 2446$ kcal. The remaining 498 kcal $(2944 - 2446)$ to make the 40% protein diet isocaloric with the 50% protein diet must come from corn starch: 498 kcal/3700 kcal = x g/1000 g, or 3700 x = 498,000 and x = 498,000/3700 = 134.6 g of corn starch (Table 5.7). The two diets are now isocaloric, but the 40% protein diet contains only 911.3 g o ingredients $(666.7 + 60 + 50 + 134.6)$. Thus $1000 - 911.3 = 88.7$ of cellulose must be added to complete the diet (Table 5.7).

The 20 and 30% protein diets are calculated in the same manner.

Tables 5.7 and 5.8 give data for each diet in terms of the weight of each ingredient and caloric content, respectively. Practical diets can be produced in the same way, although the process is complicated because a wide variety of more complex ingredients must be included and the contribution to vitamins and minerals made by the various grains and animal by-products must be taken into consideration. In the example case we have ignored the vitamins, minerals, and other contributions to total dietary composition of the menhaden meal, other than protein, although these other components may be significant.

LITERATURE CITED

Andrews, J. W. 1977. Protein requirements. In R. R. Stickney and R. T. Lovell (Eds.), *Nutrition and feeding of channel catfish.* Southern Cooperative Series Bulletin 218, pp. 10-13.

Andrews, J. W., and Y. Matsuda. 1975. The influence of various culture conditions on the oxygen consumption of channel catfish. *Trans. Am. Fish. Soc.* **104:** 322-327.

Andrews, J. W., and J. W. Page. 1974. Growth factors in the fish meal component of catfish diets. *J. Nutr.,* **104:** 1091-1096.

Andrews, J. W., and J. W. Page. 1975. The effects of frequency of feeding on culture of catfish. *Trans. Am. Fish. Soc.* **104:** 317-321.

Andrews, J. W., and R. R. Stickney. 1972. Interactions of feeding rates and environmental temperature on growth, food conversion, and body composition of channel catfish. *Trans. Am. Fish. Soc.* **101:** 94-99.

Arnold, C. R., T. D. Williams, W. A. Fable, J. L. Lasswell, and W. H. Bailey. 1977. Laboratory methods for spawning and rearing spotted sea trout. *Proc. Southeast. Assoc. Fish Wildl. Agencies* **31:** 437-440.

Bardach, J. E., J. H. Ryther, and W. O. McLarney. 1972. *Aquaculture.* Wiley-Interscience, New York. 868 p.

Bayne, D. R., D. Dunseth, and C. G. Ramirios. 1976. Supplemental feeds containing coffee pulp for rearing *Tilapia* in Central America. *Aquaculture,* **7:** 133-146.

Boyd, C. E. 1976. Nitrogen fertilizer effects on production of *Tilapia* in ponds fertilized with phosphorus and potassium. *Aquaculture,* **7:** 385-390.

Castell, J. D., D. J. Lee, and R. O. Sinnhuber. 1972a. Essential fatty acids in the diet of rainbow trout (*Salmo gairdneri*): Lipid metabolism and fatty acid composition. *J. Nutr.* **102:** 93-99.

Castell, J. D., R. O. Sinnhuber, J. H. Wales, and D. J. Lee. 1972b. Essential fatty acids in the diet of rainbow trout (*Salmo gairdneri*): Growth, feed conversion and some gross deficiency symptoms. *J. Nutr.* **102:** 77-86.

Cowey, C. B., and J. R. Sargent. 1977. Lipid nutrition in fish. *Comp. Biochem. Physiol.* **57B:** 269-273.

Cowey, C. B., J. W. Adron, A. Blair, and J. Pope. 1970. The growth of O-group plaice on artificial diets containing different levels of protein. *Helg. Wiss. Meeresunters.* **20:** 602-609.

Cowey, C. B., J. A. Pope, J. W. Adron, and A. Blair. 1971. Studies on the nutrition of marine flatfish. Growth of the plaice *Pleuronectes platessa* on diets containing proteins derived from plants and other sources. *Mar. Biol.* **10:** 145-153.

Cowey, C. B., J. A. Pope, J. W. Adron, and A. Blair. 1972. Studies on the nutrition of marine flatfish. The protein requirement of plaice (*Pleuronectes platessa*). *Brit. J. Nutr.* **28:** 447-456.

Cruz, E. M. 1975. Determination of the nutrient digestibility in various classes of natural and purified feed materials for channel catfish. Ph.D. dissertation, Auburn University, Auburn, Ala. 82 p.

Davis, A. T., and R. R. Stickney. 1978. Growth responses of *Tilapia aurea* to dietary protein quality and quantity. *Trans. Am. Fish. Soc.* **107:** 479-483.

Davis, E. M., G. L. Rumsey, and J. G. Nickum. 1976. Egg-processing wastes as a replacement protein source in salmonid diets. *Prog. Fish-Cult.* **38:** 20-22.

Dupree, H. K. 1966. *Vitamins essential for growth of channel catfish.* U.S. Bureau of Sport Fisheries and Wildlife, Technical Paper 7. 12 p.

Dupree, H. K. 1968. *Influence of corn oil and beef tallow on the growth of fingerling channel catfish.* U.S. Bureau of Sport Fisheries and Wildlife, Technical Paper 27. 12 p.

Dupree, H. K. 1977. Vitamin requirements. In R. R. Stickney and R. T. Lovell (Eds.). *Nutrition and feeding of channel catfish.* Southern Cooperative Series Bulletin 218, pp. 26-29.

Dupree, H. K., and J. E. Halver. 1970. Amino acids essential for the growth of channel catfish, *Ictalurus punctatus. Trans. Am. Fish. Soc.* **99:** 90-92.

Everhart, W. H., A. E. Eipper, and W. D. Youngs. 1975. *Principles of fishery science.* Cornell University Press, Ithaca, N.Y. 288 p.

Friedman, L., and S. I. Shibko. 1972. Nonnutrient components of the diet. In J. E. Halver (Ed.), *Fish nutrition.* Academic Press, New York, pp. 182-254.

Gorham, P. R. 1964. Toxic algae. In D. F. Jackson (Ed.), *Algae and man.* Plenum Press, New York, pp. 307-336.

Halver, J. E. (Ed.). 1972. *Fish nutrition.* Academic Press, New York. 713 p.

Hanson, J. A., and H. L. Goodwin. 1977. *Shrimp and prawn farming in the Western hemisphere.* Dowden, Hutchinson & Ross, Stroudsburg, Pa. 439 p.

Harrison, C. 1975. The essential amino acids of *Mytilus californianus. The Veliger,* **18:** 189-193.

Hickling, C. F. 1962. *Fish culture.* Faber and Faber, London. 295 p.

Hickman, C. P. 1968. Glomerular filtration and urine flow in the euryhaline southern flounder, *Paralichthys lethostigma,* in seawater. *Can. J. Zool.* **46:** 427-437.

Hublou, W. F. 1963. Oregon Pellets. *Prog. Fish-Cult.* **23:** 175-180.

Ketola, H. G. 1975. Mineral supplementation of diets containing soybean meal as a source of protein for rainbow trout. *Prog. Fish-Cult.* **37:** 73-75.

Kitamikado, M., T. Morishita, and S. Tachino. 1964. Digestibility of dietary protein in rainbow trout. I. Digestibility of several dietary proteins. *Bull. Jap. Soc. Sci. Fish.* **30:** 46-49.

Leary, D. F., and R. T. Lovell. 1975. Value of fiber in production-type diets for channel catfish. *Trans. Am. Fish. Soc.* **104:** 328-332.

Lee, D. J., and R. O. Sinnhuber. 1972. Lipid requirements. In J. E. Halver (Ed.), *Fish nutrition.* Academic Press, New York, pp. 145-180.

Lee, D. J., J. N. Roehm, T. C. Yu, and R. O. Sinnhuber. 1967. Effects of ω3 fatty acids on the growth rate of rainbow trout, *Salmo gairdneri. J. Nutr.* **92:** 93–98.

Lockwood, A. P. M. 1967. *Aspects of the physiology of crustacea.* Freeman, San Francisco. 328 p.

Lovell, R. T. 1977a. Digestibility of nutrients in feedstuffs for catfish. In R. R. Stickney and R. T. Lovell (Eds.), *Nutrition and feeding of channel catfish.* Southern Cooperative Series Bulletin 218, pp. 33–37.

Lovell, R. T. 1977b. Mineral requirements. In R. R. Stickney and R. T. Lovell (Eds.), *Nutrition and feeding of channel catfish.* Southern Cooperative Series Bulletin 218, pp. 30–32.

Lovell, R. T., and B. Sirikul. 1974. Winter feeding of channel catfish. *Proc. Southeast. Assoc. Game Fish Comm.* **28:** 208–216.

Lovell, R. T., E. E. Prather, J. Tres-Dick, and L. Chhorn. 1974. Effects of addition of fish meal to all-plant feeds on the dietary protein needs of channel catfish in ponds. *Proc. Southeast. Assoc. Game Fish Comm.* **28:** 222–228.

McGeachin. R. B. 1977. Algae fed *Artemia salina* nauplii as a food source for larval *Cynoscion nebulosus.* M.S. thesis, Texas A&M University, College Station. 39 p.

McGeachin, R. B., R. R. Stickney, and C. R. Arnold. 1977. Algae fed *Artemia salina* nauplii as a food source for larval *Cynoscion nebulosus. Proc. Southeast. Assoc. Fish Wildl. Agencies* **31:** 574–582.

Murai, T., and J. W. Andrews. 1976. Effects of frequency of feeding on growth and food conversion of channel catfish fry. *Bull. Jap. Soc. Sci. Fish.* **42:** 159–161.

National Academy of Sciences. 1973. *Nutrient requirements of trout, salmon and catfish.* National Academy of Sciences, Washington, D.C. 57 p.

National Academy of Sciences. 1977. *Nutrient requirement of warmwater fishes.* National Academy of Sciences, Washington, D.C. 78 p.

Odum, W. E. 1966. The food and feeding of the striped mullet, *Mugil cephalus,* in relation to the environment. M.S. thesis, Institute of Marine Sciences, University of Miami, Miami, Fla. 118 p.

Odum, W. E. 1968. The ecological significance of fine particle selections by the striped mullet, *Mugil cephalus. Limnol. Oceanogr.* **13:** 92–98.

O'Keefe, T. 1976. Some effects of increased dietary levels of alpha-tocopherol in channel catfish, *Ictalurus punctatus* (Rafinesque). M.S. thesis, Texas A&M University, College Station. 59 p.

Orme, L. E., and C. A. Lemm. 1973. Use of dried sludge from paper processing wastes in trout diets. *Feedstuffs,* **45** (51): 28–30.

Owen, J. M., J. W. Adron, C. Middleton, and C. B. Cowey. 1975. Elongation and desaturation of dietary fatty acids in turbot, *Scophthalmus maximus* L., and rainbow trout, *Salmo gairdneri. Lipids,* **10:** 528–531.

Phillips, A. M., Jr. 1969. Nutrition, digestion, and energy utilization. In W. S. Hoar and D. J. Randall (Eds.), *Fish physiology,* Vol. 1. Academic Press, New York, pp. 391–432.

Phillips, A. M., Jr. 1972. Calorie and energy requirements. In J. E. Halver (Ed.), *Fish nutrition.* Academic Press, New York, pp. 1–28.

Podoliak, H. A., and H. K. Holden, Jr. 1965. Distribution of dietary calcium to the skeleton and skin of fingerling brown trout. Courtland Hatchery Report 33 for the year 1964. *Fish. Res. Bull.* **28:** 64–70. New York State Conservation Department, Albany.

Podoliak, H. A., and H. K. Holden, Jr. 1966. Calcium ion regulation by fingerling brook, brown, and rainbow trout. Courtland Hatchery Report 34 for the year 1965. *Fish. Res. Bull.* **29:** 59-65. New York State Conservation Department, Albany.

Reece, D. L., D. E. Wesley, G. A. Jackson, and H. K. Dupree. 1975. A blood meal-rumen contents blend as a partial or complete substitute for fish meal in channel catfish diets. *Prog. Fish-Cult.* **37:** 15-19.

Richardson, T., A. L. Tappell, L. M. Smith, and C. R. Houle. 1962. Polyunsaturated fatty acids in mitochondria. *J. Lipid Res.* **3:** 344-350.

Robinette, H. R. 1977. Feed manufacture. In R. R. Stickney and R. T. Lovell (Eds.), *Nutrition and feeding of channel catfish*. Southern Cooperative Series Bulletin 218, pp. 44-49.

Schroeder, G. L. 1974. Use of fluid cowshed manure in fish ponds. *Bamidgeh.* **26:** 84-96.

Stickney, R. R. 1971. Effects of dietary lipids and lipid-temperature interactions on growth, food conversion, percentage lipid and fatty acid composition of channel catfish. Ph.D. dissertation, Florida State University, Tallahassee. 96 p.

Stickney, R. R. 1975. Cellulase activity in the stomachs of freshwater fishes from Texas. *Proc. Southeast. Assoc. Game Fish Comm.* **29:** 282-287.

Stickney, R. R. 1977. Lipids in channel catfish nutrition. In R. R. Stickney and R. T. Lovell (Eds.), *Nutrition and feeding of channel catfish*. Southern Cooperative Series Bulletin 218, pp. 14-18.

Stickney, R. R., and J. W. Andrews. 1971. Combined effects of dietary lipids and environmental temperature on growth, metabolism and body composition of channel catfish (*Ictalurus punctatus*). *J. Nutr.* **101:** 1703-1710.

Stickney, R. R., and J. W. Andrews. 1972. Effects of dietary lipids on growth, food conversion, lipid and fatty acid composition of channel catfish. *J. Nutr.* **102:** 249-258.

Stickney, R. R., and J. H. Hesby. 1978. Tilapia culture in ponds receiving swine waste. In R. O. Smitherman, W. L. Shelton, and J. H. Grover (Eds.), *Culture of exotic fishes symposium proceedings*, Fish Culture Section, American Fisheries Society, Auburn, Ala., pp. 90-101.

Stickney, R. R., and S. E. Shumway. 1974. Occurrence of cellulase activity in the stomachs of fishes. *J. Fish Biol.* **6:** 779-790.

Stickney, R. R., and D. B. White. 1973. Effects of salinity on the growth of *Paralichthys lethostigma* postlarvae reared under aquaculture conditions. *Proc. Southeast. Assoc. Game Fish Comm.* **27:** 532-540.

Stickney, R. R., T. Murai, and G. O. Gibbons. 1972. Rearing channel catfish fingerlings under intensive culture conditions. *Prog. Fish-Cult.* **34:** 100-102.

Stickney, R. R., H. B. Simmons, and L. O. Rowland. 1977a. Growth responses of *Tilapia aurea* to feed supplemented with dried poultry waste. *Tex. J. Sci.* **29:** 93-99.

Stickney, R. R., L. O. Rowland, and J. H. Hesby. 1977b. Water quality-*Tilapia aurea* interactions in ponds receiving swine and poultry wastes. *Proc. World Maricult. Soc.* **8:** 55-71.

Swingle, H. S. 1967. Estimation of standing crops and rates of feeding fish in ponds. *FAO Fish. Rep.* **44:** 416-423.

Swingle, H. S., B. C. Gooch, and H. R. Rabanal. 1965. Phosphate fertilization of ponds. *Proc. Southeast. Assoc. Game Fish Comm.* **17:** 213-218.

Taggart, R. B. 1974. Digestibility of carbohydrates, lipids, and proteins in channel catfish, *Ictalurus punctatus* (Rafinesque). M.S. thesis, Kansas State University, Manhattan. 57 p.

White, D. B., and R. R. Stickney. 1973. *A manual of flatfish rearing.* Georgia Marine Science Center, Technical Report Series 73-7. 36 p.

White, A., P. Handler, and E. L. Smith. 1964. *Principles of biochemistry.* McGraw-Hill, New York. 1106 p.

Wilson, R. P. 1977a. Carbohydrates in channel catfish nutrition. In R. R. Stickney and R. T. Lovell (Eds.), *Nutrition and feeding of channel catfish.* Southern Cooperative Series Bulletin 218, pp. 19-20.

Wilson, R. P. 1977b. Energy relationships in catfish diets. In R. R. Stickney and R. T. Lovell (Eds.), *Nutrition and feeding of channel catfish.* Southern Cooperative Series Bulletin 218, pp. 21-25.

Windell, J. T., R. Armstrong, and J. R. Clinebell. 1974. Substitution of brewer's single cell protein into pelleted fish feed. *Feedstuffs.* 46 (20): 22-23.

Yingst, W. L. 1978. Effects of dietary lipids on growth and body composition of channel catfish fry. M.S. thesis, Texas A&M University, College Station. 44 p.

Yu, T. C., and R. O. Sinnhuber. 1975. Effect of dietary linolenic and linoleic acids upon growth and lipid metabolism of rainbow trout (*Salmo gairdneri*). *Lipids,* 10: 63-66.

SUGGESTED ADDITIONAL READING

Andrews, J. W., and T. Murai. 1975. Studies on the vitamin C requirements of channel catfish (*Ictalurus punctatus*). *J. Nutr.* 105: 557-561.

Liang, J. K., and R. T. Lovell. 1971. Nutritional value of water hyacinth in channel catfish feeds. *Hyacinth Control. J.* 9: 40-44.

Lovell, R. T. 1973. Essentiality of vitamin C in feeds for intensively fed caged catfish. *J. Nutr.* 103: 134-138.

Mertz, E. T. 1972. The protein and amino acid needs. In J. E. Halver (Ed.), *Fish nutrition.* Academic Press, New York, pp. 106-143.

Meyers, S. P., and C. W. Brand. 1975. Experimental flake diets for fish and crustacea. *Prog. Fish-Cult.* 37: 67-72.

Meyers, S. P., D. P. Butler, and G. F. Sirine. 1971. Encapsulation a new approach to larval feeding. *Am. Fish Farmer,* July: 15 ff.

Murai, T., and J. W. Andrews. 1974. Interactions of dietary α-tocopherol, oxidized menhaden oil and ethoxyquin on channel catfish (*Ictalurus punctatus*). *J. Nutr.* 104: 1416-1431.

Murai, T., and J. W. Andrews. 1975. Pantothenic acid supplementation for channel catfish. *Trans. Am. Fish. Soc.* 104: 313-316.

Pillay, T. V. R. (Ed.). 1972. *Coastal aquaculture in the Indo-Pacific region.* Fishing News (Books), London. 497 p.

Price, K. S., Jr., W. N. Shaw, and K. S. Danberg (Eds.). 1976. *Proceedings of the first international conference on aquaculture nutrition.* University of Delaware Sea Grant Publication DEL-SG-17-76. 323 p.

Stickney, R. R., and R. T. Lovell (Eds.). *Nutrition and feeding of channel catfish.* Southern Cooperative Series Bulletin 218.

CHAPTER 6

Reproduction, Selective Breeding, and Genetics

REPRODUCTIVE STRATEGIES

Successful aquaculture is often partly dependent on the ease with which culture animals can be reproduced in captivity. Unless laboratory rearing is achieved, the culturist has little or no control on the genetic makeup of the stock, thus no attempts at improvement in such characteristics as growth rate, disease resistance, and dress-out percentage can be made through selective breeding. Most species being cultured successfully around the world are similar to the extent that their life cycles may be completed, and reproduction of the organisms will occur, under culture conditions. Possibly the most notable exception is the milkfish (*Chanos chanos*), which is reared in large numbers in the Philippines, Indonesia, and elsewhere. Although a program is underway to develop the techniques for controlled reproduction of milkfish, juveniles now must be obtained from wild populations. The dependence on wild stock effectively prevents the development of superior stocks through selective breeding. A similar dependence on wild gravid penaeid shrimp has been largely responsible for preventing the establishment of economically feasible culture of those animals in the United States.

Animals exhibit a variety of reproductive strategies; however virtually all the important aquaculture species rely on some form of sexual reproduction. Some species are hermaphroditic, but most are dioecious. Even the hermaphroditic species rarely fertilize themselves, but exchange gametes with other individuals in the population.

Viviparity and ovoviviparity are not typical of aquaculture species, although these types of development are found in certain aquatic animals. Most aquaculture species are oviparous, with the eggs being laid and development occurring in the water in most instances. Fertilization of the eggs is generally external. Sea turtles are one of the most notable exceptions to the general rule of both external fertilization and egg development in the water.

The reproductive physiology of various species of vertebrates and invertebrates has been extensively studied, and despite the apparent role of various endocrine glands in sexual activity, the pituitary gland and the gonads seem to be the most important. In fishes, hypophysectomy results in blockage of both ovulation and spermatogenesis in all cases examined to date, except for the Agnatha (Hoar, 1969). The primary interest in hormones by aquaculturists has been their use in spawning induction and sex reversal. Hormones such as carp pituitary and human chorionic gonadotropin (HCG) can be injected into ovulating fishes as a means of stimulating spawning. In the crustaceans the development of eggs and the prespawning molt in penaeids can be induced through eyestalk ablation techniques (SEAFDEC, 1976; Hanson and Goodwin, 1977).

Fish and invertebrates often can be induced to spawn merely through the manipulation of temperature and/or photoperiod. For example, channel catfish will develop gametes if exposed to an artificial winter, then brought back to normal spawning temperature, although details of the best lower temperature to use and the duration of exposure are not available. In the case of channel catfish photoperiod may not be critical; however for a variety of other species, especially marine fishes, spawning has been achieved in the laboratory only when both temperature and photoperiod were carefully controlled. The spotted sea trout, *Cynoscion nebulosus,* is a good example (Arnold et al., 1977).

The techniques utilized by Arnold et al. (1977) have also been employed by the same workers in the spawning of red drum (*Sciaenops ocellata*) and flounder (*Paralichthys* sp.). It is of particular interest that hormone injections were not utilized as a part of the spawning program, yet multiple spawns of red drum and other species were obtained, resulting in literally millions of eggs from a handful of females (C. R. Arnold, personal communication).

The aquaculturist may take an active role in preparing the animals for spawning and ensuring that the proper conditions are established, or the animals may be left to their own devices. In some cases (e.g., *Tilapia* spp.) it is possible to stock adult animals in ponds and collect the offspring following hatching of the eggs. In others (e.g., *Ictalurus punctatus* and most marine species), special conditions must be provided for the adults to ensure against resorption of eggs, spawning of eggs that are not subsequently fertilized, or spawning of eggs that have little chance for survival because of adverse environmental conditions. Finally, techniques can be utilized whereby specific individuals are paired, and each phase of reproduction is orchestrated by the aquaculturist. It is only at the last and highest level of sophistication that the aquaculturist begins to have a degree of genetic control through selective breeding.

The mere recognition of sex among aquaculture animals can be difficult in some instances, particularly with respect to fishes that have been utilized for culture. Small channel catfish males and females are particularly difficult to differentiate except during the spawning season, and even then mistakes are often made by persons who have had a good deal of experience. Marine fishes of aquaculture interest are usually difficult to sex, even as the spawning season approaches. In some species sexual recognition is based on distension of the female abdomen with eggs as spawning time nears. Any fish not exhibiting that characteristic is assumed to be a male, provided it is large enough and old enough to spawn. Significant size differences between the sexes are sometimes apparant. Male flounders (*Paralichthys* sp.) are often quite small compared to the females, and culturists who have attempted to select large animals from wild populations for brood stock sometimes end up with all females. In most species of tilapia the male grows more rapidly than the female, although the ultimate disparity at the end of the growth curve is not as distinct as that in flounders.

Some species of aquaculture animals do demonstrate significant sexual differences, especially during the spawning season. For example, *Tilapia mossambica* males are distinct in color when ripe, although the sexes may be similar throughout the rest of the year. The sexes of various tropical fishes are readily distinguishable by body shape and/or color, but as a general rule, aquaculture organisms are similar with respect to these characteristics.

Most crustaceans of interest to aquaculturists are readily separable into males and females. Some warmwater species, such as the blue crab, Florida lobster, and freshwater shrimp, carry their eggs externally; thus the females can be readily identified during the spawning season after the eggs have been extruded. In most cases other external characteristics exist that allow the culturist to quickly and surely separate the sexes of invertebrates throughout the year. For example, in the blue crab, *Callinectes sapidus*, the abdomen of juvenile females is triangular, with the apex located anterior to the base. The abdomen becomes rounded in adult females. In males the abdomen is narrow in the lateral aspect and long in the anterior-posterior aspect (Figure 6.1). From the dorsal surface the sexes are indistinguishable. Male marine and freshwater shrimp have an organ called the petasma, which is located on the first pair of pleopods and can be seen as an accessory organ protruding from those appendages. The petasma is used to transfer the spermatophore to the female, where it is attached to the thelycum, an organ located between and just anterior to the most posterior pair of walking legs.

Dorsal

Ventral

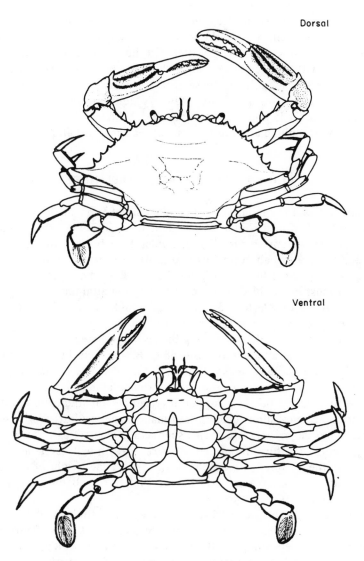

Figure 6.1 Dorsal and ventral views of a male blue crab, *Callinectes sapidus*. Original drawings by Daniel Perlmutter.

The actual methods involved in spawning various aquaculture species are highly diverse. Since it would not be possible to describe each in detail, sections on spawning have been written with the object of providing general information with respect to several of the most popular warmwater aquaculture species. In addition, these sections demonstrate the diversity of techniques and approaches being made by aquaculturists in the area of breeding. First, it is important to understand the importance of fecundity in spawning technique and the subsequent rearing of fry and larvae.

EGG SIZE AND RELATIVE FECUNDITY

The amount of time required to rear an animal from birth to market is largely a function of its initial size, although the ultimate size at slaughter is also an important factor. For example, the absolute weight increase required to take a calf from birth to market is much greater than that of a channel catfish; however the number of times the weight of the newborn calf must be doubled between birth and the attainment of its market size (approximately 225 to 275 kg) is considerably smaller than that of the catfish in growing from a few milligrams to 0.5 kg. Chickens are often used as an example in demonstrating the accomplishments that are possible with domestication and selective breeding. Broilers can be produced from eggs in as little as 6 to 8 weeks. Part of the secret lies in the fact that since the chicken lays a large egg, the newly hatched animal is extremely large relative to a newly hatched fish or invertebrate of aquaculture interest. Significant accomplishments have been made by aquaculturists with respect to increasing the growth rate of fish such as the channel catfish above that which occurs in nature; however the 6 week catfish will never be a reality as long as the fry obtained are only a fraction of the weight of a newborn chick, and there is no reason to believe that larger channel catfish eggs will be produced in the future.

On the average channel catfish females lay approximately 6600 eggs per kilogram of body weight (Clemens and Sneed, 1957). This figure varies to some extent depending on the age and general condition of the adult female, but it is a close enough approximation for our purposes. When hatched, the 20 mg fry can be reared to market size (0.5 kg) within 6 to 10 months under the proper environmental and nutritional conditions, although many culturists do not market channel catfish younger than about 18 months (growth is held down during the first summer, the fish are overwintered as fingerlings of about 10 cm, and fed to market size the following summer). Since few adult catfish can supply the needs of the typical fish farmer, there is not a great deal of expense involved in maintaining brood stock.

The eggs of tilapia are similar in size to those of channel catfish, but the mouthbrooding species generally spawn only 75 to 250 eggs per female (Bardach *et al.,* 1972). Since tilapia are multiple spawners, whereas catfish spawn only once a year in the United States, the annual production in terms of eggs per kilogram of female body weight may be similar for these two species.

It has been suggested that fish with larger eggs could be reared more rapidly than channel catfish and tilapia, since the initial size of the fry would be larger. In light of what is known about chickens and other terrestrial forms of livestock, this appears to be a reasonable hypothesis. There remains the problem of finding a species of fish (or invertebrate) that produces large eggs. One such is the gafftopsail catfish, *Bagre marinus,* in the family Ariidae. This marine fish is widely accepted as human food, unlike its relative, *Arius felis,* the sea catfish, which is held in low esteem in the coastal regions of the United States where it occurs. The female *B. marinus* lays eggs more than 1.0 cm in diameter. These are picked up by the male and incubated in his mouth until yolk-sac absorption is complete, after which the fry leave the protection of the male to feed. The process of incubation was first described by Gudger (1918). For a period of time after the fry begin to feed, they return to the mouth of the adult male if danger threatens. Because of the protection afforded the developing eggs and fry, and because the newly hatched fish weigh several grams, survival is generally excellent as long as parental care is provided. However the factors that ensure a high survival percentage in *B. marinus* also effectively rule it out as a candidate for aquaculture.

Male *Bagre marinus* are limited in the number of eggs that each can carry in the mouth and buccal cavity, and this condition is reflected in the low fecundity of the female. Whereas most fish species spawn hundreds, thousands, or even millions of eggs annually, the gafftopsail catfish spawns only a few (generally less than 50). Because of low fecundity, an aquaculturist would be forced to maintain an extremely large number of brood fish. The expense of providing space, water, and feed for their maintenance would prohibit culture on economic grounds.

Bagre marinus is an exception among marine fishes in that it lays very large eggs. Most marine species broadcast large numbers of small eggs into the water, where the eggs are fertilized, then abandoned by the parents to develop and hatch untended. The eggs of many marine fishes are laid in the open sea, and the larval animals must find their way into an estuary to have the proper conditions for growth and survival. The odds against the survival of an individual are great, but the probability that each spawning pair will leave sufficient offspring to replace themselves is good because fecundities are high.

Warmwater marine species under consideration for aquaculture in the United States have high fecundities. The minute eggs produced by *Sciaenops ocellata, Pogonias chromis, Paralichthys lethostigma,* and *P. dentatus* may number in the hundreds of thousands per female. Though only a few brood fish are required to maintain a large aquaculture facility, elaborate facilities are often needed to rear the fry and living organisms to feed them (Chapter 5). Once these species have reached the size of catfish fry and can be placed on prepared diets, their growth is comparable in many cases with aquaculture species having larger eggs. The production of large numbers of eggs per female is also typical of marine invertebrates, where problems in larval rearing may be even more pronounced than in the fishes.

SPAWNING TECHNIQUES

Channel Catfish

Much of the channel catfish brook stock utilized by catfish farmers in the United States is highly inbred within hatcheries. In addition, most people who attempt to establish a brookstock population obtain their fish from another culturist. If traced back far enough, it is quite likely that the majority of the *Ictalurus punctatus* presently in culture came from a relatively small initial population of wild fish. Many culturists believe that wild channel catfish are difficult to spawn, and they depend on domesticated fish for their brood stock. Furthermore, wild strains of catfish in the United States have been contaminated to some extent with hatchery fish as a result of stocking programs and accidental releases; thus the probability of obtaining truly wild strains is limited, especially in the southern United States.

Channel catfish weighing in excess of 20 kg have been captured in nature; however most of the brood stock utilized by catfish breeders range from 0.9 to 4.5 kg (Martin, 1967). Relatively small brood animals are used primarily because larger fish are difficult to handle. In addition, the fecundity of female catfish decreases to some extent as the fish grow larger, and the hatchability may be reduced in eggs obtained from large (thus old) fish. Channel catfish females between 0.5 and 1.8 kg average about 8800 eggs per kilogram, whereas fish larger than 1.8 kg average about 6600 eggs per kilogram of body weight (Clemens and Sneed, 1957).

Channel catfish spawn in the late spring and summer, depending on the strain of fish and the geographic region in which they are found. Generally in the continental United States spawning does not occur much before May and is usually concluded by early August, although exceptions sometimes occur. Spawning temperature is between 21 and 29 C (Clemens and Sneed, 1957).

Following spawning, brood fish are often maintained in ponds at densities of about 375 fish per hectare (Nelson, 1960), but higher densities have also been reported (Martin, 1967). During the nonspawning season brood stock are fed commercial catfish rations. It has been recommended that minnows and crayfish also be available to brood fish (Canfield, 1947). During the late winter and early spring, when gonads are developing, many culturists provide beef liver, heart, and other fresh meat products, although the actual need for such feeds has not been established when brood stock are maintained on complete diets. The theory is that mineral and vitamin requirements of brood fish are considerably greater during gonadal development than at other times of the year and that fresh organ meats can supply the demand better than commercial feeds. We have achieved excellent results in spawning channel catfish after feeding organ meats or complete diets and providing forage in the form of fathead minnows, *Pimephales promelas.*

During most of the year it is difficult to separate the sexes in a group of channel catfish; however during spawning season certain secondary sexual characteristics do become apparent. The females develop a well-rounded abdomen because of oocyte production. The ovaries become soft and palpable, and the genital pore becomes raised and inflamed (Clemens and Sneed, 1957). Male channel catfish generally have heads somewhat wider than the rest of the body (the opposite is generally true among females), dark pigment develops under the lower jaw and on the abdomen, and the genital papilla becomes well formed and tubular (Clemens and Sneed, 1957).

Prior to spawning, the brood stock may be seined from holding ponds for selection and stocking into spawning ponds, pens, or aquaria (Figure 6.2). Fish should be handled gently during the spawning season and should not be kept out of water for very long, but special precautions are not generally required and handling does not seem to interfere with subsequent spawning success.

Pond Spawning. Captive spawning of channel catfish may date back as far as the 1890s (Martin, 1967), when it was discovered that the species will readily spawn in ponds if some type of nest is made available. In nature, channel catfish often spawn under logs or in depressions along stream banks. Providing some type of artificial nest in a pond is usually a requirement for spawning, since there are often no natural nesting sites available.

Figure 6.2 Brood stock may be seined from holding ponds for selection and stocking into ponds, pens, or aquaria. The fish being held is representative of the maximum size preferred by culturists.

Artificial spawning nests of various types have been used, including but not restricted to milk or cream cans (Figure 6.3), nail kegs, beer kegs, metal drums of various sizes (e.g., grease drums), and ceramic or concrete drain tiles. Commercially manufactured spawning containers are also available. The number of nests to be placed in a spawning pond relative to the number of brood fish stocked will vary depending on whether the eggs are to be allowed to hatch in the pond or will be removed and hatched indoors. In the former case each nest will be occupied longer between spawns, and more nests should be placed in the pond. Usually there are fewer nests available than there are pairs of brood fish.

Figure 6.3 Milk cans make excellent catfish spawning containers. Though increasingly rare and expensive, their basic configuration remains among the most desirable.

In the open pond method of spawning brood fish are often stocked at densities of from about 60 to 375 fish per hectare (Martin, 1967). Some culturists prefer to stock a slightly higher proportion of females than males (e.g., four females for every three males), but others stock equal numbers of each sex. Since each male can spawn with two or more females, there is never any reason to stock more males than females.

Spawning nests are generally placed in 15 to 150 cm of water with the open end of each nest facing the middle of the pond. Nest materials that are open at both ends (e.g., drain tiles), should have one end placed against the pond bank in a manner that effectively closes it off. Following stocking of the brood fish the culturist may wish to check the nests daily; if the fish are disturbed in the act of spawning however, they may fail to reinitiate spawning activity. Thus many aquaculturists prefer to examine the nests only two or three times weekly. Nest inspection is a good idea even if the eggs are going to be allowed to hatch in the pond, since it gives the culturist an indication of spawning success and at least a rough estimate of fry production. If the eggs are to be hatched indoors, inspection at 3 day intervals is sufficient for egg collection because hatching requires at least 5 days under normal temperatures.

Some caution should be exercised when the nests are inspected. Following spawning the male catfish guards the eggs during incubation and fans them with his fins to keep a current of well-oxygenated water flowing over them. During the spawning and incubation period the male often becomes quite aggressive and may give a painful bite to the culturist who carelessly inserts a hand into a guarded nest. It is usually good practice to lift nests to the water surface and slowly empty them until the eggs can be seen. This reduces the possibility of exciting the male (or both fish, if they are in the act of spawning). If the animals become excited during spawning or incubation, the thrashing that results may disrupt the egg mass.

Pen Spawning.   In open pond spawning the aquaculturist has no control over individual pairings unless he releases only two fish into each spawning pond or uses a single male with more than one female. Because this is impractical from the standpoint of economic utilization of facilities, other means of segregating paired fish have been developed. The most widely utilized technique in this regard is pen spawning (Figure 6.4).

Pens may be constructed along the edges of a pond or in the middle. Building pens along the edge of ponds offers several advantages. The pond bank can be utilized as one side of each pen, for example, and access to the pens is simplified to a degree. The number of pens that can be placed in a given pond is variable, but in no case should the number be so large that water quality is impaired by overstocking. Maximum densities of brood fish per pond should be in line with figures presented earlier for open pond spawning.

Figure 6.4   Channel catfish spawning pens in a 0.1 ha pond.

Pens are generally constructed by stapling steel wire mesh of suitable size over a wooden frame. The mesh must be fine enough to prevent excapement of brood animals. The bottom of the wire should be sunk below the sediment surface to prevent the fish from digging under the sides of the pens. Pens of any convenient size may be constructed, but most are no larger than about 2 × 3 m.

Each pen must be provided with a suitable nest, which is stocked with a selected pair of brood fish. Care should be taken to ensure that the sexes are correctly determined, for if two males or two females are placed together not only will spawning not occur, but in the former instance there may be fighting that results in the death of one or both males. The female should be slightly smaller than the male, since the male guards the nest following spawning and fighting between the adults frequently occurs after the eggs are laid. If the male is not somewhat larger, the female may drive him off the nest or kill him, after which she will ignore the eggs (allowing them to die) or eat them. The female should be removed from the pen as soon as possible following spawning. If the eggs are removed from the nest for incubation, the pen may be restocked with a second female, or the original male may also be removed and the pen restocked with another pair of brood fish.

Aquarium Spawning. The aquarium spawning method, developed by Clemens and Sneed (1962), provides the culturist with the greatest degree of control over catfish spawning; however it is not particularly desirable in terms of producing large numbers of fry. The method is best applied when direct observation of spawning activity is desired, as in the case of a teaching laboratory or where behavioral studies are being conducted. Hormone injection of the females is required for the induction of spawning when this technique is utilized. Although not generally a requirement for successful spawning, hormone injections may also be employed in conjunction with the pen spawning technique. It is not necessary to inject males with hormones in any catfish spawning technique.

Hormones are not used to induce gonadal development, but only to initiate ovulation. Once resorption of eggs has begun (as in the case of females that are not provided with an opportunity to spawn at the time their ovaries become ripe), it is too late to reverse the process, and hormone injections will not alter the situation.

The most popular hormones for use with channel catfish are fish pituitary extracts (carp pituitary being among the most popular) and human chorionic gonadotropin (HCG). The former is not generally available commercially but must be obtained by surgically removing the pituitary gland from donor fish and making an extract with fresh or acetone-dried material.

HCG, on the other hand, is commonly available through drug suppliers; its major drawback is expense, especially if a large number of females must be injected.

Hormones are injected intraperitoneally or intramuscularly with a needle and syringe (Figure 6.5). Many culturists incorporate 10,000 IU of penicillin with the hormone injection to prevent secondary infections. Spawning generally follows the injection of about 13 mg per kilogram of body weight using pituitary extract (Clemens and Sneed, 1962) or an average of 1760 mg/kg of HCG (Sneed and Clemens, 1957).

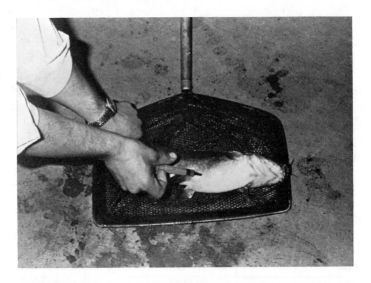

Figure 6.5 Administration of an intraperitoneal hormone injection to an adult female channel catfish.

In the aquarium spawning method fish are paried in a running water aquarium that must be large enough to permit the animals to maneuver some what. The female is injected with hormones, after which the pair is checked frequently so that they can be removed as soon as spawning is completed. If the fish do not spawn within 24 hours of the first injection, subsequent injections may be required. If spawning does not occur after three to five hormone treatments, the female should be replaced, since she may be unable to spawn. In this technique the eggs are virtually always hatched artificially.

Egg Development and Hatching. Channel catfish eggs are deposited in a yellow, adhesive mass. The male tends the eggs during incubation as previously indicated, and guards the fry until they are several days old and able to feed and seek their own protection.

Channel catfish spawn when the water temperature is between 21 and 29 C (Clemens and Sneed, 1957), and the best hatchability of eggs occurs in the range of 18 to 29 C (Martin, 1967). If the eggs are removed from the nest for artificial hatching, the temperature may be altered to some extent from that which exists in the pond or aquarium where the eggs were deposited, but drastic changes in temperature, especially outside the optimum range for hatching, should be avoided. Between 21 and 29 C, channel catfish eggs require between 5 and 10 days for hatching (Toole, 1951). The development of catfish eggs has been discussed by Murphree (1940), Clemens and Sneed (1957), and Saksena *et al.* (1961).

If the eggs are hatched by the male in a pond, the fry will remain in the nest for several days following hatching and can be removed by driving off the male and pouring the offspring into a suitable container for transfer into a fry rearing pond. Alternatively, the brood stock can be seined from the pond following spawning, leaving the fry to develop into fingerlings. A third alternative is to remove the eggs from the nests and hatch them artificially.

One advantage of the artificial hatching technique is that the spawning containers can be quickly reutilized by another pair of fish. This is particularly desirable, as previously indicated, when the pen spawning technique is employed. Also the culturist can keep an accurate account of the number of eggs hatched when artificial hatching is practiced. In addition, the culturist can be certain that the fish are well established on artificial feed before they are stocked into rearing ponds.

Channel catfish eggs can be hatched in jars receiving running water, as was once common practice in salmonid hatcheries (Canfield, 1947); however in most cases catfish eggs are hatched in troughs. The first catfish hatching trough was described by Clapp (1929); it consisted of a small raceway fitted with paddles that rotated through the water to simulate the fanning action accomplished by the adult male. The paddles, fitted on an axle, were turned by a water wheel outside the laboratory. Modern hatching trough paddles are operated by electric motors (Figure 6.6), but the frequency of thunderstorm-induced power failures during the hatching season in the southern United States suggests that reversion to water wheels might be desirable. The percentage of eggs that hatch in a trough is generally high if care is taken to prevent the establishment of fungus and bacteria on the developing egg masses (Chapter 7). In addition, fry survival is generally

Figure 6.6 Two types of channel catfish egg hatching trough. The model in the foreground has a series of paddles that move back and forth laterally through the water; the model in the background has the more traditional rotating paddles. The second hatching trough is partitioned into compartments that contain baskets of individual spawns so that the fry can be maintained in separate groups to facilitate selective breeding comparisons.

good, especially if inbreeding is not too great and if good management techniques are employed.

Fry Stocking.   Before stocking catfish fry in ponds, the ponds should be treated for predators and unwanted vegetation, and a phytoplankton bloom established. The best technique is to drain the ponds over winter or at least early in the spring and allow them to dry. They can then be treated with herbicide (either before or after filling), and a plankton bloom can be initiated early enough in the spring to ensure that it is well established before the fish are stocked. These techniques are discussed in Chapter 3.

   If fry are allowed to remain in the brood pond, that pond should be properly prepared before stocking of the brood fish, as discussed earlier. The number of brood animals stocked should be limited so that no more than 625,000 fry are produced per hectare (Martin, 1967). Normal stocking density for fry stocked into specially prepared ponds is between 125,000 and 625,000 fry per hectare (Martin, 1967). Fry can be maintained in fertilized ponds and provided supplemental feed until they reach fingerling size (several centimeters).

   When channel catfish eggs are artificially hatched, the fry may be transferred to small raceways in the laboratory through the period of yolk-sac absorption (generally less than a week following hatching), and commonly until they are well established on prepared feeds. Yolk-sac absorption is accompanied by a change in the fry from pink or orange to black. When feeding begins, the fry, which until that time have been at the bottom of the raceway, swim to the surface. A nutritionally complete diet, high in protein and energy, should be provided if optimum growth is to be achieved (see Chapter 5). At one time the yolk of hard boiled eggs, cat food, baby food, and minced liver were among the foods utilized at the onset of feeding. These have largely been replaced by trout starter rations, which provide a well-balanced diet and result in less fouling of the culture chambers than the other foodstuffs listed.

After the fry have become accustomed to prepared feed, they may be stocked into fertilized ponds at the densities outlined earlier. There is no set time for keeping fry in raceways, but the fish are usually stocked into ponds before they get longer than 2 to 4 cm.

In many parts of the United States where channel catfish are produced, fingerling fish are maintained at densities of 50,000/ha or higher throughout the first year of life. Thereafter, stocking density is reduced, with the fish being distributed into grow-out ponds during the second year of culture. In certain parts of Florida and Texas it may be possible to rear catfish from egg to market in a single growing season. The strategy would probably involve stocking fry at lower densities than are used in the two growing season technique. Ultimate production of more than 3000 kg/ha should be attainable annually in well-managed ponds, with higher densities in tanks, raceways, and cages (Chapter 2).

Striped Mullet

Normal spawning season for striped mullet, *Mugil cephalus*, along the southeast coast of the United States is between October and February, with peak activity occurring from November through January (Anderson, 1958). Since each spawning female may broadcast between 1.2 and 2.7 million eggs into the water, in aquaculture, only a few adults are required to supply sufficient stock for a large operation, provided a suitable level of survival can be maintained. Mullet eggs are fertilized immediately upon release into the water and are left untended during incubation. Depending on water temperature, mullet eggs hatch within about 48 hours of release from the female.

The first successful captive spawnings of mullet (either striped or related species) were accomplished with wild fish taken from nature after becoming ripe (Sanzo, 1936; Anderson, 1957; Yang and Kim, 1962). Mullet culture centered in the Indo-Pacific and Israel at present, was undertaken largely with wild fry until research demonstrated that captive mullet could be induced to spawn through the use of hormone injections (Tang, 1964; Liao, 1969; Yashouv, 1969). Since then considerable interest has been generated in the artificial propagation and rearing of mullet under aquaculture conditions, with most of the United States activity centered in Hawaii.

Spawning of striped mullet has been induced by the injection of ripe females with several types of hormones, including salmon pituitary (Shehadeh and Ellis, 1970; Shehadeh et al., 1973a), carp pituitary (Yashouv, 1969), HCG (Kuo et al., 1973b), and other mammalian gonadotropins (Shehadeh et al., 1973c). The initiation of ovarian development in mullet can be controlled through manipulation of temperature and photoperiod (Kuo et al., 1974a); however it appears that successful spawning is accomplished only when hormone injections are utilized. Injections must be given at the proper stage of oocyte development or the eggs will begin to deteriorate. Advantages of ovarian development through temperature and photoperiod control include the following: spawning frequency of individual fish may be increased, and the spawning season can be extended to year round (Kuo et al., 1974a).

The stage of oocyte development cannot be determined by any convenient .method of external examination of adult female mullet; instead, the eggs must be examined directly. This can be accomplished by inserting a cannula into the oviduct and withdrawing a few eggs (Shehadeh et al., 1973b). Injection of hormones is generally initiated when the oocytes are larger than 650 μ (Kuo et al., 1974b). The dosages recommended for the most commonly used hormones are discussed by Kuo et al. (1973b, 1974b).

The rearing of larval mullet has been outlined in some detail by Nash et al. (1974). Eggs obtained from hormone-injected females and fertilized by uninjected males are incubated at high density for 12 hours at 22 C in $32^0/oo$ salinity water. The eggs are then transferred to special hatching tanks at 250 eggs/1 and maintained until hatching is complete. Antibiotics are added to the water to reduce the chances for bacterial-induced mortality. Eggs can be hatched at temperatures ranging from 10 to 24 C, but 22 C is considered to be the optimum. Hatching requires about 36 hours in the temperature range 21 to 24 C, and the highest survival occurs in that range. Embryonic development of *Mugil cephalus* has been described by Liao et al. (1971) and Kuo et al. (1973a).

The newly hatched larvae average slightly more than 2.5 mm in length (Nash et al., 1974). Food is offered on the third day following hatching. A variety of natural food items have been utilized, including such animals as oyster larvae, sea urchin larvae, and *Artemia salina* (Liao et al., 1971;

Kuo *et al.*, 1973a), along with a variety of phytoplankers (Nash *et al.*, 1974). Larval survival may vary considerably depending on the type of management scheme selected; one report indicates survival in the range of 0.2 to 5% of the eggs hatched (Kuo *et al.*, 1973a). The density at which the larvae are maintained and the size of the rearing tanks appear to affect survival, as does the method by which aeration is accomplished (Nash *et al.*, 1974). Because of the extremely high fecundity of mullet, a fairly low survival rate may be acceptable, especially if the bulk of the mortality occurs before or soon after hatching when the investment in young animals is not too significant.

Panaeid Shrimp

A variety of penaeid shrimp species are commercially fished around the world, and some of these have been considered for aquaculture or are being reared by aquaculturists. Much of the emphasis in penaeid aquacultural in the United States has been aimed at the three most commonly occurring warmwater species of the southeastern Atlantic and Gulf of Mexico coasts: the white shrimp (*Penaeus setiferus*), the pink shrimp (*P. duorarum*), and the brown shrimp (*P. aztecus*). In addition, two species that occur in tropical waters off Latin America have received an increasing amount of attention recently because they seem to demonstrate certain characteristics in culture that make them appear somewhat superior to native United States species. These are *P. stylirostris* and *P. vannamei*. To date, shrimp culture in the United States has been largely experimental because of failures by investigators to close the life cycle of these animals in captivity. Apparently some recent successes in this area have been achieved by private enterprise research; thus the era of commercial shrimp farming in the United States may be rapidly approaching.

Penaeid shrimp are readily spawned in captivity and can be reared through the larval and juvenile stages to market size with a reasonable degree of success. However gonadal development and copulation do not generally occur among captive animals, and every year the culturist must obtain either gravid females or immature shrimp from nature.

The collection of gravid females is accomplished most frequently by trawling during the spawning season in areas known to have high concentrations of adult shrimp. Sufficient experience has been gained in many coastal areas of the United States to provide a fairly high probability of success in capturing gravid animals. The gravid females must have spermatophores (sacs of sperm transferred from the male to the female during copulation) attached ventrally. Since the spermatophores are easily dislodged from white shrimp during trawling and subsequent handling, particular care must be taken during these operations. For all species, trawling must be of short duration if mortality of the animals is to be kept to a minimum.

Ovarian development in penaeid shrimp can be induced by a method known as eyestalk ablation (Caillouet, 1972) in which the eyestalks of adult females are surgically removed or crushed, triggering a hormonal response. However the eggs that develop as a result of this technique are often of poor quality and few will hatch if fertilized. It may be a case of extremely rapid development, with a failure of the eggs to incorporate the levels of nutrients required by the developing embryos, or some other factor may be involved. Studies are being conducted at various laboratories to find alternative means to induce development of penaeid shrimp females at a more normal rate. Once this problem is solved, commercial shrimp culture will have made a large step toward biological and economic reality.

For spawning in the laboratory, gravid wild females with attached spermatophores are individually stocked into holding tanks of various sizes. Such shrimp usually spawn within 24 hours of capture, often at night. Following spawning, the female is removed from the container, and the eggs are allowed to incubate at about 24 C (considered to be the optimum temperature) until hatching occurs after about 12 to 16 hours (Cook and Murphy, 1969).

Each of the three commercially important penaeid shrimp native to the United States (the white, brown, and pink shrimps) go through a similar series of larval stages following hatching. These include five naupliar (nonfeeding stages), three protozoeal, three mysid, and several postlarval stages (Pearson, 1939; Heegaard, 1953; Dobkin, 1961; Cook and Murphy, 1971). Thereafter the shrimp are referred to as juveniles, having acquired the body configuration of the adults.

Larval penaeids begin feeding at the second protozoeal stage (Cook and Murphy, 1969). A variety of species of phytoplankton have been successfully utilized for feeding larval shrimp and can be produced by mass culture techniques. When the larvae reach the mysid stage they may be fed a mixture of phytoplankton and brine shrimp (*Artemia salina*) nauplii or other zooplankters. Cannibalism has often been a serious problem to penaeid shrimp culturists, although significant numbers of shrimp can still be produced from the tens to hundreds of thousands of eggs obtained from each female. Postlarval penaeid shrimp can be stocked into fertilized ponds and will accept prepared diets, some of which have been specially formulated for them. The techniques of spawning and rearing larval penaeid shrimp have been reviewed by Heinen (1976).

Some culturists have avoided establishing their own shrimp hatcheries, relying instead on natural production of wild stocks. Juvenile shrimp are obtained by trawling, trapping, or by allowing them to enter ponds on the flooding tide. A major problem associated with the latter method (practiced by certain fish culturists as well as shrimp culturists in various parts of the world) involves the predators and other undesirable species that must be eliminated from the ponds. Rose (1975) stocked marked, hatchery-reared juvenile penaeid shrimp into ponds that contained no other animals and into ponds that were stocked with wild species by tidal action. The survival of the hatchery-reared shrimp was 4 times higher when predators were not present than when predation was a threat. Ponds stocked by tidal flow contained large numbers of blue crabs (*Callinectes sapidus*) and various species of fish, along with a relatively low biomass of wild shrimp, indicating that the technique was detrimental to commercial shrimp culture. This factor is especially significant when the price of shrimp is compared with that of crabs and most fishes.

Freshwater Shrimp

Macrobrachium rosenbergii is the most popular species of freshwater shrimp now being reared by aquaculturists. Part of the popularity of *M. rosenbergii* may be due to familiarity: that species was the first for which the complete life cycle was described and controlled in the laboratory. In addition, the growth rate, maximum size, and adaptability to culture environments all seem to be optimized in *M. rosenbergii*, whereas many domestic species of *Macrobrachium* fail to measure up in one respect or another.

Goodwin and Hanson (1974) reviewed the cultural practices currently being employed with *Macrobrachium rosenbergii*, with special emphasis on the work conducted in Hawaii, where many of the pioneering studies with that species were done. In Hawaii and similar climates, *M. rosenbergii* breed throughout the year and can be maintained in ponds or tanks. In temperate climates it is often necessary to overwinter the animals indoors to prevent mortality from low temperature (the lower extreme of temperature tolerance is about 10 to 12 C in this species). Adult males will spawn at any time and do not molt once having attained maturity. Females, on the other hand, undergo a premating molt before spawning. Brood stock of both sexes should be maintained together, since the males protect the females from cannibalism during the prespawning molt.

A typical adult female (Figure 6.7) will produce up to 30,000 eggs per spawn and may spawn twice within 5 months. The eggs are extruded and attached to the pleopods of the female from 6 to 20 hours following mating and generally hatch within about 19 days.

In nature, *Macrobrachium rosenbergii* females move into brackish water to spawn, thus laboratory culture methods incorporate low salinity water as a part of the larval rearing procedure. The eggs of *M. rosenbergii* will develop properly and hatch in fresh water, but larvae produced in that medium will survive only a few days if not moved into higher salinity. Most culturists employ 12 ‰ salinity water for hatching and larval rearing. The planktonic larvae undergo 11 molts before metamorphosing into postlarvae. Feeding may begin as early as 24 to 48 hours after hatching. *Artemia salina* nauplii or other zooplanktonic species are provided as food from the onset of feeding activity.

Larval development varies as a function of temperature and salinity, with 29 C and 12 ‰ appearing to be near optimum. Under those conditions postlarvae can be produced in 35 to 40 days. Postlarval *Macro-*

Figure 6.7 An adult female *Macrobrachium rosenbergii*. Eggs are carried on the ventral surface of the animal, approximately in the location delimited by the thumbs of the individual holding the shrimp. This specimen is a small adult and may reach a maximum weight of approximately 0.5 kg.

brachium rosenbergii become benthic and, in nature, would begin to migrate upstream. Therefore, in culture, salinity is gradually reduced to that of fresh water after the postlarval stage has been reached. The remainder of the life cycle is spent in fresh water, although gravid females being utilized as brood stock are intermittently placed in salt water to spawn.

Juvenile *Macrobrachium rosenbergii* can be reared under intensive or extensive culture conditions and will feed on manufactured diets of various formulations. The complete nutritional requirements of freshwater shrimp have not been elaborated, but diets of acceptable quality have been developed.

Crayfish

The red swamp crayfish (*Procambarus clarkii*) and the white river crayfish (*P. acutus*) of Louisiana breed during May and June (Avault, 1972; LaCaze, 1976), after which the female may burrow into the pond bottom. If there is open water in the pond, burrowing will take place near the water's edge (LaCaze, 1976). Pond culturists generally drain their ponds during June to force the females to burrow. Crayfish burrows may be very complex structures, 1 m or more deep. At the surface they may be merely plugged with a mud cap, or a mud chimney may protrude several centimeters above the pond bottom (Avault, 1972).

Each burrow may be occupied by a single female or she may be joined by a male. In the latter case mating activity may continue inside the burrow (LaCaze, 1976). During mating the male deposits sperm in a receptable organ on the female. The eggs are not extruded and fertilized until September, during which time they are carried on the swimmerets of the female and held in place by a sticky substance called glair (LaCaze, 1976). Hatching takes place within about 2 or 3 weeks following extrusion and fertilization for the red swamp crayfish. White river crayfish eggs require from 3 to 8 days longer for hatching (Avault, 1972; LaCaze, 1976). LaCaze (1976) reported that the average number of young produced (not all the eggs develop and hatch properly) is approximately 400 for the red swamp crayfish, with maximum production being about 700 juveniles. The white river crayfish may produce somewhat fewer young on the average.

Hatching peaks during October, at which time crayfish culturists should reflood their ponds. Once sufficient water is present, the young crayfish that have been clinging to the abdomen of the female become free swimming and leave the burrow to enter the pond. The young are approximately 1 cm long at this time. If sufficient water is not present, some of the young crayfish may be released inside the burrows, where crowding and lack of food prevent their growth. In some cases the females leave the burrows and attempt to move overland to find water. Desiccation and predation often take a toll during such migratiions (LaCaze, 1976) and it is certainly not in the best interest of a crayfish culturist to have the brood stock and their offspring leave the ponds in search of more desirable environments.

Major problems that occur during the growth period of the young include oxygen depletions and overcrowding. Both these can be solved by proper culture system management (Avault et al., 1974).

Oysters

Oyster culture, long successful in Europe and the Orient (particularly in Japan), has been relatively less so in the United States, even though interest in oyster culture in this country goes back to the colonial period. Presently there is interest in the aquaculture of oysters on the Atlantic, Pacific, and Gulf of Mexico coasts, although total aquacultural production remains insignificant compared with that obtained from wild stocks. Three species have received the attention of United States culturists, although two of them, Crassostrea gigas (the Japenese oyster) and Ostrea edulis (the European oyster) are coldwater forms and are not discussed here.

Crassostrea virginica, the American oyster, occurs along the Atlantic and Gulf of Mexico coasts of the United States and grows best in warm water. The biology of this species has been discussed in detail by Galtsoff (1964). *C. virginica* requires a temperature range of 21 to 27 C for spawning and can be conditioned to spawn within about 6 weeks in the winter if exposed to a temperature range of 23 to 24 C (Hidu et al. 1969), even though spawning normally occurs in the spring. The most common method of spawning induction involves thermal and chemical stimulation (Loosanoff and Davis, 1963) in which ripe adults are exposed to a rapid rise in water temperature to 30 C and to a sperm suspension from a sacrificed male. Each female may release several million eggs following this treatment.

Fertilized eggs are removed from the water with a fine mesh sieve and transferred to well-aerated, filtered seawater. At 30 C the eggs will hatch into veliger larvae within 48 hours (Landers, 1968). The larvae are provided with either natural phytoplankton (commonly obtained through the use of a continuous centrifuge receiving water from a natural seawater source high in primary productivity) or artificially produced phytoplankton obtained by mass culture techniques (Loosanoff and Davis, 1963). The water is changed daily during the period of larval development.

At a temperature of 30 C metamorphosis can be expected in as little as 10 days if proper phytoplankton levels are maintained (Landers, 1968). Extremely dense or dilute phytoplankton levels will cause an increase in the time for setting of the larvae. The setting larvae, called spat, are generally placed in shallow tanks containing cultch material. Cultch material for oyster setting is often composed of empty oyster or clam shells, although a variety of other materials have been successfully employed. Depending on the expected level of mortality, the larvae may be placed in the setting tanks at densities of from 10 to 50 spat per oyster shell cultch (Landers, 1968).

Spat-laden cultch can be spread over a pond bottom, distributed in a leased area in an estuary, reared in culture tanks, or suspended in the water column from rafts or other structures. The latter method, often called string culture, provides protection of the oysters from various benthic predators such as oyster drills and has received at least some attention from United States aquaculturists (Shaw, 1960, 1962, 1968; Marshall, 1968, 1969; Linton, 1968; May, 1968, 1969). The spreading of oysters in leased estuarine areas is the least intensive form of culture and is very closely related to traditional commercial oyster fishing in which oyster shell is often returned to the sea following shucking so that the shell can provide natural cultch. The difference is that the culturist establishes a hatchery and plants cultch that has already been seeded with spat.

Cultchless oyster production has also been developed by culturists. In this technique the spat are settled in containers in the laboratory and ultimately are transferred to trays in which they are allowed to grow. When the oysters become large enough, they can be moved into natural areas in an estuary, or the whole system may be maintained indoors. A large supply of phytoplankton-rich water must be available in either case.

Of tangential importance to aquaculture is the common practice of collecting oysters from polluted waters and removing them to clean areas. The oysters are allowed to remain in uncontaminated water for several days, during which time the pollutants are expelled from the animals. Subsequently these oysters may be placed on the retail market. Culture facilities are often readily adaptable for use in this purging process.

The history of development of oyster culture throughout the world has been described by Loosanoff (1969). The techniques utilized for American oyster culture have been adopted in large part by workers on the west coast of the United States working with the Pacific oyster. Breese and Malouf (1975) described culture techniques employed for Pacific oysters, along with the nutrient media and techniques employed in the culture of algae for feeding the young oysters.

Sea Turtles

The green sea turtle, *Chelonia mydas*, is the only species presently being cultured for food anywhere in the world, and the only culturist is a commercial enterprise, located in the Cayman Islands. Green sea turtles occur, in general, between latitudes 35°N and 35°S (Ingle and Walton Smith, 1949). The young are carnivorous and will consume a variety of natural animal material as well as prepared diets (Stickney et al., 1973). However after the first year of life, green sea turtles become increasingly herbivorous and prefer marine grasses (Ingle and Walton Smith, 1949).

The breeding behavior of green sea turtles has been summarized by Ingle and Walton Smith (1949). Adults gather off the laying beaches in the spring for copulation just before and during laying. It has long been thought that eggs laid during any one year were fertilized by sperm stored from a prior mating, possibly one that occurred up to 3 or 4 years earlier (sometimes the duration between visits to the nesting beach for females). This theory is well supported by observations made by Witham (1970) on captive green sea turtles.

The female leaves the water, digs a nest in the sand, and deposits as many as 200 eggs. The eggs are then covered and the female retreats to the sea. She may return at intervals of about 2 weeks and repeat this behavior several times during a single spawning season. Incubation varies depending on temperature, but generally requires nearly 2 months. Hatchlings dig themselves out of the nests and scurry to the water. Significant

losses of young turtles inside the nests, on the way to the water, and thereafter, can be attributed to birds, predatory mammals, various types of
marine life, and man.

Attempts to mate and spawn green sea turtles in captivity are being
made by interested parties and, increasingly, positive results are being
obtained. The commercial turtle venture in the Cayman Islands was originally totally dependent on gathering eggs from natural nesting beaches
to supply its needs (Stewart, 1977), but this source is being eliminated
through the imposition of protective laws by countries in which the turtles
nest. Conservationists in the United States and elsewhere are very much
interested in protecting green sea turtles, and clashes between aquaculturists and conservationists over the protection issue have already occurred.
Many conservationist groups wish to make the culture of these animals
illegal and have promoted strict laws prohibiting the importation or consumption of green sea turtles in the United States, no matter what the
source of the animals. There is no doubt that the green sea turtle is en-
dangered in many parts of the world, and every effort should be made
to protect this animal and especially to preserve its nesting beaches until
suitable sized populations are reestablished. Both the green sea turtle,
Chelonia mydas, and its relative the loggerhead sea turtle, *Caretta caretta*
(Figure 6.8), should be reintroduced into regions from which they have
been eliminated. Such reintroductions may be aided by the utilization
of aquacultured stock, and indeed, preliminary work along those lines
has been initiated by the National Marine Fisheries Service at their Galveston,
Texas, laboratory, where loggerhead sea turtle hatchlings were maintained
for several months prior to release into the Gulf of Mexico during 1977
(Jim McVey, personal communication). The establishment of turtle mariculture operations utilizing their own brood stock and artificial nesting
beaches could provide turtles not only for human consumption but also
for restocking programs.

SPECIALIZED REPRODUCTIVE STRATEGIES

In most cases aquaculturists attempt to obtain the maximum reproduction
from available brood stock and are not concerned about overpopulation
and stunting because most cultured species are marketed before reaching
adulthood. Also of little concern with most species is escapement from
the culture facilities, since animals under culture are native to the region
in which they are being reared. However in some cases reproduction at a
small, premarket size can occur, and aquaculturists have become increasingly
interested in rearing exotic species that could disrupt the natural environ-

Figure 6.8 Juvenile marine turtles; a green sea turtle, *Chelonia mydas* (left) and a logger-head, *Caretta caretta*.

ment if released inadvertently. Both problems are exemplified by various species of tilapia, and the disruptive effects of accidentally introduced grass carp have brought considerable protest. Thus, this section discusses methods of controlling overpopulation and escapement of reproductively active tilapia, and the technique of gynogenetic production of all-female populations of grass carp.

Tilapia

In many cases of overpopulation with tilapia, stunting is so severe that few, if any, marketable fish are produced even in climates featuring extended growing seasons. *Tilapia aurea,* a species that often overpopulates ponds (Figure 6.9), is a mouthbrooder. The male of this species constructs a nest in which the female deposits her eggs. Following fertilization, the female picks up the eggs (generally 500 or fewer) in her mouth and carries them until they hatch. The fry remain in close association with the female for several days following hatching and take refuge in her mouth if danger threatens. The net result is a high rate of survival, a trait that is desirable in culture species.

Another superficially excellent quality of *T. aurea* is that multiple spawns can be expected from each female brood fish annually. Other species of tilapia also exhibit this characteristic and spawn at a small size. For example,

Figure 6.9 Dip net filled with small *Tilapia* sp. typical of fish that might be captured at the end of the growing season from an overpopulated pond in which the adults were allowed to spawn freely. Nearly all these fish are of breeding size, though few are more than about 8 cm long.

T. mossambica have been known to spawn 6 to 11 times in a single year (Chimits, 1955), and some species are reported to have spawned at a length only slightly exceeding 10 cm (Atz, 1954). Thus the offspring of a female spawning for the first time in a given spring may be spawning with her later progeny before the summer is over. Overpopulation and stunting are unavoidable if uncontrolled spawning is allowed.

A variety of methods have been utilized to overcome the stunting problem that so often occurs in tilapia production. In Central and South America, where *Tilapia aurea* and other species have become popular in recent years, attempts have been made to control overpopulation by stocking carnivores, which consume small fish entering the system. The problem with this approach, as generally recognized by fishery managers, is that the maintenance of proper predator-prey ratios, even for only one growing season, can be very difficult.

Other methods of maintaining stable tilapia populations include the use of cage culture, monosex hybridization, and sex reversal through exposure to hormones. Cage culture tends to restrict or eliminate spawning because no nest can be constructed and if the female does lay eggs, they generally fall through the cage bottom and are lost. This method is generally successful, but instances of reproduction in caged tilapia have been reported (Bardach et al., 1972; Pagan-Font, 1975).

Male tilapia generally grow more rapidly than the females (Chimits, 1955; Lowe-McConnell, 1958; Avault and Shell, 1968), especially when the fish approach adulthood and the females begin to divert large amounts of food energy to egg production. Thus the production of all-male populations of tilapia not only solves the overpopulation problem but also leads to more rapidly growing animals. Crosses between such species as *Tilapia mossambica* × *T. aurea* and *T. nilotica* × *T. aurea* produce high percentages of male offspring (Avault and Shell, 1968; Pruginin et al., 1975). In some cases all the eggs produced develop into males. The major drawback of hybridization is that extra facilities are needed to maintain two or more species of brood animals, compared to those required for maintaining only one species. The brood fish must be selected on the basis of sex (not easily distinguishable in some species of tilapia), and following spawning the adults must be returned to the proper ponds if mixing of species is to be avoided.

A technique has been devised by which all *Tilapia aurea* (and presumably other species) spawned under normal conditions can be transformed into males (Guerrero, 1975). Fry are collected from spawning tanks or ponds and are fed a standard fish diet treated with 60 μg/ml diet of 17 α-ethynyltestosterone. The hormone-treated feed is offered for about 3 weeks, after which it may no longer be effective. Female sex reversal is nearly 100% effective. Following the treatment (which is usually conducted in tanks or raceways), the fish can be stocked into ponds or other rearing chambers. Acceptance of prepared diets, including those treated with hormones, has not been a problem in feeding tilapia fry.

Thus techniques have been developed by which overpopulation of tilapia can be achieved without inordinate difficulty or expense. Monosex hybridization and the production of all-male populations through hormone treatment also have the advantage that if escapement from an aquaculture facility does occur, the chances of establishment of a self-sustaining wild population of tilapia are reduced. Cage culture does not prevent escapement in cases where cages are damaged, although it does restrict the chance for establishing wild populations by at least severely reducing the likelihood of reproduction during the normal grow-out period.

In most parts of the United States the intolerance of tilapia to low temperature prohibits the establishment of permanent populations. However in portions of Florida and Texas, as well as in the heated water of power plant cooling reservoirs, extremely large populations can become established

(Germany and Noble, 1977). When this occurs there is both observational and experimental evidence to indicate that spawning of such species as largemouth bass (*Micropterus salmoides*) is reduced or eliminated, probably as a function of competition for nesting space (Noble *et al.*, 1975).

Grass Carp

Fear that release of grass carp, *Ctenopharyngodon idella* (Figure 6.10) into the natural waters of the United States as a biological weed control organism will result in the competitive exclusion of native, more desirable fishes has led to the banning of grass carp in more than 30 states. Thus the grass carp, widely cultured around the world, cannot presently be considered to be a viable aquaculture candidate in most states unless it can be guaranteed either that the animals will not reproduce in the wild or that monosex fish can be produced exclusively for aquatic vegetation control or aquaculture. After such assurances are obtained, many state regulations will have to be revised to allow even research level activities and demonstrations of the usefulness of grass carp to be conducted.

Figure 6.10   The grass carp, *Ctenopharyngodon idella*, is widely utilized in aquatic vegetation control and aquaculture in much of the world, but its future in the United States remains in doubt.

Grass carp may never become a popular aquaculture species in the United States, but some of the work on production of monosex populations that has been done with this fish is significant because of its potential for incorporation into culture strategies applied to other species. In particular, studies aimed at the gynogenetic development of grass carp fry, conducted at the U.S. Fish and Wildlife Service, Fish Farming Experiment Station in Stuttgart, Arkansas, have provided evidence that this procedure has some merit.

Gynogenesis is the development of an ovum following sperm penetration, but without fusion of the gametes. In other words, a ripe egg is penetrated by a spermatozoan, but the genetic material contained in the sperm cell is not incorporated into the nucleus of the egg, even though the egg is stimulated to develop. In most cases the larvae formed are haploid (contain only half the normal complement of chromosomes) and die a few days after development has begun. However in a few instances an ovum is diploid (contains the normal number of chromosomes found in somatic cells) and develops into a female fish identical to its parent. The latter case is one of gynogenetic reproduction. The idea behind artificial gynogenesis is to promote the development of diploid eggs and stimulate them to divide and produce embryos.

Two ways in which haploid ova become diploid, thus can undergo gynogenetic development, have been proposed, as discussed by Stanley and Sneed (1974). In the first, the polar body that is normally released during the second meiotic or reduction division in gamete formation, recombines with the oocyte to double the haploid number of chromosomes. Successive mitotic cell divisions produce a diploid embryo. In the second mechanism the polar body is released normally during the second meiotic division and a haploid egg is formed. Upon proper stimulation, the chromosomes in the ovum are replicated as in normal mitosis, but incomplete cell division occurs. The two sets of chromosomes are incorporated into a single nucleus, and subsequent cell divisions produce an embryo with the diploid chromosome number.

Ova can be induced to develop in the absence of sperm cells (Stanley and Sneed, 1974). A needle dipped in the serum or whole blood of a fish may be used to prick the eggs and often stimulates them to divide. Alternatively, weak electrical currents may be passed through the eggs with the same result. For development that follows the definition of gynogenesis more explicitly, however, sperm from distantly related species may be utilized to stimulate the eggs, or irradiated sperm from the same species may be used. Sufficient radiation (e.g., X-ray) is utilized to denature the DNA while not destroying sperm cell motility.

In each of the cases above, gynogenetic births are rare because of the extremely low number of eggs that have diploid chromosome numbers. The incidence of diploid ova can be increased through temperature shock and other means. A rapid reduction in temperature of several degrees may induce the production of diploid ova, but the total number of gynogenetic fish obtained may be only a small percentage of the total number of eggs

available. Before this technique becomes practical for the average aquaculturist, it must be refined to the point where a reasonably high percentage
of gynogenetic fish can be produced from each female in the brood population.

SELECTIVE BREEDING AND GENETICS

Most attempts at stock improvement in commercial aquaculture have
involved selective breeding with little knowledge of genetic consequences.
As a result, a great deal of inbreeding has occurred, and in some instances
the percentage of abnormally developed individuals has increased rapidly
in recent years. This has occurred, for example, in the channel catfish
business. Moreover many exotic species, such as *Macrobrachium rosenbergii*
and *Tilapia aurea,* have been spread throughout the warmwater areas of
the United States as a result of the importation of only a few individuals
from native populations elsewhere in the world. Thus inbreeding has been
inevitable in many cases, and the consequences may not yet be apparent.

Rather than concern themselves with the problems associated with inbreeding, commercial fish culturists may instead attempt to select brood
stock for rapid growth, high food conversion efficiency, high dress-out
percentage, or other desirable characteristics. It is generally assumed that
such traits are controlled by dominant genes and that more than one gene
is responsible for each of them. The influence of environment on such
characters has been ignored in most cases, although it may be very important.
The wide range of environmental conditions that might be encountered by
a culture species could mask genetic influences on certain traits and make
selection difficult. For example, in a carefully controlled experiment, a
group of fish or invertebrates might be reared at 30 C from egg to adulthood. If the most rapidly growing individuals were mated, their offspring
might also be expected to grow rapidly under the same experimental conditions. However if animals of the F_1 generation were to be placed into a
commercial outdoor aquaculture facility where the temperature fluctuated
diurnally and seasonally, the animals might grow even less rapidly than
those produced from random matings. Superficially, the development of
laboratory strains of fish or invertebrates for aquacultural research seems
desirable, but because of constantly changing environmental conditions
in most commercial culture systems and geographical differences in climate
and water quality among laboratories as well as commercial operations,
it is difficult to conceive of the development of an aquaculture strain
corresponding to the laboratory white rat.

For selective breeding to be successful, there must be a degree of variance in one or more characteristics that are considered beneficial by the culturist. In many cases such variance is difficult to ascertain, let alone quantify. For example, dress-out percentage or percentage of fat deposited in the body cavity of a fish, may vary several percent within a given population, although morphologically the fishes may be indistinguishable externally. If the animals must be sacrificed and dressed to ascertain differences, it will be difficult to conduct selective breeding experiments! One way around this apparent dilemma is to mate a wide variety of individuals that present little in the way of identifiably distinguishing traits and examine the responses of the offspring. Such studies require a great deal of time as well as extensive facilities, and realistically, there are no laboratories in the United States capable of handling such large-scale genetics studies.

The portion of total variance within a population of animals attributable to certain kinds of genetic effects is known as heritability. This concept is developed in detail by such authors as Dobzhansky (1970) and Strickberger (1976). For purposes of this discussion, a heritability of 0.0 indicates that there was no correlation between the parents and the offspring with respect to the trait under study, whereas a heritability of 1.0 indicates that the offspring were identical to the parents for a particular trait. When heritability is low, large-scale breeding experiments involving many pairings and probably extending over several generations may be required to improve the stock. Most of the information available on heritability in agricultural livestock comes from the cattle and swine literature, and examination of studies performed on terrestrial animals may provide insight into the difficulties and potentials that lie ahead for aquaculturists. To date, warmwater aquacultural genetics can provide little information on heritability.

Because of the lack of information, most warmwater culturists base their selection of brood stock on rapid growth or some other characteristic that may or may not have a high level of heritability but is readily identifiable to the culturist. In many cases little or no improvement of the stock results from such selections, but they represent the state of the art insofar as many aquaculture species are concerned. Selection of channel catfish has been based on growth for many generations, although there has been little or no improvement obtained.

Inbreeding, the mating of closely related individuals, may result in a general reduction in the performance of the offspring, a decreased rate of survival of eggs and larvae, an increase in the frequency of deformity, and an increase in the variability of various quantitative characters. As previously indicated, a great deal of inbreeding has occurred in exotic species that have been brought into the United States. In addition, native species in aquaculture, such as the channel catfish, are experiencing high levels of inbreeding because culturists have failed to obtain new strains of animals to add into their brood stock populations. The incidence of such anomalies as fused caudal vertebrae and an increase in the frequency of albinism are just two examples of problems that have occurred because of inbreeding among channel catfish. Inbreeding leads to an increase in homozygosity (the presence of only one type of allele for one or more traits), and although this may bring improvement in certain characteristics, problems often result, and the practice should generally be avoided except when conducted by trained geneticists who have specific goals in mind and are able to manipulate large stocks of animals.

Heterosis, or hybrid vigor, is the response shown by organisms that are mated with distantly related individuals of the same species. This is known as outcrossing and can lead to an increase in heterozygosity, with resulting improvements in performance, increased survival of eggs, and so forth—the opposite of inbreeding. Hybrid vigor usually disappears in the second generation because heterozygosity is decreased. Thus it is necessary to maintain two or more inbred lines of brood animals that when crossed, will produce vigorous offspring for stocking in grow-out culture chambers. Ideally, the culturist would prefer to have separate inbred lines that were homozygous for certain traits, with one being homozygous dominant and the other homozygous recessive for each trait. The parents might have genotypes as follows:

$$\text{male} \; = \; aa \quad BB \quad cc \quad DD \quad ee$$
$$\text{female} = AA \quad bb \quad CC \quad dd \quad EE$$

When the adults depicted are mated, the offspring would be heterozygous for each gene:

$$F_1 = aA \quad bB \quad cC \quad dD \quad eE$$

If two inbred lines of fish are maintained and crosses are utilized for rearing purposes, the problem still will not be solved. Eventually it becomes necessary to replace brood animals and the gene pool will, of necessity, be altered whether the new brood stock are obtained by further inbreeding within lines or by selection among fish produced in some other manner. Strictly random mating, another way in which inbreeding can be reduced, calls for a fairly large population of brood animals, and even then a detectable level of inbreeding occurs. Eventually inbreeding becomes so severe that new brood stock strains must be acquired. Kincaid (1976) suggested that salmonid culturists maintain brood stocks of no fewer than 25 pairs and recommended 50 to 100 pairs as being more desirable.

A third scheme for reducing inbreeding involves the maintenance of three lines of brood animals and the use of rotational line mating (Kincaid, 1976, 1977). The initial mating lines generally are established by separating a strain that will produce progeny having the desired traits into three groups, identified as lines A, B, and C. During each breeding season the males from line A are mated with females from line B, males from line B with females from line C, and males from line C with females from line A (Figure 6.11). The offspring from these matings are maintained together if the parents are to be utilized in subsequent matings, or offspring are maintained separately in sufficient numbers to replace the brood stock during years when that is desired. The extent of inbreeding in the rotational line mating scheme is somewhat less than that which might occur from random matings, and this scheme solves the brood stock replacement problem, which exists when only two breeding lines are maintained.

One problem associated with the rotational line mating scheme is maintenance of separate populations of brood animals. To ensure that only the desired crosses occur, it is almost essential that each line be maintained separately (i.e., in different culture chambers). Needless to say, one of the

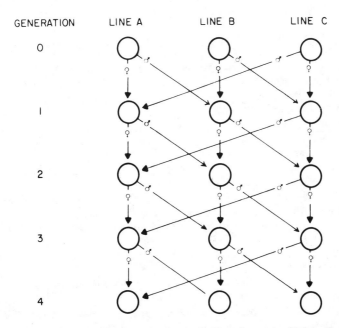

Figure 6.11 Rotational line mating scheme for minimizing the extent of inbreeding within a population of animals. Adapted from Kincaid, (1976, 1977).

most important considerations in the employment of this mating scheme is the maintenance of adequate records, to prevent mistakes in crosses.

A great deal of the research conducted by aquaculturists has been related to environmental requirements, nutritional requirements, methods of disease control, and ways in which to design and manage culture systems for optimum production. Comparatively, the level of intensity of effort in genetics and selective breeding has been somewhat below that of those other areas of research, although programs with various species have been developed in recent years. Because genetic improvement depends on long-term experimentation with large numbers of individuals, advances in knowledge tend to be slower than in other subject areas, but advances can be expected to occur if sufficient time and energy are expended by the research community. After all the other problems associated with the culture of any species have been solved, the only remaining way to improve that animal from an aquacultural standpoint is through genetics. This field will become increasingly important to aquaculturists in the future.

LITERATURE CITED

Anderson, W. W. 1957. Early development, spawning, growth and occurrence of the silver mullet (*Mugil curema*) along the south Atlantic coast of the United States. *Fish. Bull.* **57**: 397–498.

Anderson, W. W. 1958. Larval development, growth, and spawning of striped mullet (*Mugil cephalus*) along the south Atlantic coast of the United States. *Fish. Full.* **58**: 501–519.

Arnold, C. R., T. D. Williams, W. A. Fable, J. L. Lasswell, and W. H. Bailey. 1977. Laboratory methods for spawning and rearing spotted sea trout. *Proc. Southeast Assoc. Fish Wildl. Agencies*, **31**: 437-440.

Atz, J. W. 1954. The peregrinating *Tilapia*. *Anim. King.* **57**: 148–155.

Avault, J. W., Jr. 1972. Crawfish farming in the United States. In S. Abrahamsson (Ed.), *Freshwater crayfish; Papers from the First International Symposium on Freshwater Crayfish*, Lund, Austria, Studenlitt, pp. 239–250.

Avault, J. W., Jr., and E. W. Shell. 1968. Preliminary studies with the hybrid *Tilapia nilotica* × *Tilapia mossambica*. *FAO Fish. Rep.* **44**: 237–242.

Avault, J. W., Jr., L. W. de la Bretonne, and J. V. Huner. 1974. Two major problems in culturing crayfish in ponds: Oxygen depletion and overcrowding. In J. W. Avault, Jr. (Ed.), *Freshwater crayfish; Papers from the Second International Symposium on Freshwater Crayfish*. Louisiana State University, Baton Rouge, pp. 139–144.

Bardach, J. E., J. H. Ryther, and W. O. McLarney. 1972. *Aquaculture*. Wiley-Interscience, New York. 868 p.

Breese, W. P., and R. E. Malouf. 1975. *Hatchery manual for the Pacific oyster*. Oregon State University Sea Grant Program Publication ORESU-H-75-002. Corvallis. 22 p.

Caillouet, C. W., Jr. 1972. Ovarian maturation induced by eyestalk ablation in pink shrimp, *Penaeus duorarum* Burkenroad. *Proc. World Maricult. Soc.* **3**: 205–225.

Canfield, H. L. 1947. Artificial propagation of those channel cats. Prog. Fish-cult. **9**: 27-30.

Chimits, P. 1955. Tilapia and its culture. A preliminary bibliography. *FAO Fish. Bull.* **8**: 1-23.

Clapp, A. 1929. Some experiments in rearing channel catfish. *Trans. Am. Fish. Soc.* **59**: 114-117.

Clemens, H. P., and K. E. Sneed. 1957. *The spawning behavior of the channel catfish, Ictalurus punctatus.* U.S. Department of the Interior, Special Scientific Report—Fisheries, No. 219. 11 p.

Clemens, H. P., and K. E. Sneed. 1962. *Bioassay and the use of pituitary materials to spawn warmwater fishes.* U.S. Fish and Wildlife Service Resources Report 61. 30 p.

Cook, H. L., and M. A. Murphy. 1969. The culture of larval penaeid shrimp. *Trans. Am. Fish. Soc.* **98**: 751-754.

Cook, H. L., and M. A. Murphy. 1971. Early development stages of the brown shrimp, *Penaeus aztecus* Ives, reared in the laboratory. *Fish. Bull.* **69**: 223-239.

Dobkin, S. 1961. Early developmental stages of the pink shrimp, *Penaeus duorarum*, from Florida waters. *Fish. Bull.* **61**: 321-349.

Dobzhansky, T. 1970. *Genetics of the evolutionary process.* Columbia University Press, New York. 505 p.

Futch, C. R. 1976. Biology of striped mullet. In: J. C. Cato, and W. E. McCullough (Eds.), *Economics, biology and food technology of mullet.* Florida Sea Grant Program Report 15, pp. 65-69.

Galtsoff, P. S. 1964. The American oyster *Crassostrea virginica* Gmelin. *Fish. Bull.* **64**: 1-480.

Germany, R. D., and R. L. Noble. 1977. Population dynamics of *Tilapia aurea* in Trinidad Lake, Texas. *Proc. Southeast. Assoc. Fish Wildl. Agencies,* **31**: 412-417.

Goodwin, H. L., and J. A. Hanson. 1974. *The aquaculture of freshwater prawns (Macrobrachium species).* The Oceanic Institute, Waimanolo, Hawaii. 95 p.

Gudger, E. W. 1918. *Oral gestation in the gaff-topsail catfish Felichthys felis.* Papers from the Department of Marine Biology of the Carnegie Institute of Washington, Vol. 12, pp. 25-52.

Guerrero, R. D. III. 1975. Use of androgens for the production of all-male *Tilapia aurea* (Steindachner). *Trans. Am. Fish. Soc.* **104**: 342-348.

Hanson, J. A., and H. L. Goodwin (Eds.). 1977. *Shrimp and prawn farming in the Western hemisphere.* Dowden, Hutchinson & Ross, Stroudsburg, Pa. 439 p.

Heegaard, P. E. 1953. Observation on spawning and larval history of the shrimp, *Penaeus setiferus* (L.). *Publ. Inst. Mar. Sci., Univ. Tex.* **3**: 73-105.

Heinen, J. M. 1976. An introduction to culture methods for larval and postlarval penaeid shrimp. *Proc. World Maricult. Soc.* **7**: 333-344.

Hidu, H., K. G. Drobeck, E. A. Dunnington, Jr., W. Roosenburg, and R. L. Beckett. 1969. Oyster hatcheries for the Chesapeake Bay region. Univ. Md. Natur. Resour. Inst. Spec. Rep. No. 2, 18 p.

Hoar, W. S. 1969. Reproduction. In: W. S. Hoar and D. J. Randall (Eds.), *Fish physiology*, Vol. 3. Academic Press, New York, pp. 1-72.

Ingle, R. M., and F. G. Walton Smith. 1949. *Sea turtles and the turtle industry.* University of Miami Press, Miami. 107 p.

Kincaid, H. L. 1976. Inbreeding in salmonids. In T. Y. Nosho and W. K. Hershberger (Eds.), *Salmonid genetics; Status and role in mariculture.* University of Washington Sea Grant Report WSG WO 76-2. University of Washington, Seattle, pp. 33-37.

Kincaid, H. L. 1977. Rotational line crossing: An approach to the reduction of inbreeding accumulation in trout brood stocks. *Prog. Fish-Cult.* **39**: 179-181.

Kuo, C.-m., Z. H. Shehadeh, and K. K. Milisen. 1973a. A preliminary report on the development, growth and survival of laboratory reared larvae of the grey mullet, *Mugil cephalus* L. *J. Fish Biol.* **5**: 459-470.

Kuo, C.-m., Z. H. Shehadeh, and C. E. Nash. 1973b. Induced spawning of captive grey mullet *(Mugil cephalus* L.) females by injection of human chorionic gonadotropin (HCG). *Aquaculture,* **1**: 429-432.

Kuo, C.-m., C. E. Nash, and Z. H. Shehadeh. 1974a. The effects of temperature and photoperiod on ovarian development in captive grey mullet *(Mugil cephalus* L.). *Aquaculture,* **3**: 25-43.

Kuo, C.-m., C. E. Nash, and Z. H. Shehadeh. 1974b. A procedural guide to induced spawning in grey mullet *(Mugil cephalus* L.). *Aquaculture,* **3**: 1-14.

LaCaze, C. 1976. *Crawfish farming,* rev. ed. Louisiana Wildlife and Fishery Commission, Fishery Bulletin 7. 27 p.

Landers, W. S. 1968. Oyster hatcheries in the northeast. In T. L. Linton (Ed.), *Proceedings of the oyster culture workshop, July 11-13.* Marine Fisheries Division, Georgia Game and Fish Commission, Brunswick, pp. 35-40.

Liao, I. C. 1969. Artificial propagation of grey mullet, *Mugil cephalus* Linnaeus. *Chin.-Am. Joint Comm. Rural Reconstr. Fish. Ser.* **8**: 10-20.

Liao, I. C., Y. J. Lu, T. L. Huang, and M. C. Lin. 1971. Experiments on induced breeding of the grey mullet, *Mugil cephalus* Linnaeus. *Aquaculture,* **1**: 15-34.

Linton, T. L. 1968. Feasibility studies of raft-culturing oysters in Georgia. In T. L. Linton (Ed.), *Proceedings of the oyster culture workshop, July 11-13.* Marine Fisheries Division, Georgia Game and Fish Commission, Brunswick, pp. 69-73.

Loosanoff, V. L. 1969. Development of shellfish culture techniques. In K. S. Price, Jr., and D. L. Maurer (Eds.), *Proceedings of the conference on artificial propagation of commercially valuable shellfish—oysters.* University of Delaware, Newark, pp. 9-40.

Loosanoff, V. L., and H. C. Davis. 1963. Rearing of bivalve mollusks. In F. S. Russel (Ed.), *Advances in marine biology,* Vol. 1. Academic Press, London, pp. 1-136.

Lowe-McConnell, R. H. 1958. Observations on the biology of *Tilapia nilotica* Linné in east Africa waters. *Rev. Zool. Bot. Afr.* **57**: 131-170.

Marshall, H. L. 1968. Three-dimensional oyster culture research in North Carolina. In T. L. Linton (Ed.), *Proceedings of the oyster culture workship, July 11-13.* Marine Fisheries Division, Georgia Game and Fish Commission, Brunswick, pp. 62-66.

Marshall, H. L. 1969. *Development and evaluation of new cultch materials and techniques for three-dimensional oyster culture.* Division of the Commission on Sports Fisheries, North Carolina Department of Conservation and Development, Special Scientific Report 17. 34 p.

Martin, M. 1967. Techniques of catfish fingerling production. In *Proceedings of the commercial fish farming conference, Texas A & M University, February 1—2.* Texas A & M University, College Station, pp. 13-22.

May, E. B. 1968. Raft culture of oysters in Alabama. In T. L. Linton (Ed.), *Proceedings of the oyster culture workshop, July 11-13*. Marine Fisheries Division, Georgia Game and Fish Commission, Brunswick, pp. 76-77.

May, E. B. 1969. Feasibility of off bottom oyster culture in Alabama. *Ala. Mar. Res. Bull.* **3**: 1-14.

Murphree, J. M. 1940. *Channel catfish propagation*. Privately printed by T. J. Rennick. 24 p.

Nash, C. E., C.-m. Kuo, and S. C. McConnel. 1974. Operations procedures for rearing larvae of the grey mullet (*Mugil cephalus L.*). *Aquaculture*, **3**: 15-24.

Nelson, B. 1960. Spawning of channel catfish by use of hormone. *Proc. Southeast. Assoc. Game Fish Comm.* **14**: 145-148.

Noble, R. L., R. D. Germany, and C. R. Hall. 1975. Interactions of blue tilapia and large-mouth bass in a power plant cooling reservoir. *Proc. Southeast. Assoc. Game Fish Comm.* **29**: 247-251.

Pagon-Font, F. A. 1975. Cage culture as a mechanical method for controlling reproduction in *Tilapia aurea*. *Aquaculture*, **6**: 243-247.

Pearson, J. C. 1939. The early life histories of some American Penaeidae, chiefly the commercial shrimp, *Penaeus setiferus* (Linn.). *Bull. U.S. Bur. Fish.* **49**: 1-73.

Pruginin, Y., S. Rothbard, G. Wohlfarth, A. Havlevy, R. Moav, and G. Hulata. 1975. All male broods of *Tilapia nilotica* × *T. aurea* hybrids. *Aquaculture*, **6**: 11-22.

Rose, C. D. 1975. Extensive culture of penaeid shrimp in Louisiana salt-marsh impoundments. *Trans. Am. Fish. Soc.* **104**: 296-307.

Saksena, V. P., K. Yamamoto, and C. D. Riggs. 1961. Early development of the channel catfish. *Prog. Fish-Cult.* **23**: 156-161.

Sanzo, L. 1936. Contributi alla conescenza dello sviluppo ambrionaria e post-ambrionario nei Mugilidi. *Mem. R. Com. Talassogr. Ital.* **230**: 1-11.

SEAFDEC (Southeast Asian Fisheries Development Center). 1976. *Annual Report 1976.* Aquaculture Department, Southeast Asian Fisheries Development Center, Tigbauan, Iloilo, Philippines. 83 p.

Shaw, W. N. 1960. A fiberglass raft for growing oysters off the bottom. *Prog. Fish-Cult.* **22**: 154.

Shaw, W. N. 1962. Raft culture of oysters in Massachusetts. *Fish. Bull.* **61**: 481-495.

Shaw, W. N. 1968. Raft culture of oysters in the United States. In T. L. Linton (Ed.), *Proceedings of the oyster culture workshop, July 11-13*. Marine Fisheries Division, Georgia Game and Fish Commission, Brunswick, pp. 5-31.

Shehadeh, Z. H., and J. N. Ellis. 1970. Induced spawning of the striped mullet *Mugil cephalus* L. *J. Fish Biol.* **2**: 355-360.

Shehadeh, Z. H., C.-m. Kuo, and K. K. Milisen. 1973a. Induced spawning of grey mullet *Mugil cephalus* L. with fractionated salmon pituitary extract. *J. Fish Biol.* **5**: 471-478.

Shehadeh, Z. H., C.-m. Kuo, and K. K. Milisen. 1973b. Validation of an *in vivo* method for monitoring ovarian development in the grey mullet (*Mugil cephalus* L.) *J. Fish Biol.* **5**: 489-496.

Shehadeh, Z. H., W. D. Madden, and T. P. Dohl. 1973c. The effect of exogenous hormone treatment on spermiation and vitellogenesis in the grey mullet, *Mugil cephalus* L. *J. Fish Biol.* **5**: 479-487.

Sneed, K. E., and H. P. Clemens. 1959. The use of human chorionic gonadotropin to spawn warm-water fishes. *Prog. Fish-Cult.* **21**: 117-120.

Stanley, J. G., and K. E. Sneed. 1974. Artificial gynogenesis and its application in genetics and selective breeding of fishes. In J. H. S. Blaxter (Ed.), *The early life history of fish.* Springer-Verlag, New York, pp. 527-536.

Stewart, S. R. 1977. A feasibility study of commercial green sea turtle mariculture. M. S. thesis, Texas A & M University, College Station. 187 p.

Stickney, R. R., D. B. White, and D. Perlmutter. 1973. Growth of green and loggerhead sea turtles in Georgia on natural and artificial diets. *Bull. Ga. Acad. Sci.* **31**: 37-44.

Strickberger, M. W. 1976. *Genetics.* Macmillan, New York. 914 p.

Tang, Y. A. 1964. Induced spawning of striped mullet by hormone injection. *Jap. J. Ichthyol.* **12**: 23-28.

Toole, M. 1951. Channel catfish culture in Texas. *Prog. Fish-Cult.* **13**: 3-10.

Witham, R. 1970. Breeding of a pair of pen-reared green turtles. *Quart. J. Fla. Acad. Sci.* **33**: 288-290.

Yang, W. T., and U. B. Kimm. 1962. A preliminary report on the artificial culture of grey mullet in Korea. *Indo-Pacific Fish. Coun.* **9**: 62-70.

Yashouv, A. 1969. Preliminary report on induced spawning of *M. cephalus* (L.) reared in captivity in freshwater ponds. *Bamidgeh,* **21**: 19-24.

SUGGESTED ADDITIONAL READING

Price, K. S., Jr., and D. L. Maurer (Eds.). 1969. *Proceedings of the conference on artificial propagation of commercially valuable shellfish—oysters.* University of Delaware, Newark. 212p.

CHAPTER 7
Disease and Parasitism

MORTALITY IN AQUACULTURE

Natural mortality as a result of old age occurs only in aquacultural animals that are utilized as brood stock, since marketable individuals are generally sold relatively early in the normal life cycle of the species (often before reaching sexual maturity). Brood stock for many species is not maintained long enough for natural mortality to occur because the animals often demonstrate a reduction in fecundity with increasing age and in some cases become too large to handle easily (e.g., channel catfish). Other species experience a high percentage of mortality following spawning (e.g., certain members of the family Salmonidae). Mortality from causes other than old age does affect aquaculture, as has been discussed in earlier chapters. Some of these causes are deterioration of water quality, nutritional imbalance, improperly stored feed, poaching, toxicants, and predation.

Each of the sources of mortality cited above can be of significance in aquaculture, but more concern is generally expressed for losses resulting from diseases and parasitism than to those attributed to all other causes except degraded water quality. Diseases and parasite epizootics do not, of course, occur in all aquaculture facilities each year, although most culturists experience losses to these causes at various times. Only about 5% of commercial fish farmers working in the United States have severe problems in a given year (Meyer, 1967). Even so, the early detection and treatment of diseases and parasites is often crucial if a crop is to be saved from decimation.

The primary disease and parasite problems in aquaculture animals relate to viral, bacterial, fungal, and protozoan epizootics. Parasitic nematodes, trematodes, and cestodes are commonly found in aquaculture animals, but seldom are they present in concentrations sufficient to cause significant problems (although exceptions do occur).

It would be extremely rare to find any population of aquatic organisms that was completely free of any type of disease or parasitic organism. Microbiological and microscopic examination of nearly any fish or invertebrate generally reveals the presence of at least some type of potentially pathogenic organism. Infection is common in animals of all types, and indeed, infected individuals in any given population of organisms may be a nearly universal characteristic. However it does not necessarily follow that a disease epizootic will occur just because there are infected individuals present. As previously indicated, subacute chronic infections are generally not a problem because most species have natural defense mechanisms (including phagocytosis and immune responses) that guard against the development of an epizootic, as long as the population is not stressed. Severe epizootics can and often do occur following stress. The time that passes between the onset of stress and the outbreak of a disease or parasitic epizootic often ranges from 24 hours to 14 days, depending on the incubation period for the pathogen that eventually surfaces as the most significant problem. (This time frame will become more apparent when individual diseases are discussed later in this chapter.)

Aquaculture animals usually do not experience severe disease or parasite problems if the water system is well managed and the organisms are not stressed. Exceptions to this general tenet do occur, but in most cases good management practices are rewarded by healthy animals. Sometimes aquaculture organisms become stressed in spite of the best efforts of the culturist. For example, during the spring and fall, when water temperatures are often changing quite rapidly in outdoor water systems, outbreaks of parasitic protozoans or infections from bacterial pathogens may occur even under the best managerial scheme. Not only is the natural fluctuation in water temperature a stress factor, but in many cases the disease or parasitic organisms that become epizootic find optimum conditions for their growth within a particular temperature range that may exist only seasonally for relatively short periods. Since under most aquaculture strategies it is not economical for the culturist to control changes in ambient water temperature in outdoor facilities, the potential problems that exist during critical periods must be recognized and the culture animals carefully observed for any indications that an epizootic has been initiated.

Some aquaculturists try to avoid the use of any chemical treatment of their animals unless absolutely necessary, whereas others routinely use prophylactic treatments to reduce the chances of an epizootic, even though the odds against a severe problem may be great. A third approach involves the use of prophylactic treatment only when the evidence indicates that an epizootic is likely to occur—for example, during periods of rapid temperature change as a result of seasonal fluctuations, during the addition of large amounts of water of different temperature, after stress induced by deteriorating water quality, after the dilution of estuarine water by high amounts of precipitation, during handling or transportation, or following the development of any other type of stressful condition. It is virtually impossible to avoid the use of treatment drugs and chemicals entirely, but it is important to limit the dependence on such therapeutic substances to instances when they are required for the maintenance of aquatic animal health.

If an epizootic does occur, its severity may depend on a number of factors. One of the primary ones is, of course, the physical condition of the animals prior to the outbreak, and it is known that degraded physical condition will result from any type of stress to which the animals are subjected. The severity of an epizootic will be increased if a secondary infection accompanies the primary one. Bacterial infections are common following the establishment of a parasite epizootic. In many instances it is common to find a wide array of disease and parasitic problems occurring simultaneously.

The use of chemicals to control epizootics is an additional source of stress, since in many cases the amount of chemical required to kill a disease or parasitic organism is similar to the level that will cause mortality in the host. Great care must be taken to ensure that toxic levels of treatment chemicals do not occur, while providing a dosage sufficient to eradicate the pathogenic organisms. It is also important to realize that the effects on aquaculture species of many therapeutic agents have not been adequately evaluated, nor have the effects of such water quality parameters as salinity, temperature, alkalinity, and turbidity been completely examined with respect to many of the chemicals presently utilized for aquatic animal disease and parasite control.

Control of diseases and parasites depends to a great extent on the stage at which the problem is detected. Early detection is very important, as is proper diagnosis. If the first sign of an epizootic is the appearance of numerous dead animals, it may be too late to take effective action to save the remainder of the crop. It is a wise practice to treat initially only small numbers of animals, to determine the effectiveness of a treatment before exposing the whole crop to that particular therapeutic agent. If the disease or parasitic outbreak has progressed to the stage that massive losses are occurring, however, there is no time to test various types of treatment, even though selection of the wrong one in the existing circumstances may lead to increased mortality rather than reducing the impact of the problem. To be able to recognize a problem early, it is important that the culturist have frequent visual contact with the animals.

SANITATION

When an epizootic is detected, every attempt should be made to contain it. The careless transfer of equipment from one pond to another without treating that equipment with antiseptics can spread a localized disease or parasite to every pond, tank, or other type of culture chamber on an aquaculture facility. Such items as dip nets, seines, feed pails, and even the bodies and clothing of personnel who work around the aquaculture animals should be sanitized after exposure to organisms, regardless of whether there is known to be a disease or parasitic problem present. Soap and water are generally sufficient for people and their clothing, but other agents are usually more desirable for equipment sanitation. Between uses, nets can be placed in disinfectant solutions of chlorine, formalin, merthiolate, and various other commercially available preparations, or separate pieces of equipment may be assigned to each culture chamber. If strong disinfectant solutions are utilized on nets or other gear, it is important to flush the equipment thoroughly with water before use, since the residual chemicals (chlorine, formaldehyde, etc.) could be strong enough to kill exposed culture animals. The equipment should be rinsed with uncontaminated water.

Large items such as seines can be soaked in large vats of disinfectant, although this may not always be practical. Alternatively, seines may be rinsed with uncontaminated water, then dried in the sun between uses. Caution should be taken with monofilament nylon nets, since they are subject to degradation when exposed to direct sunlight. A spray treatment with chlorine solution or some other type of disinfectant may also be effective.

TREATMENT METHODOLOGY

When an epizootic does occur and chemical treatment is indicated, or when the onset of an epizootic is anticipated and it appears that prophylactic treatment might be beneficial, the appropriate chemical must be selected and properly applied. Some drugs may be mixed with the feed during formulation, added to the pellets of feed as a surface coating (often by dissolving the chemical in a lipid carrier and spraying the feed with the mixture), given in the form of an injection, or used as a bath. Other treatment chemicals are effective only when added to the water.

When treating for a disease or parasitic infestation, the first step is to isolate the affected individuals or culture chambers. In most instances this is relatively easy because water from one culture chamber usually does not flow into others. As discussed earlier, care should always be taken to avoid contamination of healthy animals with nets and other gear that have been in contact with diseased animals and not properly sanitized.

Once the affected organisms have been quarantined, the physical environment should be adjusted for the treatment of choice. For example, if a static water bath treatment is to be used in a pond, inflow water should be turned off. In running water bath treatments, the flow rate should be adjusted to ensure that the chemical will be present long enough to act on the parasite or disease organism but not so long that the culture animals are further stressed.

In some cases it is possible to effect treatment through the manipulation of the culture environment by altering either a physical or a chemical factor. Certain parasites like *Ichthyophthirius multifiliis* can be controlled by increasing or reducing temperature or by adding salt to the water (increasing salinity). This protozoan also responds to various treatments with harsh chemicals, but the two methods mentioned are often effective and place less stress on the culture animals than do drugs.

Another nonchemical means of treatment consists of the interruption of some link in the life cycle of the parasite. Various trematodes that infect fish have rather complex life cycles, usually involving one or more intermediate hosts, one of which is often a snail. If snails can be eradicated from the culture chambers, the life cycle of the parasite will be broken and further infestation will be curtailed. Similarly, wading birds of many kinds, in addition to being direct predators on aquaculture animals, carry parasites that can be transferred to fish. It is illegal to destroy migratory wading birds, but they can be discouraged by various "scarecrow" devices or, in the case of outdoor tanks and raceways, by placement of hardware cloth or larger mesh wire covers over the culture chambers.

Treatment techniques vary to some extent depending on the type of culture system, particularly when a treatment chemical is to be added directly to the water. Differences among water systems with respect to disease treatment were discussed in Chapter 2. Recall that care must be taken in treating closed recirculating water systems to avoid destruction of the microflora associated with the biofilter. Treatment of open raceways or tanks may be accomplished with or without first reducing or stopping the flow of water, although different amounts of chemical are required to effect control in the two instances. In any intensive water system, provision for aeration must be made when water flow is curtailed for any length of time. Cages may be treated in a variety of ways (Chapter 2), and ponds can be treated by adding chemicals to the water under static conditions. Pond levels may have to be reduced to some extent to conserve on chemical quantity (which can present a major expense to the culturist). When drugs are added to feed, no special precautions are usually required. Despite fear that antibiotics will harm the microflora of biological filters, the experience of some workers seems to indicate that these concerns may not be warranted. Definitive research is lacking on the subject.

Treating large areas of water can be difficult. Small ponds (e.g., those 0.5 ha or less) generally can be treated from the bank. Chemicals are usually dissolved or diluted in water and dispersed as evenly as possible over the surface of the water by broadcasting them in some manner. Buckets, hand-operated sprayers of the type used for applying herbicide and pesticide, and long-handled dippers have been effectively utilized to treat small ponds.

It is often difficult to reach all areas of large ponds from the bank. In those cases chemicals may be diluted with water and poured into the wake of an outboard motor. The chemical should be added slowly enough that all portions of the pond can be covered before the indicated amount of chemical has been used up. If the pond has relatively deep water in one or more areas, more chemical should be utilized there to ensure even distribution. Aircraft may be employed to spray treatment chemicals over very large ponds, although the culturist should be certain that the spray tanks and nozzles of crop-dusting planes have been thoroughly cleaned of any residual herbicide or pesticide.

Topical treatments are sometimes effective but are often impractical in large aquaculture operations because of the logistical problems involved in capturing infected animals from a large basically healthy population and applying a chemical to local infections. In most instances a whole culture chamber will receive treatment if a disease or parasite has been found to affect any portion of the animals in that chamber.

Bath treatments are of three general types. Pond treatments discussed earlier can be thought of as long-term or indeterminate baths, since dissipation of the chemical is not by dilution but through degradation, which may require several days or even weeks. Extended baths require several hours and are conducted in static water that must be replaced with new water after treatment is complete. Such baths may be conducted in ponds or smaller culture chambers. Aeration is necessary in raceways, tanks, and cages if extended baths are utilized, since water flow must be terminated for the duration of the bath.

Short-term treatments of large numbers of animals can be accomplished through dips or flushes. Dip treatment usually involves capturing the animals and dipping them, in groups or individually, in one or more treatment chemicals. Dips may last from a few seconds to a few minutes and usually involve fairly strong concentrations of chemicals. Following dip treatment the culture animals are returned to their culture chambers. In flush treatment the chemical is added directly to running water in the culture chambers at a strength somewhat in excess of that utilized for extended or indeterminate baths and often weaker than that of dip treatment. The water is allowed to run so that the chemical is diluted and eventually completely removed.

As already noted, oral treatment of diseases and parasites consists of the application of chemicals to the feed. This can be a very effective means of treatment, and indeed, certain diseases can be treated in no other way. This method does require that the feed be consumed if the drug is to be effective, however, and seriously ill animals will not feed, thus usually succumb to the disease. Fishes in the early stages of the disease, on the other hand, may continue to feed long enough for the treatment to be effective.

A final method by which drugs can be introduced to culture animals is through injection. This method, like topical application of drugs, calls for the capture of each individual animal, followed by inoculation. Although the approach is suitable for small numbers of animals, a large commercial aquaculturist of fish or invertebrates would find it too time-consuming and expensive.

A new area of prophylaxis is being developed for fishes that may have a revolutionary effect on commercial culture of at least some organisms. Waterborne vaccines are under development for certain bacterial infections of channel catfish (Donald Lewis, personal communication). The whole area of immunology of fish is currently receiving long-needed attention by aquatic animal health specialists, and perhaps one day an array of vaccines will be available that can be used in bath treatments at an affordable price.

CHEMICAL CONTROL

The use of drugs, herbicides, pesticides, fish toxicants, and a variety of other chemicals on food fish is regulated (and presently undergoing review) by the U.S. Food and Drug Administration (FDA) and the U.S. Environmental Protection Agency (EPA). With respect to chemicals cleared for use against diseases and parasites of food fishes, until recently the FDA had approved only sodium chloride and vinegar, neither of which has been widely utilized except on a very few parasites. In addition to these two chemicals, about 100 substances generally regarded as safe for human consumption are available for use without clearance. That list includes such items as sugar, baking soda, pepper, and other food ingredients that have little or no application in disease and parasite control. FDA approval is required on most chemicals utilized on food animals to ensure that the consumer is not threatened by the drugs.

A variety of chemicals have found use in aquaculture, although most still have not received official approval from federal authorities, and some, like malachite green, may never be cleared for use on food fish. Table 7.1 lists the chemicals that had been approved as of February 1976. Other chemicals are being screened by federal and private laboratories, but too little information on them is available to make clearance determinations.

Table 7.1   Therapeutic Agents Cleared for Use on Food Fish as of February 1976 (Meyer et al., 1976)

Compound	Primary Uses
Vinegar (acetic acid)	Useful on some external parasites
Sodium chloride	Useful on some external parasites of freshwater fishes
Oxytetracycline (Terramycin)	Antibiotic that is effective against a variety of pathogenic bacteria
Sulfamerazine	Sulfonimide bacterial control agent

For a chemical to be cleared, extensive scientific studies must be conducted. The government can undertake those studies, but it is usually left up to drug and chemical companies to provide data on which clearance decisions are based. When there are potential sales for large amounts of a particular product that has not been cleared, the manufacturer may find it in his interest to seek clearance (a process that may require millions of dollars and several years). Aquaculture in the United States is presently not a large enough industry to stimulate the level of intensive study and monetary investment required for federal clearance of very many drugs, although the same is not true in the case of terrestrial livestock. The clearance of the antibiotic oxytetracycline (Terramycin) early in the 1970s was considered a major advancement for aquaculture disease control.

Even though a drug or chemical has been officially approved for use in aquaculture, the substance should never be used unless there is a clear need. Routine application can lead to more serious problems in culture system management than the assumption of a conservative posture that features treatment only as a last resort. Some of the reasons for this view are as follows: (1) drugs are expensive and their continuous use may mean the difference between profit and loss of an aquaculture venture, (2) the constant use of antibiotics can lead to the development of resistant strains of bacteria, (3) phytoplankton and zooplankton may be killed, leading to water quality deterioration and limitation of natural food following routine chemical usage, (4) the more chemicals are used, the more likely it is that a miscalculation in treatment dosage will occur, leading to an overdose and possibly mortality, (5) biofilter efficiency may be impaired or destroyed by chemicals added to closed recirculating water systems, and (6) the injudicious use of chemicals can have a damaging effect on the environment as well as on the person handling them.

CALCULATION OF TREATMENT LEVELS

Before the amount of chemical to be added to a pond, tank, or raceway can be calculated, the volume of the culture chamber must be determined. In circular tanks this is readily accomplished by multiplying the depth of the chamber by πr^2. Similarly, in square or rectangular tanks or raceways, the volume can be calculated by multiplying length by width by depth. The calculation of pond volumes is somewhat more involved, especially for round or irregularly shaped ponds. Calculation of pond volumes and determination of treatment levels has been presented in a simple manner by Meyer (1968). In large ponds it is often sufficient to multiply the surface area by mean depth to determine volume. However in small ponds, the volume lost as a function of bank slope can be significant, leading to the addition of excessive amounts of chemicals. In such cases the volume should be calculated more precisely by subtracting the portion lost under the slopes of the levees from the volume assumed from total surface area. This calculation is simplified if all pond banks have the same slope. In most ponds great precision in measurement is not required because slight errors will not appreciably influence the final treatment level of a chemical, and there is usually some latitude between therapeutic and lethal doses of drugs presently in use.

Most ponds constructed in the United States are laid out in acres rather than hectares; thus volume is often measured in acre-feet. However treatment levels of chemicals are usually presented in milligrams per liter (mg/l) or parts per million (ppm). Thus it is often convenient to convert pond volumes to the metric system for calculation of treatment levels. The conversion units presented in Appendix 1 can be helpful in working with one system, then the other.

As an example of a treatment level calculation, assume a 0.5 acre pond with a mean depth of 3 ft is going to be treated with a chemical in the amount of 5 mg/l. How many kilograms of the chemical must be added to obtain the proper dosage?

(0.5 acre)(3 ft)	= 1.5 acre-ft
(1.5 acre-ft)(43,560 ft^3/acre-ft)	= 65,340 ft^3
(65,340 ft^3)(7.5 gal/ft^3)	$\doteq$ 490,050 gal
(490,050 gal)(3.8 l/gal)	= 1,862,190 l
(1,862,190 l)(5 mg/l)	= 9,310,950 mg
9,310,950 mg	= <u>9.3 kg</u>

In fairly large bodies of water it is generally sufficient to figure chemicals to the nearest 0.1 kg, although in aquaria, small culture chambers, or small containers utilized for short-duration dip treatments, chemical dosages must be measured to the nearest 0.1 g or even the nearest milligram. Common sense should be used when chemicals are being measured. For example, if 50 kg of a substance will be added to a pond, it would be foolish to measure to the nearest milligram. On the other hand, if a small tank is to receive 10 mg of a chemical, a lethal dose might result if the scale utilized for measurement was accurate only to the nearest gram.

The example just presented can be simplified to some extent if the transition from English to metric units is made in an earlier step:

$$(0.5 \text{ acre})(3 \text{ ft}) = 1.5 \text{ acre-ft}$$
$$(1.5 \text{ acre-ft})(1233.51 \text{ m}^3/\text{acre-ft}) = 1850.27 \text{ m}^3$$
$$(1850.27 \text{ m}^3)(1000 \text{ l/m}^3) = 1,850,270 \text{ l}$$
$$(1,850,270 \text{ l})(5 \text{ mg/l}) = 9,251,350 \text{ mg}$$
$$9,251,350 \text{ mg} = \underline{9.3 \text{ kg}}$$

The precise number of milligrams derived differed considerably in the two methods of calculation (9,310,950 vs. 9,251,350), yet rounding off to the nearest 0.1 kg resulted in the same dosages in both cases (9.3 kg). The difference between the two answers at the milligram level is attributable to the imprecision of the conversion factors, which allows inherent errors to be carried throughout the calculations. In the treatment of large volumes of water, minor differences are unimportant in terms of the effectiveness of treatment or the toxicity due to overdose. Of more serious consequence are mathematical errors that may result in gross over- or undertreatment.

If ponds are constructed on the basis of hectares, or if the culturist converts acreage to hectares for purposes of calculating dosages, the procedure is even further simplified. For example, if an aquaculturist wishes to add 5 mg/l of a chemical to a pond that is 0.25 ha in area and has an average depth of 0.5 m, the following calculation could be made:

$$(0.25 \text{ ha})(0.5 \text{ m})(10,000 \text{ m}^2/\text{ha}) = 1250 \text{ m}^3$$
$$(1250 \text{ m}^3)(1000 \text{ l/m}^3) = 1,250,000 \text{ l}$$
$$(1,250,000 \text{ l})(5 \text{ mg/l}) = 6,250,000 \text{ mg}$$
$$6,250,000 \text{ mg} = \underline{6.3 \text{ kg}}$$

EXAMINATION OF ANIMALS FOR DISEASE AND PARASITES

Bacterial characterization in aquatic animals, as in terrestrial organisms, requires isolation of the bacteria from infected tissue, culture of the material collected in one or more types of nutrient medium, and additional procedures such as staining. Once the bacteria have been identified, colonies can be challenged with various antibiotics to determine sensitivity prior to treatment. Because of the ease with which contamination can occur, good sterile technique is required throughout the necropsy, incubation of cultures, and subsequent examination of culture results.

All materials that come in contact with the cultures must be sterilized. Personnel conducting bacteriological examinations should wear clean laboratory-type clothing, wash their hands before and after each examination, and not smoke or drink in the laboratory. Extreme caution must be taken to ensure that any bacteria that grow on the culture media were obtained during necropsy, not introduced through contamination.

Amlacher (1970) stressed that fishes should be examined and cultured for external bacteria before the animals are opened or otherwise cut in any way. External cultures can be obtained from swabs of the integument and gills. The swabs are subsequently used to streak culture plates. Care must be taken in collection of the animals, to avoid contamination. Live animals can be transported in clean ice chests or plastic bags partially filled with water. Dead or moribund animals can be placed in sterile plastic bags and transported on ice.

When bacterial cultures of internal organisms are taken, the animal should be secured to a board and the surface swabbed with alcohol or other disinfectant before any incisions are made (Amlacher, 1970). Samples from the liver, spleen, heart, and kidney can be obtained with a sterile loop, which is then used to streak the culture plates.

Most pathogenic bacteria present on fish stain gram negative. Identification is often possible through a combination of staining and examination of the shape and color of the colonies produced, along with determination of whether colonies will grow on certain kinds of culture media. For example, examination of bacteria associated with channel catfish would reveal that *Pseudomonas* sp. produce large, broad colonies of whitish grey on blood agar, whereas *Flexibacter columnaris* will not grow on blood agar.

Some aquaculturists do their own diagnostic work on bacteria and parasites, but most utilize the services available through various state and federal agencies. Laboratories operated by the U.S. Fish and Wildlife Service, state Agricultural Extension Services, and several universities will diagnose and recommend treatment for diseases and parasites of fishes and invertebrates sent to them. Such diagnostic facilities are more often than not located where aquaculture is well established, although most will provide service for persons outside their respective regions. Many diagnoses can be made by the culturist, especially those involving parasites. Bacterial characterization requires more equipment and expertise and is not an area in which many culturists become involved.

When examining fish or invertebrates for external and gill parasites, a fresh specimen should be obtained so that living parasites will be present. Not only do parasites often drop off dead animals, making the chances of finding the cause of an infestation difficult, but identification and location of parasites is facilitated when living specimens are available. If an animal must be transported for some distance before being examined, it may be necessary to preserve it in formalin. However since parasites may be distorted when formalin is utilized, making identification difficult, this procedure should be avoided when possible.

As a first step in examination of animals for parasites, gills and fins, or parts thereof, should be removed and placed under a cover glass on a microscope slide. A drop of water should be added. Body scrapings obtained with a scalpel can be handled in the same manner. Low power examination of the slide often serves to permit recognition of helminth, crustacean, and large protozoan parasites. High power may be required for small protozoans.

When the blood of fishes is examined for protozoans, the method of Strout (1962) seems to work well. A few drops of blood are placed in a vial and allowed to clot. Then a drop of the clear serum is pipetted onto a microscope slide and examined under high power. In another method a drop of whole blood is placed on a microscope slide, allowed to dry, and stained with either Giemsa's or Wright's stain (Kudo, 1954) prior to examination under the microscope.

Fish blood can be obtained in several ways. For fingerlings it is usually best to sever the tail across the caudal peduncle and allow the blood to drip into a collection vial. In larger fish, sacrifice is not usually necessary. A needle and syringe can be utilized to obtain blood by either heart puncture or caudal puncture. Obtaining blood by heart puncture requires more practice than does caudal puncture, but both methods provide sufficient quantities of blood without permanently damaging the fish. In the caudal puncture technique the fish is placed with the ventral side uppermost and the needle is inserted into the midline of the caudal peduncle until it comes in contact with the vertebral column. The needle is withdrawn slightly and blood can be aspirated into the syringe.

When the external examination for parasites is complete, the animal may be opened and the internal organs (especially the intestines, mesenteries, liver, gonads, kidney, gall bladder, and urinary bladder) examined. Small pieces of tissue can be placed on microscope slides in the manner indicated earlier. Most of the pathogenic parasites that plague aquaculturists occur externally, although a few internal ones lead to frequent or at least occasional problems.

COMMON DISEASES AND PARASITES

A great variety of diseases and parasites occur in aquatic organisms, and undoubtedly the list will continue to expand for many years to come. It is not possible to examine in detail all those that have been identified, nor even to discuss in detail each pathogenic organism that has been a problem to warmwater aquaculturists in the past. Some groups of organisms hold wide varieties of bacteria and parasites in common, and others have pathogenic organisms to which they alone are particularly susceptible or to which they appear to be immune. The same is true within species. In addition to pathogenic viruses, bacteria, and parasites, a series of nutritional diseases has been recognized in aquaculture animals. This book cannot hope to examine in detail each pathogen or nutritional disease with respect to every species of aquaculture organism now under consideration in the warmwater regions of the United States. Instead, this section demonstrates the types of disease and parasitic organisms that might be encountered in selected aquatic animals.

Of the warmwater species presently under culture for human food in the United States, perhaps more is known about the channel catfish, *Ictalurus punctatus,* than any other. Considerable data have also been collected on oyster and shrimp diseases. Thus this discussion is based primarily on those organisms. Similar diseases and parasites can be expected in conjunction with other species from the phyla represented by the animals indicated, although the methods of treatment and rates of chemical application may vary considerably from species to species and among sizes or life stages within species. Before examining some of the diseases and parasites of channel catfish, shrimp, and oysters, some comments on nutritional diseases and an unusual disease that strikes many kinds of fishes are in order.

Nutritional Fish Diseases

The comments below on nutritional fish diseases are restricted largely to channel catfish and trout, since little information is available about other species. There is virtually no information on the nutritional diseases that attack most invertebrates under consideration for culture or presently under culture in the United States.

Nutritional deficiency symptoms associated with dietary vitamin imbalances are among the best documented of all nutritionally related diseases in fish, with good data being available on both salmonids and channel catfish (Snieszko, 1972; Dupree, 1966, 1977; Lovell, 1975). Symptoms of deficiency and excesses in vitamins are similar in both types of fish, thus it can probably be assumed that other groups of fishes will demonstrate comparable disease etiology. Nearly every known vitamin has associated with it symptoms of hypovitaminosis, and the fat-soluble vitamins can also produce symptoms of hypervitaminosis (Table 5.1). No attempt has been made to deal with each vitamin in detail in this section, although two of the most important ones, vitamin C and pantothenic acid, are considered. The reader can obtain additional information on the other vitamins from Chapter 5.

Among the most serious consequences of vitamin deficiency, largely because alteration of the diet to provide that vitamin in required amounts will not reverse the problem in affected individuals, is the occurrence of broken backs in channel catfish. Fractured vertebral column, preceded by unusual spinal curvature (lordosis and scoliosis) results from a deficiency of vitamin C (ascorbic acid) and can afflict large numbers of catfish when intensive culture conditions are utilized. Pond culturists do not generally see this problem, or at least the problem does not progress until vertebral columns are actually fractured, since even in heavily stocked ponds the fish are able to obtain some vitamin C from natural foods. Proper storage and utilization of feed within a few months of purchase, even formulations heavily fortified with vitamin C, is important. Excessive losses of vitamin C activity have been reported within 16 weeks in feed stored at 20 C (Andrews and Murai, 1975), and the half-life of vitamin C in stored feeds has been reported as less than 3 months (Lovell and Lim, 1978). Lordosis and scoliosis have been observed in salmonids when vitamin C levels were low, but the broken back syndrome does not seem to be prevalent in the coldwater fishes (Snieszko, 1972).

Nutritional gill disease is also relatively common among salmonids (Snieszko, 1972) and has been known to affect channel catfish (Lovell, 1975; Dupree, 1966). The symptoms in trout include loss of appetite, congregation of fish near sources of inflowing water, clubbing and fusing of the gill filaments beginning with the distal ends, and accelerated respiration (Snieszko, 1972). Catfish exhibit reduced growth, clubbed gills, erosion of the lower jaw and fins, and mortality (Dupree, 1966; Lovell, 1975). Bacterial gill disease, caused by myxobacteria, shows similar pathology and should be ruled out through bacteriological examination before nutritional gill disease is confirmed. Nutritional gill disease results from insufficient levels of dietary pantothenic acid (Rucker et al., 1952; Halver, 1953). It has been recommended that catfish diets contain at least 250 mg of pantothenic acid per kilogram of feed (Murai and Andrews, 1975).

Vitamins are not the only elements of fish diets that can lead to nutritional diseases. Thyroid tumors, for example, appear to occur in many families of teleost fishes (MacIntyre, 1960) and are related to deficiencies in iodine (Marine and Lenhard, 1910, 1911; Marine, 1914). Other nutritionally related diseases in salmonids include liver degeneration, visceral granuloma, anemia, and pigmentation impairment (Snieszko, 1972). Such diseases may result from a variety of causes. As discussed in Chapter 5, high levels of starch in trout can cause symptoms of diabetes. This has also been reported in other fish species. Starch is generally present in much higher levels in channel catfish diets than in those of trout, but Cruz (1975) determined that the digestion of protein and fat was suppressed when high dietary levels of starch were present. Enlarged livers in channel catfish have also been associated with high levels of dietary carbohydrate (Simco and Cross, 1966).

As indicated by Lovell (1975), nutritional diseases of fish can occur when toxic materials or such chemicals as enzyme inhibitors are present in the feed; however most problems are associated with a nutritional imbalance in a complete diet for fish being reared under intensive culture conditions. The imbalance may result from an excess of a particular nutrient, or it can occur in response to the complete absence of a nutrient or its presence at a level below that required by the animals. Only complete understanding of the nutritional requirements of aquaculture organisms will allow the feed manufacturer and aquaculturist to provide adequately for the species under culture.

Swim Bladder Stress Syndrome

A series of symptoms representative of a newly recognized disease called swim bladder stress syndrome (SBSS) has recently been reported (Clary and Clary, 1978). Similar to, and probably in the past often mistaken for gas bubble disease or some type of bacterial infection, the disease is reported to occur in four phases. In the first phase the fish swim in a head-down position near the surface. The second phase is characterized by fish swimming at the surface with backs exposed to the atmosphere. In the third phase the fish intermittently swim in an erratic fashion on their sides. Finally, in the fourth phase of the disease, an excessive amount of gas is produced in the swim bladder and the fish become extremely distended in the abdominal region. Fish in the fourth stage of SBSS are buoyed at the surface of the water on their backs. We have observed both channel catfish and tilapia fry enter phase four when the fish were held in small raceways at our laboratory, and suspicions of either gas bubble disease or some type of bacterial infection, could not be substantiated.

Clary and Clary (1978) indicated that environmental stress is the cause of SBSS. The major criterion for induction of the syndrome appears to be the presence every day of low dissolved oxygen for several hours. Oxygen does not have to become depleted, but merely depressed below the desired 5 mg/l value for each of several days before the fish pass from phase one (which can occur if any additional stress occurs after only one 5 hour exposure to depressed dissolved oxygen) into the latter phases of SBSS. Algal blooms, high stocking rates, high levels of ammonia, and other stress-inducing factors may be contributory.

The syndrome can be alleviated by increasing aeration and reducing density of animals (Clary and Clary, 1978). These authors report that about 40% of the fish that enter phase four fail to recover, but the remainder and virtually all those in the first three phases of SBSS can be saved.

Since SBSS is a relatively newly recognized disease many culturists will be unaware of it when the symptoms appear. It has probably been a misdiagnosed problem around fish hatcheries for years, and the simplicity with which it can be overcome may mean a significant reduction in losses in the future as culturists become aware of the symptoms and cure.

Diseases and Parasites of Channel Catfish

Channel Catfish Virus Disease (CCVD). CCVD is the only known pathogenic virus disease that affects channel catfish. Since first reported in 1968 (Fijan, 1968; Fijan et al., 1970), CCVD has been observed in nine southern states (Plumb and Gaines, 1975). An epizootic of CCVD may result in losses as high as 95% among fry and fingerlings (Plumb, 1971a), with larger fish being either cariers or immune.

The agent responsible for CCVD has been characterized as a herpesvirus (Wolf and Darlington, 1971) that has been isolated from fingerling channel catfish during epizootics (Plumb, 1972). CCVD can be maintained in tissue culture using brown bullhead cells. It is generally suspected that channel catfish brood stock are carriers of the virus and that these fish transmit the disease by way of the reproductive cells and/or the fluids associated with reproduction (Wellborn et al., 1969; Plumb, 1971b). However the virus has not yet been isolated from suspected carrier adults. Fish infected with the virus can transmit it to other fry and fingerlings through the water (Plumb, 1972).

Commercial and laboratory stocks appear to be the primary sources of fish carrying the virus, since mortalities attributed to CCVD have not been reported from natural populations of channei catfish. Mortality may occur as early as 32 hours following infection (Plumb, 1971a), and the virus can be isolated from a variety of organs in infected fish. The highest level of viral activity has been found in the kidneys (Plumb, 1971c).

Affected fish may swim erratically or hang vertically in the water column with the head uppermost. Other symptoms sometimes but not always observed in conjunction with CCVD include distension of the abdomen with the presence of fluid in the peritoneal cavity; exophthalmia; anemia; hemorrhaging of the gills, fin bases, skin, kidneys, and other internal organs; and absence of food in the intestines (Plumb, 1971a, 1972). Epizootics may be influenced by poor culture conditions—for example, low dissolved oxygen, high temperature, crowding, secondary infections, and improper handling (Plumb, 1971a, 1973).

The symptoms of CCVD just described are among those found in a variety of other diseases, including swim bladder stress syndrome (SBSS). Fluid in the peritoneal cavity is a common symptom of internal bacterial infections, and empty intestines are not uncharacteristic of any disease that makes the fish refuse food. Before CCVD is confirmed, therefore, it is important that other maladies be ruled out. It would be imprudent to destroy brood stock only to determine later that the problem was due to SBSS rather than CCVD! Elimination of CCVD calls for the destruction of all brood stock associated with an epizootic; thus verification of this condition must not be taken lightly.

Fingerlings that survive a CCVD outbreak can be reared to market size but should not be maintained for brood stock. Special attention should be given to the shipment of fish that may be carriers of CCVD to avoid introducing the disease into formerly uncontaminated portions of the United States.

The immune response of channel catfish to CCVD has been studied (Heartwell, 1975), and it appears that a high level of immunity exists in fish that do not contract the disease during an epizootic. However there is no known cure for animals that do contract the disease, nor can fish be immunized against a CCVD epizootic. The best protection for animals not previously exposed is to isolate them from affected organisms. In addition, all equipment exposed to diseased fish should be properly disinfected, and all culture chambers, including ponds, should be sterilized after fish with CCVD have been maintained in them (Plumb, 1972). Chlorine solutions appear to be effective in sterilizing ponds in which CCVD outbreaks have occurred (Fijan et al., 1970).

Hemorrhagic Septicemia.   The gram-negative bacteria *Aeromonas hydrophila* (formerly *A. liquefaciens*) and *Pseudomonas fluorescens* have been linked with a disease of channel catfish known variously as hemorrhagic septicemia, infectuous abdominal dropsy, and red mouth disease (Bullock and McLaughlin, 1970). Furunculosis, a condition in which open sores resembling boils develop on the body of salmonids, is caused by a related bacterium, *A. salmonicida.*

The incubation period of hemorrhagic septicemia varies depending on environmental conditions and the physical condition of affected fish, but is usually not longer than 10 to 14 days. External symptoms include shallow greyish or red ulcers, inflammation around the mouth, exophthalmia, and distension of the abdomen (associated with the presence of bloody or slightly opaque fluid in the peritoneal cavity). Internally, in addition to the presence of fluid in the abdominal cavity, the kidneys may be swollen and soft, the liver pale or green, and there may be blood in the intestine (Snieszko and Bullock, 1968).

Techniques that may be employed to prevent epizootics include obtaining disease-free fish for stocking, adding antibiotics to the water in which fish are shipped, avoidance of handling fish that have been exposed to the disease, and selectively breeding for resistant strains of catfish (Snieszko and Bullock, 1968). Among these measures, only the latter has yet to be effectively employed by catfish culturists.

If an epizootic does occur, various therapeutic agents have been found effective against the disease. Among the recommended treatments are: 50 to 75 mg/kg of Chloromycetin* (chloramphenicol) or oxytetracycline in the feed (Snieszko and Bullock, 1968; Anonymous, 1970). Alternatively, nitrofurazone (Furacin) may be added to the feed for 10 days at 90 mg/kg. If the disease has progressed to the point that the fish will not feed, oxytetracycline or nitrofurazone may be added to the water (Meyer and Hoffman, 1976).

Columnaris disease. This disease is produced in channel catfish by the myxobacterium *Flexibacter columnaris* (formerly *Chondrococcus columnaris*). The disease is characterized by discoloration on the body (greyish or yellowish areas), which may develop into shallow ulcerations. The gills may also be affected, with eventual destruction of the gill lamellae and filaments (Snieszko and Ross, 1969; Meyer and Hoffman, 1976). High mortalities may result from the disease if it is not properly diagnosed and treated (Figures 7.1 and 7.2).

The bacteria may be transmitted by infected fish, carrier fish that are asymptomatic, and possibly through the water (Snieszko and Ross, 1969). Incubation may be as brief as 24 hours, depending on the conditions under which the fish are being maintained. All sizes and ages of catfish are equally susceptible (Snieszko and Ross, 1969).

As a means of preventing the disease, fish should not be overcrowded,

---

*Throughout this chapter drugs that do not appear in Table 7.1 but have been found to be effective against specific diseases or parasites are mentioned. This does not mean that they are cleared for use on food fish, although in some cases there may be no effective treatment chemical that has been officially cleared. Mention of these unregistered chemicals does not imply a recommendation for their use.

Figure 7.1 Channel catfish fingerlings afflicted with bacterial disease (a combination of *Aeromonas* and columnaris).

Figure 7.2 Bacterial infections such as columnaris disease may affect the gills, leading to erosion of the filaments and severe impairment of respiration. Original photograph by Dr. S. K. Johnson, reproduced with permission.

279

especially during warm weather. Prophylaxis includes the addition of nitro-furazone to hauling and holding tanks at 5 to 10 ppm active ingredient. Copper sulfate may be used at 1 ppm in soft water or 2 ppm in hard water to prevent, or possibly control outbreaks of columnaris disease (Snieszko and Ross, 1969).

For fish that have developed external columnaris infections, diquat may be used as a 4 day bath, with the best dosage being about 2 to 4 ppm of diquat cation concentration. Antibiotics added to the water may also be effective in controlling columnaris epizootics. Snieszko and Ross (1969) recommended 10 to 20 ppm of chlortetracycline (Aureomycin) or 5 to 10 ppm of chloramphenicol (Chloromycetin) for aquarium fishes. For systemic infections, the addition of 11 g/100 kg of dry feed of sulfamerazine or 6.6 g/100 kg of dry feed of oxytetracycline (Terramycin) may be effective when the feed is presented over a period of 10 days. The nitrofuran compound Furanase may be effective in the form of a 1 hour bath (Snieszko and Ross, 1969).

Fungal Infections.   Fungal infections of channel catfish are genrally secondary to injury or some type of disease that has led to disruption of the integument. Channel catfish eggs, however, can be directly attacked by fungi, with the initial invasion of dead eggs spreading to those that are viable.

Fungi of the genus *Saprolegnia* are primarily responsible for infections in catfish, although other genera have also been linked with outbreaks in eggs as well as in fish of all sizes (Scott and O'Bier, 1962). *Saprolegnia* spp. are universally present in freshwater environments and are important in the degradation of organic detritus. The presence of fungus on fish is readily detectable, since the animals appear to have a white cottony growth on one or more parts of the body. Eggs also show this characteristic growth.

Channel catfish eggs are particularly susceptible to fungal attack when they are hatched indoors in hatching troughs. In nature, the adult male tends the eggs and removes any that die and are attacked by fungus. This type of activity is labor intensive and generally is avoided by culturists who utilize indoor hatching facilities. The problem may be reduced to some extent by the use of well water rather than surface water, since the former is less likely to become contaminated with fungus. In any case chemical treatment is frequently required to limit losses of eggs due to fungi. Treatment generally involves dipping egg masses into fungicides two or more times weekly. Various concentrations and times of treatment with malachite green (an analine dye having fungicidal properties) have been utilized, with a 30 second treatment in 1:15,000 solution being popular. Burrows (1949) recommended a 5 ppm malachite green dip treatment for 1 hour.

Malachite green is considered to be carcinogenic, thus posing a hazard to culturists who work with or near it frequently (it is virtually impossible to work with malachite green without dyeing both skin and clothing), and it is unlikely that this drug will ever gain official clearance. Betadine (a commercially available antiseptic) will also effectively control fungus on channel catfish eggs when utilized as a 1% solution. Eggs should be dipped in this solution for 10 minutes.

Treatment of fingerling or adult channel catfish for fungi also usually involves dipping the animals in a fungicidal solution. A 10 to 30 second dip in a 1:15,000 solution of malachite green has often been used effectively. Longer treatments in lower concentrations of malachite green have also been successfully employed (reviewed by Hoffman, 1969). Clemens and Sneed (1958) determined that a 1 hour bath in 0.3 ppm malachite green was safe for channel catfish fingerlings, but suggested that a 1 ppm flush treatment might be better. Routine treatment of catfish as a prophylactic measure is neither necessary or desirable.

Protozoan Parasites. A variety of protozoans are known to parasitize freshwater fishes, including channel catfish. Because of the diversity of protozoan parasites, only some of the more common epizootic-causing species are discussed.

1. *Ichthyophthirius multifiliis.* The ciliated protozoan *Ichthyophthirius multifiliis* (Figure 7.3) is the causative agent of ichthyophthiriasis, or "ich" (Figure 7.4). *I. multifiliis* attacks the integument of various fishes, including channel catfish, and may cause a thickening of the epithelium and the production of excess mucus (Meyer and Hoffman, 1976). The parasite feeds on the epithelial tissues and fluids of the fish and can lead to high levels of mortality.

Infected fish may be found congregated at the water inflow or drain of ponds or other types of culture chambers and may attempt to rub themselves against the bottom or sides of the culture chamber, apparently to scratch or dislodge the parasites (Meyer, 1966a; Meyer and Hoffman, 1976). The presence of an epizootic is indicated by the occurrence of small white nodules or pustules on the body of the fish. In advanced cases the entire body surface of the fish may become covered with the nodules.

The life cycle of *Ichthyophthirius multifiliis* involves development of the adult parasite in the nodules on the body of the fish. The adult parasites eventually leave the fish to become free swimming for a few hours, after which they form cysts on suitable substrates. Within each cyst the parasite undergoes multiple cell division, forming numerous juveniles that finally emerge from the cyst and swim through the water, seeking a fish on which

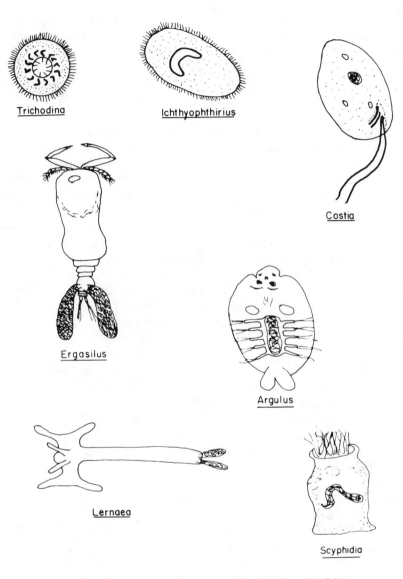

Figure 7.3 Parasites that commonly invade channel catfish.

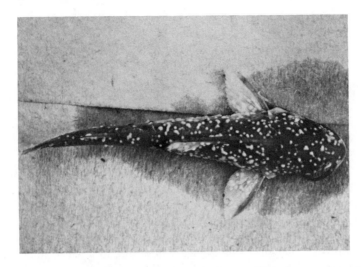

Figure 7.4 Fingerling channel catfish exhibiting white pustules on the body surface indicative of *Ichthyophthirius multifiliis*. Original photograph by Dr. S. K. Johnson, reproduced with permission.

to attach. Once established on the body of a fish, the parasite grows into the adult form and the cycle is repeated (Meyer and Hoffman, 1976).

No effective methods of treatment have been found for *Ichthyophthirius multifiliis* during the cyst stage or when the parasites are attached to the fish; however they can be destroyed during the free-swimming stages. Since the stages of development of the parasite may exist simultaneously, a single treatment is generally ineffective in controlling an epizootic. Rather, several treatments over a period of days or weeks are necessary to ensure that each individual parasite is destroyed during a vulnerable period.

Among the treatments that have been promoted are malachite green at concentrations from 0.1 to 1.25 ppm in baths (Allison, 1957; Beckert and Allison, 1964; Johnson, 1961), formalin at 15 ppm in ponds (Allison, 1957) or 200 to 250 ppm as 1 hour baths (Meyer and Collar, 1964; Davis, 1953), combinations of formalin and malachite green (Leteux and Meyer, 1972), copper sulfate at 0.5 ppm weekly in ponds (Meyer, 1966a; Leteux and Meyer, 1972), and salt water (Allen and Avault, 1970). Johnson (1976) tested a variety of chemicals on channel catfish infected with *I. multifiliis* and found that new infestations occurred with all the types of treatment mentioned above except sodium chloride at 0.2% (2°/oo). The formalin and malachite green mixture was not tested by Johnson. Other effective treatments included chloramine T, quinine sulfate, and quinine bisulfate (Johnson, 1976).

The optimum temperature range for ichthyophthiriasis is from 21 to 24 C, and one effective treatment is to raise the water temperature of culture tanks to 29 to 32 C (Meyer, 1966a; Johnson, 1976). Because of the rather restricted temperature range for the development of epizootics, ichthyophthiriasis is most commonly encountered during the spring and fall when water temperatures are changing.

2. *Trichodina* spp. Among the most common external parasites to affect channel catfish are various species in the genus *Trichodina* (Figure 7.3). The parasite may occur on the body, fins, and gills of infected fish. Under magnification *Trichodina* spp. appear as circular transparent animals with an internal disc structure that contains hooks. The animal is a protozoan with cilia located completely around the circumference of the cell. The cilia may not be readily apparent under the microscope (Davis, 1953; Meyer, 1966b). The aboral surface of the organism contains an adhesive disk, although the animals are often found to be in rapid motion on the surface of the host (Davis, 1947, 1953).

Trichodiniasis is characterized by the appearance of irregular white blotches on the head and dorsal surface of infected fish. There may also be fraying of the fins and a loss of appetite associated with an epizootic (Davis, 1947, 1953). Fish may produce excessive amounts of mucus and will flash (Hoffman and Lom, 1967). Epidermal necrosis may also occur (Davis, 1947).

Trichodina epizootics can be treated with dips of 30°/oo salt water, 1:500 solutions of acetic acid, or 1:4000 solutions of formalin (Davis, 1953). In ponds, the addition of sufficient formalin to make a 15 ppm solution may be effective (Amlacher, 1970).

3. *Costia* spp. Among the flagellated protozoans that attack fish, those in the genus *Costia* create some of the most frequent problems. *Costia* spp. may attack the gills and external surfaces of fish (Rogers and Gaines, 1975). As the result of excessive mucus production, affected fish display a characteristic greyish-white or bluish cast in the film that covers the external surface. Epidermal necrosis may result from *Costia* epizootics (Amlacher, 1970). Costiasis may produce high levels of mortality and is especially serious when it occurs on fingerling catfish (Davis, 1953).

Depending on the species, *Costia* is essentially an oval or pear-shaped organism (Figure 7.3) that moves rapidly when viewed under the microscope (Davis, 1953). Control of this parasite has been effected with 1:500 acetic acid dips (Davis, 1953) and 1:4000 formalin solutions (Fish, 1940), as well as methylene blue, sodium chloride, potassium permanganate, and copper sulfate (Amlacher, 1970).

4. *Myxosporidians.* Parasites responsible for such afflictions as the whirling disease of salmonids have received a great deal of attention by fish disease specialists. However other myxosporidian parasites may be important to warmwater fish culturists. One genus, *Henneguya*, may lead to epizootics resulting in high levels of mortality (McCraren *et al.*, 1975).

A number of species of *Henneguya* have been described in conjunction with channel catfish and other species. Infections often appear as white cysts on the skin, gills, barbels, adipose fins, and various other tissues, including internal ones. Spores of *Henneguya* resemble spermatozoans. The scrapings from viable cysts can be examined microscopically for their characteristic spores.

Some forms of *Henneguya* can lead to epizootics resulting in high levels of mortality. The literature does not provide information on effective treatment, although methods utilized for the treatment of other myxosporidians that attack salmonids might be employed. For internal infections, treatment may not be feasible with any recognized therapeutic agent.

Helminth Parasites. Various parasitic worms are commonly found, but in most cases infestations are not severe enough to cause significant problems to catfish producers. However the crowded conditions in aquaculture that seem to favor other disease and parasitic organisms can also lead to epizootics by helminths. Among the flatworms (Platyhelminthes), members of the classes Trematoda and Cestoda are exclusively parasitic. Certain species within the spiny-headed (Acanthocephala) and roundworm (Annelida) phyla are also parasitic. In channel catfish most of the severe helminth problems are caused by monogenetic trematodes.

The distinction between monogenetic and digenetic trematodes varies to some extent depending on which authority is consulted. Chandler and Read (1961) state that monogenetic trematodes differ from digenetic species in that the former do not undergo asexual development during any stage of the life cycle, whereas asexual development does occur in the digenetic species. Villee *et al.*, 1963) divide the two on the basis of number of hosts; that is, monogenetic trematodes have life cycles involving a single host, whereas digenetic species utilize two or more hosts during their lives. In reality, both definitions are valid, since species that utilize only one host (Monogenea) reproduce only sexually, and species having two or more hosts (Digenea) also employ asexual reproduction at some point in the life cycle.

The life cycle of digenetic trematodes often involves the presence of the adult parasite in the body of a fish. Eggs are released from the adult trematode and enter the water, where they hatch into larvae. The larval trematodes enter a gastropod mollusk and undergo further development.

The later larval stages of the parasite may then enter a mayfly naiad and become encysted within the body of that host. A fish eating the mayfly becomes infected when the cyst ruptures and releases larval parasites, which then develop into adults (Anonymous, 1970). Alternatively, fish may serve as an intermediate, rather than the final host for digenetic tremvatodes. Typical species of digenetic trematodes that infect channel catfish include *Alloglossidium corti,* which is found in the intestines, and *Clinostomum marginata,* which occurs in cysts in the flesh of the fish (Meyer, 1966b). The latter is referred to as the yellow grub of fishes and reaches adulthood only when consumed by the proper bird host. This parasite normally is not a problem if fish-eating birds are kept away from culture chambers. The life cycle of various digenetic trematodes can also be interrupted by eliminating snails, which often serve as intermediate hosts.

The most common monogenetic trematodes infecting catfish are *Gyrodactylus* (found primarily on the body and fins, but also occasionally on the gills) and *Cleidodiscus* (found only on the gills). These parasites have been treated effectively with 25 ppm formalin in ponds or 250 ppm formalin as a 1 hour bath (Meyer, 1966b). Dylox, potassium permanganate at 5 ppm, and potassium dichromate at 20 ppm have also been utilized effectively (Meyer, 1966b; Anonymous, 1970).

Copepods.   A variety of parasitic copepods have proved troublesome to aquaculturists. Many species have become highly modified in their physical appearance, with these modifications adapting them for their parasitic existence. Among the species that commonly infect channel catfish, *Ergasilus* is found on the gills, whereas *Argulus* (known as the fish louse) is found on the surface of the body and may resemble scales (Meyer, 1966b). Both species (Figure 7.3) feed on body fluids.

Ergasilus looks similar to the free-living cyclopoid copepods, except that one pair of antennae are developed into stout hooks and the mouth parts are designed for biting (Bowen, 1966). Respiratory impairment, anemia, and poor growth are among the symptoms of serious *Ergasilus* infection. Moreover the disruption of the tissues associated with the parasite often leads to the outbreak of bacterial and other types of secondary infection. Suggested treatment for *Ergasilus* is 0.25 ppm Dylox weekly in ponds (Anonymous, 1970).

Argulus grossly resembles the horseshoe crab (*Limulus polyphemus*) in many respects, although the parasite never attains a size larger than a few millimeters. Epizootics of *Argulus* are often initiated when parasitized fish are released into culture facilities that were previously free of the parasite (Bowen and Putz, 1966). Secondary infections associated with *Argulus* frequently occur. Heavily infested fish may swim erratically and will flash.

Weight loss is a common symptom (Bowen and Putz, 1966). Because of the size of the parasite and its location on the surface of the body, diagnosis is generally relatively simple. A variety of chemicals have been utilized in the treatment of *Argulus* (reviewed by Bowen and Putz, 1966), but none seems to be any more effective than the 0.25 ppm Dylox treatment recommended for *Ergasilus*.

A third species of parasitic copepod, *Lernaea cyprinacae* (Figure 7.3), often called the anchor parasite, has been known to attack channel catfish when that species is reared in conjunction with scaled fishes (Meyer, 1966b). Once attached to a fish, the head of the parasite is modified and literally becomes anchored in the flesh of the host. Dylox at 0.25 ppm has been utilized effectively in the treatment of this copepod also (Meyer, 1966b; Anonymous, 1970).

Other Common Fish Diseases

Vibriosis. The symptoms of hemorrhagic septicemia occur in various marine and estuarine fishes, and the bacterial pathogen of primary responsibility for the disease is *Vibrio anguillarum*. *Vibrio* spp. have also been found in freshwater hemorrhagic septicemia infections (Ross *et al.*, 1968), although such occurrences do not seem to be common. One species of *Vibrio* that may cause disease not only in fish but in humans, is *V. parahaemolyticus* (Bullock and McLaughlin, 1970). In the case of humans, *V. parahaemolyticus* leads to a form of food poisoning; thus it is important to avoid epizootics of this bacterium in cultured fishes, even though most North Americans cook fish well before consuming it.

Aquarium studies in which *Vibrio* epizootics occurred have demonstrated that antibiotics, either injected into the fish or mixed with the feed, are effective forms of treatment (Oppenheimer, 1962; Farrin *et al.*, 1957). The experience of aquaculturists with *Vibrio* and related bacterial epizootics is limited to an extent by the low level of marine and estuarine fish culture in the United States. However *Vibrio* has led to significant problems in such nations as Japan where marine fish culture is extensive (Sindermann, 1970c).

Euryhaline marine fishes may be highly susceptible to attack by bacteria when reared in fresh water. For example, the red drum, *Sciaenops ocellata*, has been known to exhibit chronic *Aeromonas* infections when reared at relatively high density in ponds and raceways in fresh water (Robert B. McGeachin, unpublished data). Affected fish were also unusually susceptible to secondary invasion by fungus (probably *Saprolegnia*). Routine treatments with nitrofurazone (2 or 3 times weekly) were required for the bacterial infection. The fungus was effectively treated with malachite green.

Recurrence was rapid in both cases if intermittent treatment was discontinued.

Viral Disease. Perhaps the most common and best known viral disease affecting marine and estuarine fishes is lymphocystis. This disease also occurs in fresh water but has not been a significant problem in warmwater fish culture except under saline water conditions. Lymphocystis generally occurs as whitish nodules on the fins, head, and sometimes the body of the fish. The nodules are caused by the increase in size and encapsulation of connective tissue cells (Sindermann, 1970c).

The viral origin of lymphocystis was suspected long before the etiology of the disease was provided by Walker (1962) and Walker and Wolf (1962). Lymphocystis is not generally fatal unless it forms around the mouth of the fish to an extent that greatly interferes with the ingestion of food, but the disease is highly contagious (Sindermann, 1970c). Under culture conditions lymphocystis spreads rapidly and can have an adverse effect on consumer acceptance of some species. For example, flounders sold with the head and fins intact would be hard to market if those areas were infested with lymphocystis nodules (Figure 7.5). The disease has been reported from tank

Figure 7.5 *Paralichthys* sp. with lymphocystis nodules prominent around the mouth and on the caudal fin.

cultured flounders of the genus *Paralichthys* (Stickney and White, 1974) and appeared to enter the culture facility in the water since the problem did not recur after a UV sterilization device was added to the water treatment system.

There is no known cure for lymphocystis. Fish affected with the disease should be destroyed to prevent the infection of other animals. Tanks, ponds, or other types of culture chambers should be sterilized with chlorine solutions following removal of the fish.

Parasitic Infections. Fish other than channel catfish that are of aquaculture interest are susceptible to a variety of parasite problems, but few have been of serious consequence in warmwater food fish production. The incidence of parasitic epizootics will undoubtedly increase concurrent with the development of fish mariculture and as new freshwater fishes are more widely introduced into United States aquaculture. Parasite problems have been faced in many countries outside the United States however (Sindermann, 1970a, 1970c), and the solutions developed in those countries deserve careful study because they can serve as the basis for treatment of future epizootics here.

Diseases and Parasites of Shrimp

Shrimp culture in the United States is in its infancy, yet a variety of diseases and parasites have been identified in both marine and freshwater species. Although this discussion is confined to the diseases and parasites of penaeid shrimps, the reader should be aware that similar problems occur in freshwater shrimp and crayfish (Johnson, 1977, 1978). Many of the infections that have been reported were observed in natural populations or were found to be incidental to cultured shrimp. This does not mean that extensive epizootics will not occur in the future, but it indicates that disease and parasite treatment for penaeid shrimp is not well developed. Many diseases and parasites have been described, but there are few data available relative to the effectiveness of particular therapeutic agents. It may be assumed that related diseases and parasitic infestations in fish and crustaceans can be treated in a similar fashion, although this may not always be true, especially when the toxicity of a treatment chemical may greatly differ between vertebrate and invertebrate culture organisms.

Bacteria may invade the body fluids and exoskeleton of penaeid shrimp. *Vibrio* spp. (Lewis, 1974; Vanderzant *et al.*, 1970) and *Pseudomonas* spp. are the bacterial agents most likely to be found in shrimp (Johnson, 1975). Other types of bacteria attack the exoskeleton (Cook and Lofton, 1973) and may lead to blackening of the infected area (Figure 7.6). Such exoskeleton invaders are not dangerous to humans who consume contaminated shrimp, but the results of their presence are somewhat unsightly and could lead to avoidance of such animals by the consumer if the shrimp were sold on ice with the exoskeleton intact.

Larval marine crustaceans are susceptible to fungi of the genus *Lage-*

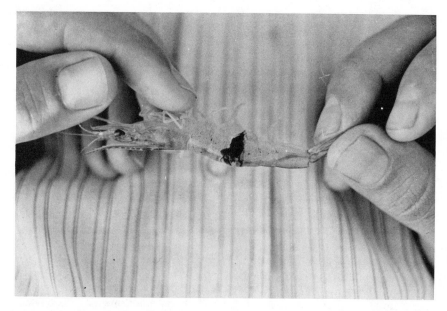

Figure 7.6 Bacterial invasions of the exoskeleton may lead to extensive discoloration in cultured penaeid shrimp. Photograph used with permission of Dr. S. K. Johnson.

nidium, and because of the high levels of mortality that have been known to occur in various species of crustacea, including penaenid shrimps, it is believed that consequences in aquaculture might be serious (Bland *et al.,* 1976).

Fungal infections of larval crustacea have been successfully treated by combinations of malachite green and at least one other chemical (Bland *et al.,* 1976).

Only a limited amount of information is available on fungal infections of juvenile and larger shrimp (Lightner and Fontaine, 1973; Overstreet, 1973), although infestations have occurred and it has been predicted that significant problems can be expected when shrimp aquaculture is practiced more widely (Johnson, 1975). Whether epizootics become common probably depends on whether the marine fungi responsible are able to invade healthy tissue or must await some type of primary infection before they can become established.

A variety of protozoan parasites have been found living as ectocommensals (organisms living on the surface of a host organism but causing no apparent damage to the host) and are not considered to be a problem in most cases. Internal parasitic protozoans of shrimp may be pathogenic, however. The most common parasitic protozoans of shrimp are microsporidians, which cause a disease called cotton shrimp or milk shrimp. The infection may occur throughout the musculature of the animal, or it may be limited to certain organs (Johnson, 1975). Affected animals have a white hue that is quite distinct from that of healthy individuals. The affected tissues contain large numbers of spores. In most cases infected animals are discarded. In nature the problem is limited to a very small percentage of the catch, but it is conceivable that epizootics could occur under the crowded conditions that exist in aquaculture.

Although the life cycles of such common microsporidians as *Nosema, Pleistophora,* and *Thelohania* are not known with any precision, a theoretical model has been presented by Johnson (1975) in which the spores of the protozoan parasites are ingested by shrimp, which then develop the infection. Presumably infected shrimp lose spores into the water, leading to the infection of other individuals.

Shrimp are subject to helminth parasitism, including the invasion of trematodes, cestodes, and nematodes, although none have been known to cause massive mortalities. Nematodes have been commonly discovered in natural shrimp populations but have not caused severe problems in cultured shrimp to date (Johnson, 1975). In the case of some parasitic nematodes, such as *Thynnascaris* sp., penaeid shrimp may act as an intermediate host (Johnson, 1975).

Freshwater shrimp (*Macrobrachium* spp.) and crayfish (primarily *Cambarus* spp., *Orconectes* spp., and *Procambarus* spp.) are susceptible to a variety of diseases and parasites similar to those of penaeid shrimp. A variety of bacteria, fungi, protozoans, helminths, isopods, ostracods, and other organisms have been known to affect freshwater shrimp and crayfish (Johnson, 1977). As in the case of the penaeids, aquaculturists have relatively little experience in disease control with these organisms.

Crustacean Shell Disease

Various crustaceans have been known to contract a form of exoskeleton necrosis commonly called shell disease. The history of observations of the malady and an indication of species that have been known to contract the disease were reviewed by Rosen (1970).

Commercial crustacean species in both marine and fresh waters are subject to the disease. In the United States it is of particular interest that the blue crab, *Callinectes sapidus,* and the American lobster, *Homarus americanus,* are among the species of North American crustaceans most likely to contract the disease. Neither is presently being produced successfully in commercial aquaculture, but both are or have been under study and may hold potential as aquaculture candidates in the future.

The syndrome of shell disease involves development of necrotic lesions on the exoskeleton. The lesions may first appear as small round, reddish-brown depressions that eventually form irregular sites of erosion, which become enlarged with time. In some cases extensive portions of the exoskeleton become involved (Rosen, 1970).

The specific organisms responsible for crustacean shell disease have yet to be firmly established, but chitinoclastic bacteria are strongly implicated (Rosen, 1970). Such bacteria are ubiquitous in aquatic environments (Lear, 1963) and probably attack the crustacean following disruption of the natural defense mechanisms found in the epicuticle of the exoskeleton (Rosen, 1970).

Mortality may occur within several months after the onset of infection; however if ecdysis occurs before the animal dies, the disease will be shed with the old exoskeleton (Rosen, 1970). Apparently the infection is limited to the exoskeleton and does not become established in underlying soft tissues. The disease does appear to be highly contagious (Rosen, 1970), especially when animals are crowded, as is virtually always the case in aquaculture.

Diseases and Predators of Oysters

Oyster Fungus Disease.   One of the most serious problems in the culture of oysters in warm water is caused by the fungus *Labyrinthomyxa marina* (= *Dermocystidium marinum*). The problem seems to be directly related to the planting of oysters at high densities (Mackin, 1961). The disease does not affect oysters until they are a few months old, and mortality increases directly with the age and size of the animals (Sindermann, 1970c). Menzel and Hopkins (1955) concluded that the fungus causes a reduction in growth rate, followed by cessation of growth, and finally death.

Labyrinthomyxa marina is limited by cool water and low salinity, thus it is a problem only during part of the year in such areas as Chesapeake Bay but may cause serious damage throughout the year in the Gulf of Mexico. A *Labyrinthomyxa*-like outbreak in oysters (*Crassostrea virginica*) resulted in the death of approximately 30 to 34 million oysters (90 to 99% of the population) in Pearl Harbor, Hawaii, in 1972, under apparently normal environmental conditions (Kern et al., 1973). Aside from decreasing temperature and salinity, there is no known cure for *L. marina* epizootics. Early attempts at selectively breeding oysters for resistance to the disease were not successful (Bardach et al., 1972).

The histopathology of oyster fungus disease has been studied by Mackin (1961). Inflammation characterizes the early stages of infection, followed by fibrosis and extensive lysis of the tissues. All tissues may be involved

under conditions of high temperatures and salinities, but the most extensive damage occurs in the connective tissues, the adductor muscle, the
digestive epithelium, and the blood vessels.

Oyster Predators. Though neither diseases nor parasites, certain predators
of oysters deserve recognition because of the significant levels of mortality
they can impose on natural or cultured populations. The most important
predators of oysters are starfish, annelids, predacious gastropods, crabs,
fishes, boring sponges, and cephalopods (Lunz, 1940; Hopkins, 1956;
Sindermann, 1970b).

Among the most serious of the oyster predators are the gastropod mollusks called oyster drills. Two species, *Urosalpinx cinerea* and *Eupleura
caudata,* represent significant problems to *Crassostrea virginica.* These
pests occur from Cape Cod to Florida along the east coast of the United
States, and sparingly along the Gulf coast (Galtsoff *et al.,* 1937). Oyster
drills are able to bore circular holes in the shells of oysters and other mollusks with a special rasping organ called the radula, which contains thousands of minute chitinous teeth arranged in rows (Galtsoff *et al.,* 1937).
Once the shell of the oyster has been penetrated, the drill consumes its
host. Details on oyster drill proboscis structure and function were determined by Carriker (1943). The early stages of *U. cinerea* develop optimally
at between 20 and 25 C (Ganaros, 1958).

Methods to rid areas of oyster drills and control their populations have
included trapping (Nelson, 1931) and the use of various chemicals (Loosanoff *et al.,* 1960; MacKenzie, 1970). The rearing of oysters on strings or
in baskets off the substrate is another effective means of controlling this
problem. Divers have also been assigned to collect the oyster drills, but
this is expensive and time-consuming.

Starfish also lead to high levels of destruction in oyster beds along the
Atlantic coast of the United States. The problem is caused by *Asterias
forbesi,* which in laboratory studies were observed to kill as many as five
1 year old oysters each in a single day (Loosanoff and Engle, 1942). These
authors reported that lime was effective in killing the starfish and controlling reinfestation for considerable periods. Starfish can also be ignored
when oysters are reared off the bottom. Starfish cannot be cut into pieces
as a method of destroying them, since each piece often will regenerate into
a new predator. Formaldehyde injections by divers will kill starfish, but
again the process requires considerable time and expense.

A final problem among many others affecting oysters is the mudworm
(Polydora websteri), which forms unsightly black areas on the inner faces
of oyster shells (making infected oysters unacceptable for the half-shell
trade) and making the shells brittle (MacKenzie and Shearer, 1961). Heav-

ily infested oysters may become weakened and eventually die (Roughley, 1922, 1925). Oysters can be protected from mudworms to some extent if they are reared off the bottom (Loosanoff and Engle, 1943).

Parasites and Commensals Affecting Invertebrates

Many invertebrates of aquaculture interest can be infected with a variety of parasites; however as was true in the case of shrimp, epizootics leading to mass mortalities do not generally occur. Once again, this may be largely a function of the inexperience of United States aquaculturists in the rearing of marine invertebrates. As aquaculture develops, the incidence and severity of parasitic attacks may increase, since expanding culture increases the chances for obtaining stock that is infected in some manner.

Among the ectocommensal organisms that occur in conjunction with aquaculture invertebrates, the barnacle deserves mention. Barnacles are commonly found attached to the carapace of crabs, shrimp, and other invertebrates (they have also been found on such vertebrates as the green sea turtle). In most cases barnacles present no problem with respect to survival, growth, or marketability of aquaculture organisms. In addition, these commensals are shed during ecdysis. It is conceivable that barnacles could develop in places on the host's body where impaired vision or ambulation would result, but such occurrences appear to be rare.

LITERATURE CITED

Allen, K. O., and J. W. Avault, Jr. 1970. Effects of brackish water on ichthyophthiriasis of channel catfish. *Prog. Fish-Cult.* **32:** 227-230.

Allison, R. 1957. Some new results in the treatment of ponds to control some external parasites of fish. *Prog. Fish-Cult.* **19:** 58-63.

Amlacher, E. 1970. *Textbook of fish diseases.* T.F.H. Publications, Jersey City, N.J. 302 p.

Andrews, J. W., and T. Murai. 1975. Studies on the vitamin C requirements of channel catfish (*Ictalurus punctatus*). *J. Nutr.* **105:** 557-561.

Anonymous, 1970. *Report to the fish farmers.* U.S. Bureau of Sport Fisheries and Wildlife, Resource Publication 83. 124 p.

Bardach, J. E., J. H. Ryther, and W. O. McLarney. 1972. *Aquaculture.* Wiley-Interscience, New York. 868 p.

Beckert, H., and R. Allison. 1964. Some host responses of white catfish to *Ichthyophthirius multifiliis* Fouquet. *Proc. Southeast. Assoc. Game Fish Comm.* **18:** 438-441.

Bland, C. E., D. G. Ruch, B. R. Salser, and D. V. Lightner. 1976. Chemical control of *Lagenidium,* a fungal pathogen of marine crustacea. *Proc. World Maricult. Soc.* **7:** 445-472.

Bowen, J. T. 1966. *Parasites of freshwater fish. IV. Miscellaneous 4. Parasitic copepods*

Ergasilus, Achtheres, and Salmincola. U.S. Bureau of Sport Fisheries and Wildlife, Fish Disease Leaflet No. 4. 4 p.

Bowen, J. T., and R. E. Putz. 1966. *Parasites of freshwater fish. IV. Miscellaneous. 3. Parasitic copepod Argulus.* U.S. Bureau of Sport Fisheries and Wildlife, Fish Disease Leaflet 3. 4 p.

Bullock, G. L., and J. J. A. McLaughlin. 1970. Advances in knowledge concerning bacteria pathogenic to fishes (1954–1968). In S. F. Snieszko (Ed.), *A symposium on diseases of fishes and shellfishes.* American Fisheries Society, Special Publication 5. American Fisheries Society, Washington, D.C., pp. 231–242.

Burrows, R. E. 1949. Prophylactic treatment for control of fungus (*Saprolegnia parasitica*) on salmon eggs. *Prog. Fish-Cult.* **11:** 97–103.

Carriker, M. R. 1943. On the structure and function of the proboscis in the common oyster drill, *Urosalpinx cinerea Say. J. Morphol.* **73:** 441–506.

Chandler, A. C., and C. P. Read. 1961. *Introduction to parasitology.* Wiley, New York. 822 p.

Clary, J. R., and S. D. Clary. 1978. Swim bladder stress syndrome. *Salmonid,* March–April: 8–9.

Clemens, H. P., and K. E. Sneed. 1958. The chemical control of some diseases and parasites of channel catfish. *Prog. Fish-Cult.* **20:** 8–15.

Cook, D. W., and S. R. Lofton. 1973. Chitinoclastic bacteria associated with shell disease in *Penaeus* shrimp and the blue crab (*Callinectes sapidus*). *J. Wildl. Dis.* **9:** 154–158.

Cruz, E. M. 1975. Determination of nutrient digestibility in various classes of natural and purified feed materials for channel catfish. Ph.D. dissertation, Auburn University, Auburn, Ala. 90 p.

Davis, H. S. 1947. Studies of the protozoan parasites of fresh-water fishes. *Fish. Bull.* **41:** 1–29.

Davis, H. S. 1953. *Culture and diseases of game fishes.* University of California Press, Berkeley. 332 p.

Dupree, H. K. 1966. *Vitamins essential for growth of channel catfish.* U.S. Bureau of Sport Fish and Wildlife, Technical Paper 7. 12 p.

Dupree, H. K. 1977. Vitamin requirements. In R. R. Stickney and R. T. Lovell (Eds.), *Nutrition and feeding of channel catfish.* Southern Cooperative Series Bulletin 218, pp. 26–29.

Farrin, A. E., L. W. Scattergood, and C. J. Sindermann. 1957. Maintenance of immature sea herring in captivity. *Prog. Fish-Cult.* **19:** 183–189.

Fijan, N. 1968. Progress report on acute mortality of channel catfish fingerlings caused by a virus. *Bull. Off. Int. Epiz.* **69:** 1167–1168.

Fijan, N. N., T. L. Wellborn, Jr., and J. P. Naftel. 1970. *An acute viral disease of channel catfish.* U.S. Bureau of Sport Fisheries and Wildlife, Technical Paper 43. 11 p.

Fish, F. F. 1940. Formalin for external protozoan parasites. *Prog. Fish-Cult.* (old series), **48:** 1–10.

Galtsoff, P. S., H. F. Prytherch, and J. B. Engle. 1937. *Natural history and methods of controlling the common oyster drills (Urosalpinx cinerea Say and Eupleura caudata Say).* U.S. Bureau of Fisheries, Fishery Circular 25. 24 p.

Ganaros, A. E. 1958. On development of early stages of *Urosalpinx cinerea* (Say) at constant temperatures and their tolerance to low temperatures. *Biol. Bull.* **114:** 188–195.

Halver, J. E. 1953. Fish diseases and nutrition. *Trans. Am. Fish. Soc.* **83:** 254–261.

Heartwell, C. M. III. 1975. *Immune response and antibody characterization of the channel catfish (Ictalurus punctatus) to a naturally pathogenic bacterium and virus.* U.S. Bureau of Sport Fisheries and Wildlife, Technical Paper 85. 34 p.

Hoffman, G. L. 1969. *Parasites of freshwater fish. I. Fungi.* U.S. Bureau of Sport Fisheries and Wildlife, Fish Disease Leaflet 21. 6 p.

Hoffman, G. L., and J. Lom. 1967. Observations on *Tripartiella bursiformis, Trichodina nigra* and a pathogenic trichodinid, *Trichodina fultoni. Bull. Wildl. Dis. Assoc.* **3:** 156–159.

Hopkins, S. H. 1956. *The boring sponges which attack South Carolina oysters, with notes on some associated organisms.* Contribution to the Bears Bluff Laboratory 23. 30 p.

Johnson, A. K. 1961. Ichthyophthiriasis in a recirculating closed-water hatchery. *Prog. Fish-Cult.* **20:** 129–132.

Johnson, S. K. 1975. *Handbook of shrimp diseases.* Texas A&M University Sea Grant Publication, TAMU-SG-75-603. College Station. 19 p.

Johnson, S. K. 1976. Laboratory evaluation of several chemicals as preventatives of ich disease. In *Proceedings of the 1976 fish farming conference and annual convention of the Catfish Farmers of Texas.* Texas A&M University, College Station, pp. 91–96.

Johnson, S. K. 1977. *Crawfish and freshwater shrimp diseases.* Texas A&M University Sea Grant Publication, TAMU-SG-77-605. College Station. 19 p.

Johnson, S. K. 1978. *Some disease problems in crawfish and freshwater shrimp culture.* Texas A&M University Fish Disease Diagnostic Laboratory Leaflet FDDL-S11, College Station. 5 p.

Kern, F. G., L. C. Sullivan, and M. Takata. 1973. *Labyrinthomyxa*-like organisms associated with mass mortalities of oysters, *Crassostrea virginica,* from Hawaii. *Proc. Nat. Shellfish. Assoc.* **63:** 43–46.

Kudo, R. R. 1954. *Protozoology,* 4th ed. Thomas, Springfield, Ill. 966 p.

Lear, D. W. 1963. Occurrence and significance of chitinoclastic bacteria in pelagic waters and zooplankton. In C. Oppenheimer (Ed.), *Symposium on marine technology.* Thomas, Springfield, Ill., pp. 594–610.

Leteux, F., and F. P. Meyer. 1972. Mixtures of malachite green and formalin for controlling *Ichthyophthirius* and other protozoan parasites of fish. *Prog. Fish-Cult.* **34:** 21–26.

Lewis, D. H. 1974. Response of brown shrimp to infection with *Vibrio* sp. *Proc. World Maricult. Soc.* **4:** 333–338.

Lightner, D. V., and C. T. Fontaine. 1973. A new fungus disease of the white shrimp *Penaeus setiferus. J. Invertebr. Pathol.* **22:** 94–99.

Loosanoff, V. L., and J. B. Engle. 1942. *Use of lime in controlling starfish.* U.S. Fish and Wildlife Service, Resource Report 2. 29 p.

Loosanoff, V. L., and J. B. Engle. 1943. *Polydora* in oysters suspended in the water. *Biol. Bull.* **85:** 69–78.

Loosanoff, V. L., C. L. MacKenzie, Jr., and L. W. Shearer. 1960. Use of chemicals to control shellfish predators. *Science,* **20:** 1522–1523.

Lovell, R. T. 1975. *Nutritional diseases in channel catfish.* FAO Technical Conference on Aquaculture, Kyoto, Japan. FIR:AQ/Conf/76/E.6.

Lovell, R. T., and C. Lim. 1978. Vitamin C in pond diets for channel catfish. *Trans. Am. Fish. Soc.* **107:** 321–325.

Lunz, G. R., Jr. 1940. The annelid worm, *Polydora*, as an oyster pest. *Science*, **92:** 310.

MacIntyre, P. A. 1960. Tumors of the thyroid gland in teleost fishes. *Zoologica*, **45:** 161–170.

MacKenzie, C. L. 1970. Control of oyster drills, *Eupleura caudata* and *Urosalpinx cinerea*, with the chemical Polystream. *Fish. Bull.* **68:** 285–297.

MacKenzie, C. L., Jr., and L. W. Shearer. 1961. Chemical control of *Polydora websteri* and other annelids inhabiting oyster shells. *Proc. Nat. Shellfish. Assoc.* **50:** 105–111.

Mackin, J. G. 1951. Histopathology of infection of Crassostrea virginica (Gmelin) by *Dermocystidium marinum* Mackin, Owen, and Collier. *Bull. Mar. Sci. Gulf Caribb.* **1:** 72–87.

Mackin, J. G. 1961. Mortalities of oysters. *Proc. Nat. Shellfish. Assoc.* **51:** 21–40.

McCraren, J. P., M. L. Landolt, G. L. Hoffman, and F. P. Meyer. 1975. Variation in response of channel catfish to *Henneguya* sp. infections (Protozoa: Myxosporidea). *J. Wildl. Dis.* **11:** 2–7.

Marine, D. 1914. Further observations and experiments on goitre (so-called thyroid carcinoma) in brook trout (*Salvelinus fontinalis*). III. Its prevention and cure. *J. Exp. Med.* **19:** 70–88.

Marine, D., and C. H. Lenhart. 1910. Observations and experiments on the so-called thyroid carcinoma of brook trout (*Salvelinus fontinalis*) and its relation to ordinary goitre. *J. Exp. Med.* **12:** 311–337.

Marine, D., and C. H. Lenhart. 1911. Further observation and experiments on the so-called thyroid carcinoma of the brook trout (*Salvelinus fontinalis*) and its relation to endemic goitre. *J. Exp. Med.* **13:** 455–475.

Menzel, R. W., and S. H. Hopkins. 1955. The growth of oysters parasitized by the fungus *Dermocystidium marinum* and by the trematode *Bucephalus cuculus*. *J. Parasitol.* **41:** 333–342.

Meyer, F. P. 1966a. *Parasites of freshwater fishes. II. Protozoa 3. Ichthyophthirius miltifilis.* U.S. Bureau of Sport Fisheries and Wildlife, Fish Disease Leaflet 2. 4 p.

Meyer, F. P. 1966b. *Parasites of freshwater fishes. IV. Miscellaneous. 6. Parasites of catfishes.* U.S. Bureau of Sport Fisheries and Wildlife, Fish Disease Leaflet 5. 7 p.

Meyer, F. P. 1967. The impact of diseases on fish farming. *Am. Fishes Trout News*, March–April.

Meyer, F. P. 1968. *Treatment tips, how to determine quantities for chemical treatment in fish farming.* U.S. Bureau of Sport Fisheries and Wildlife, Resource Publication 66. 17 p.

Meyer, F. P., and J. D. Collar. 1964. Description and treatment of a *Pseudomonas* infection in white catfish. *Appl. Microbiol.* **12:** 201–203.

Meyer, F. P., and G. L. Hoffman. 1976. *Parasites and diseases of warmwater fishes.* U.S. Fish and Wildlife Service, Resource Publication 127. 20 p.

Meyer, F. P., R. A. Schnick, K. B. Cumming, and B. L. Berger. 1976. Registration status of fishery chemicals, February, 1976. *Prog. Fish-Cult.* **38:** 3–7.

Murai, T., and J. W. Andrews. 1975. Pantothenic acid supplementation of diets for catfish fry. *Trans. Am. Fish. Soc.* **104:** 313–316.

Nelson, J. R. 1931. *Trapping the oyster drill.* New Jersey Agriculture Experimental Station Bulletin 523, New Brunswick. 12 p.

Oppenheimer, C. H. 1962. On marine fish diseases. In G. Borgstrom (Ed.), *Fish as food*, Vol. 2. Academic Press, New York, pp. 541–572.

Overstreet, R. M. 1973. Parasites of some penaeid shrimps with emphasis on reared hosts. *Aquaculture*, **2:** 105–140.

Plumb, J. A. 1971a. Channel catfish virus disease in southern United States. *Proc. Southeast. Assoc. Game Fish Comm.* **25:** 489–493.

Plumb, J. A. 1971b. *Channel catfish virus research at Auburn University.* Agriculture Experimental Station, Auburn University Progress Report Series 95, Auburn, Ala. 4 p.

Plumb, J. A. 1971c. Tissue distribution of channel catfish virus. *J. Wildl. Dis.* **7:** 213–216.

Plumb, J. A. 1972. *Channel catfish virus disease.* U.S. Bureau of Sport Fisheries and Wildlife Fish Disease Leaflet 18 (revised). 4 p.

Plumb, J. A. 1973. Effects of temperature on mortality of fingerling channel catfish (*Ictalurus punctatus*) experimentally infected with channel catfish virus. *J. Fish. Res. Bd. Can.* **30:** 568–570.

Plumb, J. A., and J. L. Gaines, Jr. 1975. Channel catfish virus disease. In W. E. Ribelin and G. Migaki (Eds.), *The pathology of fishes.* University of Wisconsin Press, Madison, pp. 287–302.

Rogers, W. A., and J. L. Gaines, Jr. 1975. Lesions of protozoan diseases in fish. In W. E. Ribelin and G. Migaki (Eds.), *The pathology of fishes.* University of Wisconsin Press, Madison, pp. 117–141.

Rosen, B. 1970. Shell disease of aquatic crustaceans. In S. F. Snieszko (Ed.), *A symposium on diseases of fishes and shellfishes.* American Fisheries Society Special Publication 5. American Fisheries Society, Washington, D.C., pp. 409–415.

Ross, A., J. E. Martin, and V. Bressler. 1968. *Vibrio anguillarum* from an epizootic in rainbow trout (*Salmo gairdneri*) in the U.S.A. *Bull. Off. Int. Epiz.* **69:** 1139–1148.

Roughley, T. C. 1922. *Oyster culture on the George's River, New South Wales.* Technical Education Series, Technological Museum, Sydney, Vol. 25, pp. 1–69.

Roughley, T. C. 1925. The story of the oyster. *Aust. Mus. Mag.* **2:** 1–32.

Rucker, R. R., H. E. Johnson, and G. M. Kaydas. 1952. An interim report on gill disease. *Prog. Fish-Cult.* **14:** 10–14.

Scott, W. W., and A. H. O'Bier. 1962. Aquatic fungi associated with diseased fish and fish eggs. *Prog. Fish-Cult.* **24:** 3–15.

Simco, B. A., and F. B. Cross. 1966. Factors affecting growth of channel catfish, *Ictalurus punctatus. Univ. Kans. Publ. Mus. Nat. Hist.* **17:** 1–191.

Sindermann, C. J. 1970a. Disease and parasite problems in marine aquiculture. In W. J. McNeil (Ed.), *Marine aquiculture.* Oregon State University Press, Corvallis, pp. 103–134.

Sindermann, C. J. 1970b. Predators and diseases of commercial marine mollusca of the United States. *Ann. Rep. Am. Malacol. Union,* **1970:** 35–36.

Sindermann, C. J. 1970c. *Principal diseases of marine fish and shellfish.* Academic Press, New York. 369 p.

Snieszko, S. F. 1972. Nutritional fish diseases. In J. E. Halver (Ed.), *Fish nutrition.* Academic Press, New York, pp. 403–437.

Snieszko, S. F., and G. L. Bullock. 1968. *Freshwater fish diseases caused by bacteria belonging to the genera Aeromonas and Pseudomonas.* U.S. Bureau of Sport Fisheries and Wildlife, Fish Disease Leaflet 11. 7 p.

Snieszko, S. F., and A. J. Ross. 1969. *Columnaris disease of fishes.* U.S. Bureau of Sport Fisheries and Wildlife, Fish Disease Leaflet 16, 4 p.

Stickney, R. R., and D. B. White. 1974. Lymphocystis in tank-cultured flounder. *Aquaculture,* **4:** 307–308.

Strout, R. G. 1962. A method for concentrating hemoflagellates. *J. Parasitol.* **48:** 100.

Vanderzant, C., R. Nickelson, and J. C. Parker. 1970. Isolation of *Vibrio parahaemolyticus* from Gulf coast shrimp. *J. Milk Food Technol.* **33:** 161–162.

Villee, C. A., W. F. Walker, Jr., and F. E. Smith. 1963. *General zoology.* Saunders, Philadelphia. 848 p.

Walker, R. 1962. Fine structure of lymphocystis virus of fish. *Virology,* **18:** 503–505.

Walker, R., and K. E. Wolf. 1962. Virus array in lymphocystis cells of sunfish. *Am. Zool.* **2:** 566.

Wellborn, T. L., N. N. Fijan, and J. P. Naftel. 1969. *Channel catfish virus disease.* U.S. Bureau of Sport Fisheries and Wildlife, Fish Disease Leaflet 18, 3 p.

Wolf, K., and R. W. Darlington. 1971. Channel catfish virus: A new herpesvirus of ictalurid fishes. *J. Virol.* **8:** 525–533.

SUGGESTED ADDITIONAL READING

Duijn, C. V., Jr. 1967. *Diseases of fishes.* Iliffe Books, London. 309 p.

Hoffman, G. L. 1967. *Parasites of North American freshwater fishes.* University of California Press, Berkeley. 486 p.

Meyer, F. P. 1966. *A review of the parasites and diseases of fishes in warm-water ponds in North America.* FAO World Symposium on Warm-water Pond Fish Culture, Rome. FS:IX/R3. 29 p.

Sindermann, C. J., and A. Rosenfield. 1967. Principal diseases of commercially important marine bivalve mollusca and crustacea. *Fish. Bull.* **66:** 335–385.

Snieszko, S. F. (Ed.). 1970). *A symposium on diseases of fishes and shellfishes.* American Fisheries Society Special Publication 5. American Fisheries Society, Washington, D.C. 526 p.

Wellborn, T. L., Jr., and W. A. Rogers. 1966. *A key to the common parasitic protozoans of North American fishes.* Zoology-Entomology Department Series 4, Auburn University, Auburn, Ala. 17 p.

CHAPTER 8
Harvesting, Processing, and Economics

One of the primary advantages of intensive culture is the ease with which harvesting can be accomplished in systems of virtually all types. In closed recirculating water systems as well as in open systems utilizing tanks or raceways, harvesting is often merely a matter of draining the culture chambers and dipping out the crop. In large raceways and circular tanks, the water may be partially drained and the animals herded into a relatively small volume through the use of movable screens. As the number of animals is reduced through dip netting, the volume confining those remaining can be further decreased by movement of the screens and additional lowering of the water level.

In most cases animals harvested from intensive culture systems can be loaded directly into hauling tanks for transport to the processing plant. Indoor culture operations should be designed to allow access by hauling vehicles to the immediate vicinity of each culture chamber, or some suitable method should be devised to move the harvested animals to a loading area as easily and rapidly as possible.

The harvesting of cages is also a relatively simple matter. Cages can be towed into shallow water and their contents dipped out. A gantry fitted with block and tackle and a basket will be convenient for moving the dip netted animals from the cage to a waiting transportation truck. Alternatively, the whole cage can be hoisted from the water for harvesting, but this is generally not good practice because a cage filled with harvestable animals may rupture when removed from the water. This problem is not as significant in the case of small cages (Figure 8.1), but production cages may be a cubic meter or greater in volume, containing several hundred kilograms of animals at harvest.

No large investment in harvesting equipment is usually required for intensive aquaculture: dip nets may represent the most significant item of expense. Since the mesh size, handle length, and frame dimensions

Figure 8.1 Experimental cages (approximately 1.0 m high and 0.5 m in diameter) small enough to be lifted from the water for harvesting.

of dip nets are highly variable, the culturist may order nets that fit the needs of the culture system being managed. Nets of various size mesh. are useful for handling larvae, fry, juvenile, and adult animals. In general, relatively large mesh is used for harvesting, although it should not be so large that marginal sized, but marketable individuals escape or become gilled or otherwise entangled. For channel catfish it is desirable that the nets be tarred. Commercial petroleum-based substances specifically designed to coat nets are available, or nets may be ordered pretarred. Because tarring is messy, the purchase of precoated nets is encouraged. When tar is utilized to coat the webbing of the net, catfish spines are less likely to become fouled. Untarred nets often entangle a large number of fish. Removal of catfish from such nets is time-consuming, can be harmful to the animals, and often results in injury to the culturist when the spines enter the fingers.

Animals representing a variety of sizes may be found in a given culture chamber at the end of the growing season. If animals of submarketable size are present, it may be necessary to separate them from the remainder of the crop so that the smaller individuals can be retained on feed until they reach a market size. Separation can be done by hand, but this method requires a good eye and will slow the harvest process considerably when large numbers of animals must be inspected one by one. Graders have been developed that automatically separate aquatic animals by size. Most graders consist of wooden or metal slats set in the bottom of a table or basket. The slats are set a specific distance apart to retain animals above a certain size. Smaller individuals fall through. Graders are useful not only in separating marketable from submarketable animals, but also when separating fingerlings into various size groups for stocking. The graders for the latter process have slats much closer together than those used for harvest and come in various sizes. Some graders are made to float so that netted animals can be dropped into them while the grader is in the culture chamber. Submarketable animals escape back into the pond, tank, cage, or raceway, and the marketable individuals remain within the grader, from which they are moved to the transportation truck. Commercial graders are available or they can be rather simply constructed by the culturist.

HARVESTING EXTENSIVE CULTURE SYSTEMS

The harvesting of ponds is greatly facilitated when construction has resulted in regular shapes with properly sloped banks, easy access to at least one side by vehicle, large drain lines that allow rapid and complete emptying, and the incorporation of a harvest basin near the drain. Many ponds lack one or more of these features. Each is important, but the regularity of pond shape is of least concern. In some instances it is more convenient to lay out a series of ponds to conform to natural variations in terrain than to greatly alter the site to accommodate square or rectangular ponds. It is important that pond bottoms be smooth an clean. If culture ponds do not have drains, the ponds can be dewatered by pumping.

In most instances harvesting is not preceded by complete drainage of ponds, but the level is reduced to some extent (e.g., by half) to concentrate the animals. Thereafter seining is undertaken to capture the majority of the fish. A full pond requires longer seine hauls and the operation is often less efficient than when water volume is reduced (Figure 8.2). Small ponds can be seined by hand, whereas tractors or trucks may be required to pull

Figure 8.2 Seining a full pond can be difficult. If the seine crew attempts to walk at the top of the levee, there will be little control of the net and little leverage with which to pull. If the crew enters the water, the animals may be able to escape around the ends of the seine.

seines in large ponds. Once the majority of the fish have been removed by the initial seining operation, the pond may be further drained and the operation repeated. If a harvest basin is present, final drainage of the pond will concentrate the fish in that area, where they may be captured by dip netting.

Commercial catfish farmers frequently practice a technique known as multiple harvesting. In this strategy marketable fish are removed periodically throughout the growing season, with submarketable individuals replaced. The ponds are not drained when this technique is utilized. Fingerlings may be added to replace harvested fish, or the bulk of the fish in the pond may be harvested before restocking occurs.

Seines should be approximately 1.5 times the width of the pond, to ensure that they will bow out during harvesting operations (Figure 8.2). For additional capacity, seines can be equipped with a bag (Figure 8.3). Besides increasing seine capacity, bags may reduce escapement. The depth of the seine should be at least twice the depth of the water to be harvested, and even three times water depth is not too much for catfish, where it is extremely important to keep the net on the bottom. Seine mesh should be small enough to retain all harvestable animals, but it should not be smaller, since reduction in mesh size will increase the resistance of the

Figure 8.3 Seines having a bag (indicated by the line of floats behind the body of the seine) may be somewhat more efficient than nets without this feature and are useful when large biomasses of animals are to be harvested.

net in the water and make it more difficult to pull. In addition, the cost of netting increases with decreasing mesh size.

The upper line (float or head line) of a seine is usually fairly standard in configuration, consisting of a rope to which cork, plastic, Styrofoam, or other buoyant material (visible in Figures 8.2 and 8.3) is attached at intervals. Bottom lines, on the other hand, may vary to some extent. The bottom line (or lead line) is traditionally made from rope to which lead weights are attached at intervals. At present it is possible to purchase lead lines of rope within which a lead core has been implanted. In ponds with muddy bottoms the traditional lead line is often inefficient (having a tendency to dig into and drag through the mud, allowing the seine to roll up, with consequent loss of the culture animals), and a mud line should be used. Mud lines are composed of a number of relatively small diameter ropes, loosely tied together. The ropes are made from a material that readily absorbs water (e.g., cotton). Mud lines tend to maintain contact with the bottom without digging in or lifting off the sediments; thus escapement under the seine is reduced. Such lines will fray and wear rapidly in ponds having firm sediments.

Harvesting is the most labor-intensive phase of a typical aquaculture operation. Several persons are required on seining crews, even if trucks or tractors are utilized to do the bulk of the physical work. The aquaculturist may be required to add to the normal work force during harvest operations, and the added expense should be considered to be an operating expense.

Once captured, the animals must be loaded into waiting trucks before mortalities occur. Much of the operation may be by hand labor, or certain kinds of equipment may be utilized to assist in the operation. The mechanics of harvest and number of times harvesting is required in a given year depend on the species under culture, the market being served, and the geographical area in which the culture facilities lie.

LIVE-HAULING

Whether an aquaculture crop is to be sent off to another commercial or private fish farm, recreational lake, local restaurant, wholesale food market, or large processing plant, it is necessary to ship the animals in good condition. For many species the product is acceptable only when shipped alive. The most notable exception involves shrimp, which are traditionally headed and hauled on ice. Oysters can be hauled in burlap bags but are still alive when they reach the market. Crabs and lobsters are hauled on ice and are also alive, though dormant, when they reach the processor or restaurant. Fish hauling differs somewhat from that for the species just mentioned, since fish must be hauled in water. Hauling is usually accomplished by placing the animals in live-hauling tanks carried by or permanently attached to trucks. Once concentrated in harvest basins or by seining, aquaculture animals can be removed from the pond or held in live cars (pens constructed of netting or wire and placed in the pond to hold animals in high concentrations). In both cases the animals are eventually transferred to the hauling tank, either by the use of baskets or by lifting the entire live car from the water with the aid of a suitable mechanical device.

Virtually all aquaculturists need some type of hauling tank as a part of their normal operations. In many cases this need is satisfied with a tank that will fit the bed of a pickup truck. Such tanks are generally about 0.5 m in depth, with the outside dimensions equal to the available space on the truck bed. Live-hauling tanks can be constructed of wood, metal, or fiberglass, with the latter being desirable because of its light weight and durability (Figure 8.4). Such tanks are commercially available, or the culturist may wish to construct one of unique design or configuration.

A live-hauling tank is not merely a watertight box in which to haul aquatic animals. It must provide the proper environmental conditions to

Figure 8.4 Hauling tanks, such as this commercially available, truck-mounted fiberglass model, are a necessary part of nearly every aquaculture operation.

maintain the organisms in good condition while they are being transported or otherwise held. Because of the high concentration of animals that may be contained in hauling tanks, some type of aeration must be provided, even if the animals are to be held only for a short period. Mechanical agitators connected to the 12 V electrical system of the truck are popular. This type of aerator is generally mounted through holes of the proper size in the top of the hauling tank (Figure 8.4). Aeration may also be provided through airstones supplied from a tank of compressed air or oxygen. Such tanks must be properly secured according to state law to avoid the risk of damage to the valve, which could result in extreme danger. A cracked valve on a compressed gas bottle may turn that cylinder into a missile.

Commercial live-haulers have trucks weighing several tons. Such vehicles are often equipped with air compressors to provide constant aeration; others rely on agitators, or compressed air or oxygen from bottles. For long-distance hauling the water in such trucks may be circulated, filtered, and even refrigerated to provide a more suitable environment for the animals being transported.

Tanks of any appreciable size, even those that are carried on the bed of a pickup truck, should have baffles in them, or they should be compartmentalized into chambers of not much more than 1 m^2. Both the presence of baffles and the compartmentalization of transportation tanks tend to dampen the oscillation of water that results when the truck accelerates, brakes, or turns. When a great volume of water shifts inside a large undampened hauling tank, the truck becomes hard to control.

Refrigeration is not economical or convenient in conjunction with pickup truck sized transportation tanks. Because of the relatively small volumes of water being hauled, however, it is generally economical to place ice in the hauling tank if the water temperature gets too high. If the water remains at or below about 30 C for warmwater fishes, ice may not be necessary, although fewer losses will be incurred, even with warmwater species, if the temperature is reduced to about 25 C. Only a few kilograms of ice is required to achieve that temperature. A thermometer should be carried in the truck so that the temperature of the water can be checked every 2 to 4 hours and more ice added as necessary.

Fish may sometimes be hauled under the influence of an anesthetic, although the use of such drugs on fish headed for processing has not been approved by the United States government. Many species become excited when handled, confined in small volumes of water at high densities, and hauled for long distances. If the fish are to be restocked into other water bodies, serious losses can be anticipated as a result of such stresses. In such cases anesthetics, and even the addition of antibiotics to the water, may be helpful.

A considerable amount of research has been conducted on the fish anesthetics tricaine methanesulfonate (MS-222) and quinaldine. Hunn and Allen (1974) found that successful anesthesia was achieved with channel catfish at an MS-222 concentration of 150 mg/l or quinaldine at 30 mg/l. Bell (1964) reviewed the literature on fish anesthetics and reported that brief exposures to MS-222 of 68 to 111 mg/l were often recommended; the most useful range of that chemical was 40 to 80 mg/l. Brood catfish can be successfully anesthetized at an MS-222 concentration of 111 mg/l (Crawford and Hulsey, 1963). Studies with midrange and coldwater fishes provide similar data. Schoettger and Steucke (1970a) reported that 100 to 150 mg/l of MS-222 or 10 to 20 mg/l of quinaldine were effective for anesthetizing northern pike (*Esox lucius*), muskellunge (*E. masquinongy*), and walleye (*Stizostedion vitreum*). The same authors (Schoettger and Steucke, 1970b) tried a mixture of MS-222 and quinaldine on rainbow trout (*Salmo gairdneri*) and northern pike and found that the best results were obtained when the mixture contained 20 to 30 mg/l of MS-222 and 5 mg/l of quinaldine. Striped bass (*Morone saxatilis*) can be anesthetized for spawning at concentrations of either 21 mg/l of MS-222 or 2 mg/l of quinaldine (Bonn et el., 1976). Allen and Harman (1970) cautioned that the pH of MS-222 solutions must be checked before placing fish in them; if the pH has changed too much from that of the ambient water, the solutions must be buffered. In an experiment these authors made up a solution of 150 mg/l of MS-222 in water with an initial hardness of 12 mg/l and pH 8. The resulting solution had a pH of 4.5, which would be lethal or damaging to exposed fish.

Other anesthetics have been utilized successfully. Bardach et al. (1972) reported that Chinese carps are often transported in 6.7 to 7.7 μg/l of sodium barbital or 1 to 4 g/l or urethane when water temperatures are within the range of 25.5 to 32 C. Sodium bicarbonate (NaHCO₃) has also been used effectively as an anesthetic on brook trout (*Salvelinus fontinalis*) and common carp (*Cyprinus carpio*) by Booke et al. (1978). They suggested that pH-controlled release of carbon dioxide was the cause of anesthesia achieved when 642 mg/l of sodium bicarbonate was present in water at pH 6.5.

It is possible that various antibiotics could be effective in providing some degree of control over bacterial disease problems of transported fishes. Sniesko and Bullock (1968) reported that mixtures of penicillin and streptomycin at 10 to 50 mg/l are effective against *Aeromonas* and *Pseudomonas* during live-hauling and that Acriflavin at about 10 mg/l may also be effective.

Live-hauling tanks can be utilized to transport fry and fingerling fish as well as marketable animals, but because of the small size of fish fry and invertebrate larvae, it is often more simple to ship them in polyethylene bags, partially filled with water and topped off with oxygen from a compressed gas bottle. Following saturation of the water and air space with oxygen, the bag is closed and sealed. When the bags are placed in insulated boxes, they can be shipped long distances safely. It is generally safe to leave fish fry and invertebrate postlarvae or larvae under those conditions for at least 24 hours, or longer depending on the number of animals in each polyethylene bag. Several thousand catfish fry or shrimp postlarvae can be shipped in only a few such bags. Again, small amounts of ice can be used to prevent overheating in hot weather, but the insulated box should help maintain the water at or near the temperature that existed when the bags were filled, until the animals arrive at their destination. To reduce the production of metabolites, animals should not be fed before live-hauling, regardless of the method of hauling to be utilized.

Large aquaculture enterprises that move great quantities of animals throughout the year may require a heavy truck designed especially for transporting aquatic organisms, but the average aquaculturist often can manage with a small transportation tank and a pickup truck (Figure 8.4). When the small volume culturist must move large numbers of animals to a distant processing plant or retailer, shipment may be contracted to a commercial hauling firm. In catfish farming regions of the southern United States, commercial live-haulers are available, but this is not the case in all areas. As increased demand for these services is demonstrated in the future, live-haulers will begin to branch out into newly profitable areas.

SPECIALIZED CULTURE STRATEGIES

Not all aquaculture animals are reared in culture chambers that can be drained or chambers that hold the animals in a small volume of water, even though the container itself is placed in a large water body (e.g., cage culture). The extensive culture of oysters, mussels, and clams may be conducted in leased estuarine or offshore areas that cannot be drained (although the animals may be exposed intermittently by tidal action). String culture in such areas does provide a means of rapid and relatively easy harvest, since the animals can be collected merely by pulling the strings from the water. Most mollusk culture, however, is done by seeding large areas of estuarine bottom with young animals and allowing them to grow under the influence of natural productivity. Thereafter, they can be dredged using well-established commercial harvesting techniques.

The culture of mollusks intertidally does provide the advantage of harvesting a drained area since the crop can be collected during low tide. However interidal animals may grow somewhat more slowly than subtidal ones because of cessation of feeding while exposed to air. Areas with relatively great tide ranges and those with small tide ranges but very gently sloping beaches, may have large expanses of exposed bottom during spring low tides. In some parts of the United States, notably in the Gulf of Mexico, seasonal wind patterns may control tidal period and amplitude, particularly in areas where the natural tidal range is very small. In many of the Texas bays, for example, the prevailing southerly winds of summer tend to keep the water well up on the shoreline for extended periods (several days or even weeks), whereas during the winter frequent shifts of wind to the north push the bay waters out, to expose extensive areas of bay bottom for similarly long periods. Since oysters and other mollusks grow most rapidly during the summer, intertidal culture in such areas might result in rapid growth, and harvest would be facilitated in the winter when the animals are frequently exposed to the air. One danger in relying on this approach would be "winterkills" of oysters due to low temperatures during periods when the animals are exposed.

Another specialized case, and one that is not widely seen in conjunction with aquaculture in the United States, is the fencing off of large areas of ocean or even reservoirs to form enclosures for rearing motile aquatic animals. Fish and shrimp, for example, could be reared in relatively shallow marine areas surrounded by netting of an appropriate mesh size. Such areas would have to be rather large, so that boats with trawls or other types of collecting gear could enter for harvesting. Various problems could be expected in conjunction with such operations. Netting often tends to foul, leading not only to interference with the free exchange of water between the enclosure and the surrounding environment, but also to the eventual partial sinking or even rending to the net. Any tear in the net offers a convenient means of escape for the culture animals and a point of entry for predators. Firms that have attempted culture in fenced areas often hire SCUBA divers to inspect and repair the netting. SCUBA divers can also be employed to rid the enclosure of predators, but this is expensive.

Large net or wire enclosures in fresh water have not been utilized to date in the United States, although it is certain that fouling would be a less serious problem than in the marine environment. As in salt water, predation may be significant, as well as competition for supplemental feed between the culture organisms and any wild animals that were inadvertently allowed to remain within the enclosure during construction or entered thereafter.

FEE FISHING

Fee fishing operations may be independent of other aquacultural activities, or they may be integrated into the production of food animals for wholesale. Fee fishing provides the sportsman with an opportuntiy to catch high quality aquatic animals and ensures a reasonable degree of success. This concept has been widely employed by salmonid and catfish producers but has not yet been developed for most marine fish or aquacultured invertebrates, although fee fishing operations of those types could become popular in the future.

In most fee fishing operations one or more ponds on an aquaculture facility are opened to the public upon payment of an entry fee. Alternatively, the fisherman may pay for the fish captured on the basis of weight. Generally management imposes both an entry fee and a charge per fish caught, based on weight.

In most cases the aquaculturist stocks the fee fishing ponds with large numbers of marketable fish, reared elsewhere on that or another facility. As fish are removed by the sportsmen, the culturist adds to the stock in the fish ponds. When the culturist charges by the fish or by weight of fish caught, it is advantageous to stock sufficient numbers of animals to give the customer a positive fishing experience; that is, the customer should catch fish relatively easily and in sufficient numbers that there will be no disappointment with the expenditure of time or money. This obviously helps to ensure that the customer will return.

Fee fishing operations tend to work well in the vicinity of large cities and in areas where there are few traditional fishing locations such as streams, lakes, ponds, and reservoirs. City dwellers often find it inconvenient to mount major fishing expeditions, but they might be tempted to fish regularly at a fee fishing operation located only a short distance from home.

A well-operated fee fishing business may offer a variety of services in addition to fishing itself. A bait and tackle shop may be lucrative, and if tackle is not for sale, it should be available for rent. Many fee fishing operations provide a place for the fisherman to clean the catch, or the employees of the facility will clean the fish (again, at a specified cost per unit weight or individual). Picnic areas and other recreational sidelines to fee fishing could be provided but should not detract from the main thrust of the enterprise.

Added incentive to fisherman can be provided by stocking at least a few very large fish in the fee fishing pond and offering a prize or special rates if one of them is caught. Another means of enticing business is to mark fish and give cash or prizes for the capture of marked individuals.

PROCESSING AND MARKETING

The presence of commercial processing facilities in various parts of the United States is a function of the volume of aquaculture produce reared in the immediate vicinity and the proximity of commercial fisheries producing the same type of product. In general, producers of such marine invertebrates as shrimp, crabs, lobsters, clams, and oysters should have little or no difficulty with processing, since those species can be realistically reared only in regions of the country that currently support natural populations of the same or similar species. Large commercial fisheries for edible teleosts of many species of aquaculture interest are not well developed in the warm marine waters of the United States, although considerable numbers of such organisms as striped mullet (*Mugil cephalus*) are caught and commercially processed in certain parts of the country (mullet is most popular in Florida and Hawaii). Presently being processed in moderate to large numbers are such fishes as flounder (*Paralichthys* sp. in warm water and *Pseudopleuronectes americanus* in the Atlantic off of the northeastern United States), halibut (*Hippoglossus stenolepis* along the Pacific coast of the United States), cod (*Gadus morhua* off the New England coast), red drum (*Sciaenops ocellata*), black drum (*Pogonias cromis*), and spotted sea trout (*Cynoscion nebulosus*)—each of last three occurring along the southeastern Atlantic coast of the United States and in the Gulf of Mexico. For some of these species there exist commercial fisheries, whereas for others, the fish are incidental to the catch of shrimp (*Penaeus* spp.) or other animals. Once the aquaculture of the warm water fishes listed (mullet, flounder, red drum, black drum, and spotted sea trout) has developed, if indeed that occurs, there will be suitable processing facilities available in many areas to handle the influx of cultured product.

The future production of large numbers of fishes or invertebrates through aquaculture may require the construction of additional processing facilities, either as separate entities from the aquaculture venture or in association with cultured fish production. Depending on the volume of aquacultured product available, a commercial processor may or may not wish to expand into handling cultured fishes. If commercial processing is not available, the culturist may decide to process the fish or to sell them to restaurants, fish markets, or grocery stores, which would do their own processing.

With respect to warm water aquaculture in the fresh waters of the United States, the most widely cultured foodfish at present is the channel catfish (*Ictalurus punctatus*). In 1973 the total production received at the major processing plants in the United States that handled cultured catfish was 8.8 million kg (Lovell and Ammerman, 1974). This decreased to 7.3 million kg in 1975 (Brown, 1977). The actual total weight of channel catfish processed in the United States may significantly exceed the values available; since data are obtained only from the major processing plants. Small processors are much more numerous than the handful of major processing plants, and the former, which do not report their volume to any agency from which the data can be recovered, may handle a considerable portion of the total processing volume in any given year.

Marketing has been a problem for aquaculturists in the past, although the situation has greatly improved in recent years. The United States public continues to consume fishery products at a fairly low rate compared with beef, pork, and poultry; thus part of the job of aquacultural marketing specialists is to instruct the consumer on the benefits of consuming fishery products, and in particular, those that are reared under aquaculture conditions. The Catfish Farmers of America (an organization formed to promote farm-raised catfish and to convene catfish farmers and researchers periodically to discuss various aspects of the business) have been successful to a degree in promoting their product over sometimes less-expensive imported fish. Imported catfish come largely from the Amazon River Basin in Brazil, where the fish captured do not even belong the family Ictaluridae. Nonictalurid fishes sold under the name "catfish" are also imported into the United States from other regions of the world, and nonnative catfishes make up a considerable percentage of the total amount consumed in the United States. At least part of the reason for the success of commercial catfish producers in marketing their product, even when it is more expensive than the imports, is that farm-raised fish are often of superior quality, although "muddy" tasting channel catfish are raised and sold by fish farmers. Quality control must be a primary consideration in domestic catfish production, since a consumer who has once bought off-flavor fish will not purchase catfish again; especially when the label boasts that the fish were farm-raised domestic animals.

Distinctions in flavor between wild and commercial stocks may not always be apparent. For example, commercially harvested shrimp may be indistinguishable from those produced by mariculturists, and it might be difficult to shift a buyer's preference to shrimp reared under controlled conditions rather than out in the Gulf of Mexico or in the Atlantic off the southeastern United States. Competitive prices between wild and aquacultured shrimp and other mariculture products would appear to be important for the success of aquaculture under those circumstances, at least in the early stages of the industry. As research provides information on ways in which the flavor of aquacultured animals can be controlled by diet, it may be possible to produce cultured shrimp and other species whose flavor is preferable to that of wild animals. However there is a limit to this concept, since people often purchase fishery products because of the unique flavors which each naturally possesses. Alteration of that natural flavor, even to enhance it, may result in consumer rejection.

Most of the channel catfish sold in the United States are supplied to the consumer skinned, gutted, and beheaded, either in the market or in restaurants. Frozen catfish are sometimes available, but the sale of fish in that form has not been extensive to date. Lovell and Ammerman (1974) suggested several additional potential markets for cultured catfish, including military posts, national restaurant chains, and national food store chains. The commercial aquaculturist must not forget that unless a market exists or can be developed, the produce cannot be profitably sold. Thus each culturist must also be a spokesman for the product being reared in his or her facility.

Aquaculturists sometimes contract with local restaurants or markets as outlets for cultured animals. These arrangements are especially lucrative in areas where no processing plant is available and competition is not too severe. The culturist must either provide live animals to the buyer or process them and deliver the product on ice or frozen. In most cases, the culturist delivers a specified amount of product each week or month. Depending on species and the type of culture system being employed, however, marketable animals may not be available throughout the year.

Mechanical equipment has been developed to process a variety of fish and shellfish, but much of the processing that is done today continues to be by manual labor. However maintenance of a full-time crew is often difficult in cases when harvesting is seasonal or intermittent throughout the year.

Most fish and shellfish markets in the United States are now expanding. Contrary to the dip in total live weight of processed catfish demonstrated in the figures for 1973 and 1975 cited earlier, there appears to be a trend toward increasing production of that species (Brown, 1977). Similarly, the demand for other species of fish and shellfish is high thoughout the country. Aquaculture can be expected to step in to fill some of this demand in the future, provided inflation does not drive the price of land, water, and equipment so high that aquacultural products become noncompetitive with wild stocks or imported species. Since the cost of capturing wild stocks is also increasing as costs of boat construction, equipment, and especially fuel rise, however, a pessimistic outlook may not be warranted.

A significant problem faced by catfish producers and others is that the bulk of the crop enters the market over a period of only a few months. Most of the processing of channel catfish is accomplished during the fall and early winter; thus the market is glutted during those periods and may go begging at other times. Indeed, availability of the imported product throughout the year is one reason for the wide utilization of these catfish by restaurants and even fish markets. Restaurants require fish during all seasons to operate effectively; they cannot sell catfish during certain periods only. This is not the case for all fishery products, however. Crayfish are a seasonal crop, and in Louisiana restaurants specializing in them must resort to other main dish meals during the times when crayfish are not available. Similarly, oysters are not sold in most states during the months of May through August (coinciding with the post spawning period when the glycogen reserves in the oysters are depleted and the product is of inferior quality). Shrimp, like oysters, are largely a seasonal crop, although the shrimping and oyster seasons are sufficiently long and the prices sufficiently high to make freezing of those organisms for sale during periods of unavailability economically feasible. Most catfish restaurants are specialty houses, with catfish being the primary fishery product served. Alteration in that approach to sell meals featuring whatever fishery item is seasonally available might induce more reliance on farm-raised fish, but at present the restaurant trade does not seem to be heading in that direction with respect to catfish.

Tilapia are becoming popular in many regions of the United States and, if reared under traditional aquaculture methods, will also be available for only short periods annually. The production of tilapia in power plant cooling reservoirs offers a way around this problem, and this approach has also been used for channel catfish production. Fish reared free in a power plant cooling lake are difficult to catch except during the winter, when they are concentrated in the warm water discharge area unless confined in ponds or cages. Tilapia do have firm flesh and may be best marketed as a frozen product, in which case a short harvest period from ambient temperature ponds, tanks, or raceways would be a less serious disadvantage than if the fish were sold on ice.

Aquaculturists do sometimes produce products with off-flavors, and some of the reasons for this were outlined in Chaper 5. One basic rule of aquaculture is that the producer must be interested not only in the most rapid growth at least expense of the cultured organism, but must also be concerned with consumer acceptance. One of the best means of ensuring quality control is through sound management during the growing season. To have some guarantee against the sale of off-flavor animals, samples from each culture chamber should be cooked and taste-tested for acceptability prior to sale. This is rarely done by commercial aquaculturists, but the practice might pay off if even a few loads of fish are diverted from the marketplace because of poor quality. Off-flavors do not require destruction of the crop. They are often a result of blue-green algae in the culture chambers, and the problem generally can be solved by maintaining the animals for a brief period (10 days to 2 weeks should be sufficient) in uncontaminated water. The fish can be kept at high densities in a holding trough receiving flowing well water. If the fish are fed a complete prepared ration, they should become purged of off-flavors relatively quickly, after which they can be marketed.

AQUACULTURAL ECONOMICS

Most of the economic studies of species under culture in warm water in the United States have been conducted on channel catfish, since few other species have demonstrated a profit to culturists. A considerable number of commercial aquaculture ventures have been initiated with penaeid shrimp, oysters, tilapia, and other species, but much of the information obtained by commercial producers is proprietary. In addition, many of these ventures have failed.

In first attempts to look at the economics of channel catfish culture were undertaken by Swingle (1957, 1958). Since then, several economic evaluations of channel catfish operations have been made, and one of the most recent was based on aquaculture in California (Anonymous, 1976). Brown (1977), in his book on fish farming economics around the world, devoted a considerable amount of attention to the economics of channel catfish culture in the United States. It is not possible to provide precise estimates for the various costs involved in establishing an aquaculture facility for channel catfish or any other species, since the costs of land, water, energy, feed, and other items vary considerably from one region of the country to another and are constantly changing.

The prospective culturist should attempt to determine the cost of each item required to establish the enterprise, ascertain the market price of the species to be reared, project production in terms of kilograms per hectare of water available, and draw up a budget. The initial investment will vary considerably as a function of the amount of area to be placed into production, the average size of the individual culture chambers, well depth or availability of surface water, proximity to a feed mill and market, and requirements and costs of energy. Table 8.1 lists some of the items that

Table 8.1 Partial List of Equipment and Supplies Required by a Typical Aquaculture Facility (in Addition to Land, Buildings, a Suitable Supply of Water, and Appropriate Culture Chambers).

Tractor with mowing attachments
Pickup truck
Hauling tank for pickup truck
Generator for use during power failures
Mechanical agitators or other emergency aeration devices
Dip nets
Seines
Chemicals for disease prevention and control
Chemicals for aquatic vegatation control
Dissolved oxygen meter or chemicals for Winkler oxygen determination
Appropriate chemicals and apparatus to test water pH, hardness, alkalinity, ammonia, nitrate, and salinity (the latter required only in marine and estuarine systems)
Backup pumps
Microscope to aid in identification of disease and parasitic organisms
Scales for weighing feed and animals
Feed
Feed pails or buckets, or some type of mechanical device to dispense feed into culture chambers
Fertilizer
Appropriate facilities for spawning and hatching if these activities are to be conducted.
Disinfectants for contaminated ponds, raceways, culture tanks, and equipment
Complete set of hand tools
Various plumbing supplies

should be considered as part of the initial investment for nearly any aquaculture venture. Since certain items will already be available on nearly any ranch or farm, persons involved in agriculture may be able to enter the aquaculture business with less financial difficulty than persons who are obliged to purchase the land and all the equipment at the outset of an aquaculture venture.

Parker (1976) ranked five types of culture system in order of increasing intensity as follows: ponds, cages in static water, raceways, cages in flowing water, and closed recirculating water systems. He then examined the costs of such items as energy, water, land, and disease control in relation to each of the five water systems. The graphs generated from that work showed clearly that there is no straight line relationship among types of culture system and the cost of any feature or resource associated with the operation of those systems. For example, the investment in land may be very high in terms of cost per kilogram of production if animals are to be grown in ponds, but it is relatively low when static cages are utlized. Land cost per kilogram of production increases in raceway systems and is reduced in flowing water cage systems. Closed recirculating water systems often call for the lowest investment in land. A similar pattern exists for costs of water on the basis of kilograms of product reared exclusive of energy costs; but when energy costs for each water system are examined, a very different pattern emerges. Ponds require the least amount of energy and closed recirculating water systems the most. Based on the relatively simple analysis of projected costs for various aspects of a variety of aquaculture systems presented by Parker (1976), it is clear that a general economic plan would be difficult to devise unless considerable detail about a proposed system is known.

Anderson et al. (1978) examined the economics of a poultry-tilapia production system to determine the price a poultry producer would have to obtain per kilogram of fish reared to make it worthwhile to add ponds for tilapia production to a laying hen operation. The tilapia ponds would be a means of waste treatment with the additional benefit of a secondary product. One major consideration was the known value of chicken manure as a fertilizer on pastures or terrestrial crops. A linear programming model was used to evaluate various sizes of operation (with chicken numbers held constant and the value of manure, number and sizes of ponds, and tilapia densities fluctuating). Although no operation has been established from which actual data could be obtained, sufficient research had been conducted to allow reasonable assumptions to be made. The data indicated that if the fish were to be sold for fish meal (in which case the producer would receive only a few cents per kilogram), the venture would not be economically feasible; it would be more profitable to place the manure on the land or dispose of it in a sewage treatment lagoon. If, on the other hand, a price far below the actual existing sale price but well above the fish meal price were obtained for the fish, the farmer would be justified in establishing the tilapia production operation. The higher price (more than 40¢/kg) would mean that the fish were being sold for human consumption, and that would require clearance by the United States government.

In a recent study by Shang and Fujimura (1977), the economics of establishing a *Macrobrachium rosenbergii* operation in Hawaii were evaluated for culture pond sizes ranging from a total of 0.4 to 40 ha. The cost of pond construction was placed at $4750 for a 0.4 ha pond, including design, access road, and drainage system. The annual operating costs varied from a high of $8927 for a 0.4 ha operation (single pond) to $10,160 per hectare for a 40 ha operation (equivalent to $4464/0.4 ha of water surface). These authors determined that at a price of $6.60/kg (based on reasonable wholesale prices in Hawaii), a farm would become profitable if it were at least 4 ha in area.

Aquaculture can be a profitable business, not only for the fresh water shrimp culturist in Hawaii, but for others as well. There are many successful culturists of channel catfish and other species the United States today, but there are many more who have failed because of unavoidable biological problems, economic setbacks, poor management (both fiscal and biological), and a variety of other reasons. Aquaculture, like traditional agriculture, requires a great deal of work and should not be considered a get-rich-quick enterprise.

The future of aquaculture in the United States remains in question insofar as the direction in which aquaculture will develop is uncertain, and the amount of expansion that will occur in the future is not known. The ultimate role of aquaculture will depend in large part on the intelligent use and conservation of water, the development of alternative sources of energy (if intensive culture systems are to be competitive with extensive pond systems), and a commitment from both the public and private sectors for the funds required to conduct research at the level that has been undertaken in the past with terrestrial livestock and the salmonids.

LITERATURE CITED

Allen, J. L., and P. D. Harman. 1970. Control of pH in MS-222 anesthetic solutions. *Prog. Fish-Cult.* 32: 100.

Anderson, R. G., W. L. Griffin, R. R. Stickney, and R. E. Whitson. 1978. Bioeconomic assessment of a poultry and tilapia aquaculture system. In: R. Nickelson, (Ed.), *Proceedings of the annual tropical and subtropical fisheries technological conference.* Texas A&M University, College Station: 126-141.

Anonymous, 1976. *Catfish farming in California: An economic guide*. University of California, Division of Agricultural Sciences Leaflet 2892. University of Calfornia, Berkeley. 3 p.

Bardach, J. E., J. H. Ryther, and W. O. McLarney. 1972. *Aquaculture*. Wiley-Interscience, New York. 868 p.

Bell, G. R. 1964. *A guide to the properties, characteristics, and uses of some general anesthetics for fish*. Fisheries Research Board of Canada, Bulletin 148. 4 p.

Bonn, E. W., W. M. Bailey, J. D. Bayless, K. E. Erickson, and R. E. Stevens (Eds.). 1976. *Guidelines for striped bass culture*. Southern Division, American Fisheries Society, Washington, D.C. 103 p.

Booke, H. E., B. Hollender, and G. Lutterbie. 1978. Sodium bicarbonate, an inexpensive fish anesthetic for field use. *Prog. Fish-Cult*. **40**: 11–13.

Brown, E. E. 1977. *World fish farming: Cultivation and economics*. Avi, Westport, Conn. 397 p.

Crawford, B., and A. Hulsey. 1963. Effects of MS-222 on spawning of channel catfish. *Prog. Fish-Cult*. **25**: 214.

Hunn, J. B., and J. L. Allen. 1974. Urinary excretion of quinaldine by channel catfish. *Prog. Fish-Cult*. **36**: 157–159.

Lovell, R. T., and G. R. Ammerman (Eds.). 1974. *Processing farm-raised catfish*. Southern Cooperative Series Bulletin 193. 59 p.

Parker, N. C. 1976. A comparison of intensive culture systems. In *Proceedings of the 1976 fish farming conference and annual convention of the Catfish Farmers of Texas*. Texas A&M University, College Station, pp. 110–120.

Schoettger, R. A., and E. W. Steucke, Jr. 1970a. Quinaldine and MS-222 as spawning aids for northern pike, muskellunge and walleyes. *Prog. Fish-Cult*. **32**: 199–201.

Schoettger, R. A., and E. W. Steucke, Jr. 1970b. Synergistic mixtures of MS-222 and quinaldine as anesthetics for rainbow trout and northern pike. *Prog. Fish-Cult*. **32**: 202–205.

Shang, Y. C., and T. Fujimura. 1977. The production economics of fresh water prawn (*Macrobrachium rosenbergii*) farming in Hawaii. *Aquaculture* **11**: 99–110.

Snieszko, S. F., and G. L. Bullock. 1968. *Fresh water fish diseases caused by bacteria of the genera Aeromonas and Pseudomonas*. U.S. Bureau of Sport Fisheries and Wildlife, Fish Disease Leaflet 11. 7 p.

Swingle, H. S. 1957. Preliminary results on the commercial production of channel catfish in ponds. *Proc. Southeast. Assoc. Game Fish Comm*. **10**: 160–162.

Swingle, H. S. 1958. Experiments on growing fingerling channel catfish to marketable size in ponds. *Proc. Southeast. Assoc. Game Fish Comm*. **12**: 63–72.

CHAPTER 9

Minnows, Goldfish, and Sport Fish

BACKGROUND

Historically, and continuing in the present, the numerical bulk of the aquatic organisms produced in the United States have been reared for use as bait, ornamental and pet species, or for sport fishing. The idea of bait production usually brings to mind the rearing of minnows and related species of fishes; however other organisms also fit the category of bait. Small penaeid shrimp are often utilized for bait, and the production of bait-sized shrimp under culture conditions may be somewhat simpler and less expensive than the rearing of shrimp for human consumption. The concept of rearing crayfish (primarily the red swamp crayfish, *Procambarus clarkii*) for bait has also been suggested, and the methods whereby this could be accomplished have been investigated (Huner, 1975, 1976; Huner and Avault, 1976).

Dobie *et al.* (1956) discussed 20 species of fishes that have been reared for bait commercially in the United States (Table 9.1). Of these, the golden shiner (*Notemigonus chrysoleucas*), fathead minnow (*Pimephales promelas*), and goldfish (*Carassius auratus*) are currently being reared by warmwater aquaculturists. In Arkansas alone during 1975, approximately 7000 ha of golden shiners was produced, with a value in excess of $10 million (Bailey *et al.*, 1976). In the same year those authors determined that Arkansas producers reared more than 950 ha of fathead minnows and about 100 ha of goldfish, worth over $1.1 million and $750,000, respectively. The total value of all aquacultured animals in Arkansas during 1975 was placed at $24 million (Bailey *et al.*, 1976). This total included channel catfish (*Ictalurus punctatus*) and various other food species as well as sport fish. Nearly one-half the total dollar value of production could be attributed to bait fishes.

Goldfish are reared not only for bait, but also for use as ornamentals and pets. One goldfish producer (Ozark Fisheries, Inc., in Stoutland, Missouri) estimated that it has fish in one out of every five homes in the

320

**Table 9.1   Important Bait Fishes of the United States According to
Dobie *et al.* (1956), with Species Names Updated According
to the American Fisheries Society (1970)**

Common Name	Scientific Name
Blacknose dace	*Rhinichthys atratulus*
Bluntnose minnow	*Pimephales notatus*
Brassy minnow	*Hybognathus hankinsoni*
Central mudminnow	*Umbra limi*
Common shiner	*Notropis cornutus*
Creek chub	*Semotilus atromaculatus*
Emerald shiner	*Notropis atherinoides*
Fathead minnow	*Pimephales promelas*
Finescale dace	*Phoxinus neogaeus*
Golden shiner	*Notemigonus chrysoleucas*
Goldfish	*Carassius auratus*
Hornyhead chub	*Nocomis biguttatus*
Longnose dace	*Rhinichthys cataractae*
Northern redbelly dace	*Phoxinus eos*
Pearl dace	*Semotilus marginita*
River chub	*Nocomis micropogon*
Southern redbelly dace	*Phoxinus erythrogaster*
Spotfin shiner	*Notropis spilopterus*
Stoneroller	*Campostoma anomalum*
White sucker	*Catostomus commersonnii*

United States (Anonymous, 1972). Because of the large demand for gold-fish on the nonbait market, that species is discussed under a separate subheading. The multimillion dollar annual business in tropical fishes is not included in this book, although pond culture of certain tropical species does occur in Florida.

Channel catfish culture developed in the United States as a part of the governmental program of providing eggs, fry, and fingerlings for stocking into farm ponds, community lakes, reservoirs, and other types of impoundment. At present, the largest production for such stocking programs involves various species of salmonids (trout and salmon) in cold water, with largemouth bass (*Micropterus salmoides*) and sunfishes (particularly bluegill, *Lepomis macrochirus*) in warm water. Certain midrange species are also produced in large numbers. At one time a large stocking program for marine waters, including species other than those in the family Salmonidae, was also underway, and a similar program may be developing at present.

Channel catfish culture, as has been demonstrated in the preceding eight chapters, is now one of the mainstays of warmwater aquaculture in the United States, and that species is in great demand in the food fish marketplace. Channel catfish are still produced in large numbers by private and public hatcheries for stocking into recreational fishing waters, but the techniques of culture are identical to those utilized by food fish producers. The difference is that except for the brood stock, few if any channel catfish are held in hatcheries until they reach market size (approximately 0.5 kg), since the expense is too great. Release into impoundments normally occurs when the fish are at the small fingerling size (a few centimeters long). This is also true of bass and sunfish, which may be smaller at release sizes than are catfish.

Hatchery programs in the United States go back many years, well into the nineteenth century. The United States government, through the Commission of Fish and Fisheries, Bureau of Commercial Fisheries, Fish and Wildlife Service, and National Marine Fisheries Service (the latter two exist today), have released billions of aquatic animals in natural waters. Table 9.2 lists the releases for the year ending June 30, 1897. It is of particular interest that winter flounder, *Pseudopleuronectes americanus* (listed in Table 9.2 as flatfish), were spawned for release before 1900. Successful modern spawning of pleuronectiform fishes was not accomplished until a few years ago when Shelbourne (1964) cultured plaice, *Pleuronectes platessa,* through artificial spawning. Since then, lemon sole (*Microstomus kitt*), Dover sole (*Solea solea*), and turbot (*Scophthalmus maximus*) have been spawned in captivity in cold water (Adron *et al.,* 1974), and *Paralichthys* sp. have been spawned in warm water (C. R. Arnold, Personal communication).

Large-scale stocking programs, confined for the most part to fresh water, continue today and have been very successful in providing excellent fishing for centrarchids in warmwater fish ponds, lakes, and reservoirs; for trout in lakes and streams that have cold enough waters; and for midrange species, primarily along the northern tier of states, but also in reservoirs in the middle latitudes and even some southern states. Ocean stocking programs, such as with winter flounder, did not appear to influence the size of the catch as compared with periods during which augmentation stocking did not occur. There was never any evidence that the fish released from the hatcheries returned to the fishery. In fact, it is likely that most did not survive upon release. Stocking of marine fishes has failed, with the notable exceptions of the salmon that continue to be produced in great numbers in the Pacific Northwest and in the Northeast. There is presently a great deal of interest in the rearing in captivity of loggerhead and green sea turtles (*Caretta caretta* and *Chelonia mydas,* respectively) to a size that

Table 9.2   Eggs, Fry, and Juveniles Distributed by the U.S. Commission of Fish and Fisheries During Fiscal 1897 (Brice, 1898)[a]

Common Name	Scientific Name	Number
Shad (American shad)	*Alosa sapidissima*	134,545,500
Quinnat (chinook) salmon	*Oncorhynchus tshawytscha*	32,104,049
Atlantic salmon	*Salmo salar*	2,329,809
Landlocked (Atlantic) salmon	*Salmo salar*	150,566
Silver (coho) salmon	*Oncorhynchus kisutch*	298,137
Steelhead (rainbow) trout	*Salmo gairdneri*	499,690
Loch Leven (brown) trout	*Salmo trutta*	49,709
Rainbow trout	*Salmo gairdneri*	768,123
Von Behr (brown) trout	*Salmo trutta*	23,780
Black spotted trout[b]	*Salmo mykiss*	42,200
Brook trout	*Salvelinus fontinalis*	1,359,510
Lake trout	*Salvelinus namaycush*	13,509,149
Swiss lake trout (Arctic char)	*Salvelinus alpinus*	36,082
Yellow-fin trout[b]	*Salmo mykiss macdonaldi*	7,930
Golden trout	*Salmo aguabonita*	45,000
Whitefish (lake whitefish)	*Coregonus clupeaformis*	95,049,000
Yellow perch	*Perca flavescens*	1,025
Pickerel (Northern pike)	*Esox lucius*	1,700
Striped bass	*Morone saxatilis*	450,000
Lake herring	*Coregonus artedii*	7,299,000
Black (largemouth) bass	*Micropterus salmoides*	95,358
Black (smallmouth) bass	*Micropterus dolomieui*	2,719
Crappie (white crappie)	*Pomoxis annularis*	2,125
Rock bass	*Ambloplites rupestris*	42,687
Strawberry bass	*Pomoxis sp.(?)*	3,129
Codfish (Atlantic cod)	*Gadus morhua*	98,258,000
Flatfish (winter flounder)	*Pseudopleuronectes americanus*	64,095,000
Lobster	*Homarus americanus*	115,606,065
Tautog	*Tautoga onitis*	624,000
Mackerel (Atlantic mackerel)	*Scomber scombrus*	652,000
Sea bass (black sea bass)	*Centropristis striata*	193,000
Total		568,144,042

[a] Common names in parentheses are the presently accepted names according to American Fisheries Society (1970). All scientific names have been updated to reflect current taxonomic classifications (American Fisheries Society, 1970).

[b] Species a variant of other trout (probably either brown or brook) that are no longer recognized as being distinct.

will give them an increased chance of survival over that of hatchlings. Problems with disease in culture are one of the major impediments to the success of programs of that nature (Stickney et al., 1973; J. McVey, personal communication). The United States government, through the National Marine Fisheries Service, is also initiating a program to culture such species as red drum (*Sciaenops ocellata*) for ultimate release into estuaries in an attempt to augment natural populations, which are worked by both sport and commercial fishermen. The success of such new programs remains to be evaluated. In any case the information gained will be of value to commercial aquaculturists.

BAIT MINNOWS

The techniques involved in rearing fathead minnows (*Pimephales promelas*) and golden shiners (*Notemigonus chrysoleucas*) have been outlined by several authors—for example Allan (1952), Prather et al. (1953), Dobie et al, (1956), Guidice (1968), Meyer et al. (1973), Johnson and Davis (1978), and Johnson (1978). In general, minnows are reared in ponds of the type described in Chapter 2. The environmental requirements, except for spawning, are essentially the same as those described for other freshwater fishes, that is, an adequate supply of good quality water and well-designed ponds that can be rapidly filled and completely drained are desirable, if not essential.

Smith (1968) recommended that the water supply for minnow ponds should not be less than about 95 l/min for each hectare of production. Johnson and Davis (1978) recommended a rate of 3 times that amount. Most authorities recommend wells rather than surface water supplies, and Smith (1968) stated that the ideal water supply should have a pH range of 6.0 to 8.5 and an alkalinity of at least 40 mg/l.

Johnson and Davis (1978) recommended rectangular ponds 1.2 to 1.5 m deep at one end and not less than 75 cm deep at the other. The banks should have a slope of 3:1. Those authors stated that livestock ponds may be utilized for bait minnow production but that management of such ponds is more difficult and production lower than in ponds established specifically for minnow culture.

Two methods of production have been utilized for bait minnows. In the intensive culture scheme moderate numbers of fish are stocked in fertilized ponds and provided with low level supplemental feed. This method is used primarily by culturists having limited supplies of water and farmers who produce minnows in livestock ponds (Johnson and Davis, 1978). Intensive minnow production calls for continuous water exchange and daily feeding.

Extensive minnow ponds may produce from about 100 to 300 kg/ha annually, and intensive ponds can be expected to produce between 600 and 1500 kg/ha each year (Johnson and Davis, 1978).

Plankton blooms should be established beginning 2 to 4 weeks before stocking. Nitrogen, phosphorus, and potassium in the ratio 4:4:1 should be applied at the rate of 10 kg/ha of phosphorus. As an example, 50 kg/ha of 20-20-5 fertilizer would be required to provide 10 kg/ha of phosphorus (Johnson and Davis, 1978). Fertilization should be repeated at intervals of 2 to 3 weeks or as required to maintain the phytoplankton bloom as discussed in Chapter 3.

Minnows may be fed prepared feeds similar to those used for channel catfish (Chapter 5). The feed should be finely ground (almost to a powdery consistency) and should be fed at a rate of 10 to 30 kg/ha daily (Johnson and Davis, 1978). Early minnow feeds included such ingredients as cottonseed meal, meat scraps, wheat middlings, dry dog food, cooked cornmeal, fish meal, oatmeal, bone meal, and clam meal (Dobie *et al.*, 1956). Modern commercial feeds exceed 30% protein and may contain such ingredients as soybean and other grain meals, fish meal, meat scraps, distiller's solubles, vitamins, and minerals.

Minnows are most often harvested by lift netting or seining, with seining being the preferred means of collection. A lift net is a piece of suitable mesh-sized net material attached to a boom by ropes. The boom extends from the shore over the water. The net is lowered to the bottom of the water and baited with feed. After suitable numbers of minnows have entered the area and begun to feed, the net is lifted from the water with the aid of the boom and the fish are trapped. The fish may be graded as they are removed from the lift net or seine (Chapter 8), and those too small for market can be returned to the pond (Johnson and Davis, 1978).

Minnow retailers must hold fish in good condition for sale, often for several days or even weeks following capture of the fish from a pond. Minnow holding is usually done in rectangular poured concrete or concrete block tanks, but it may also be accomplished in tanks made from galvanized metal, wood, or fiberglass (Johnson, 1978). The holding tanks should be put in a shed to protect the fish from direct sunlight and to keep down algal growth. Electric agitators are usually placed in the tanks to provide aeration. Minnows can be maintained under these conditions for up to a week without feed, but if held longer they should be fed at the rate of 0.5 to 1.0% of body weight daily (Johnson, 1978).

Diseases and parasites of bait minnows include bacteria, fungi, protozoans, and copepods. Many of the diseases and parasites reported for channel catfish (Chapter 7) are identical or similar to those that attack minnows. Chemical treatments are also essentially the same for both minnows and channel catfish, although the treatment levels of the chemicals sometimes differ. Table 9.3 gives suggested levels of use for various chemicals.

Table 9.3   Treatment Levels of Various Chemicals Utilized in the Control of Minnow Diseases and Parasites [adapted from Johnson (1978)]

Chemical	Treatment Level	Problem Treated
Sodium chloride	Less than 2°/₀₀	Protozoans, especially *Ichthyophthirius multifiliis*
Potassium permanganate	2 mg/l	Protozoans
Copper sulfate	0.5–1.0 mg/l	Protozoans
Formalin	25 mg/l	Protozoans
Methylene blue	1.0 mg/l	Protozoans
Chlortetracycline	0.1–3 mg/l	Bacteria
Oxytetracycline	1–3 mg/l	Bacteria
Nitrofurazone	1–5 mg/l	Bacteria

Golden Shiner Reproduction

Golden shiners are used as bait in the size range of about 5 to 7 cm, but the adult attains approximately 25 cm and will live up to 8 years (Altman and Irwin, 1957). Spawning is initiated in the spring when the water temperature reaches about 18 C and continues into the summer or until the water is above approximately 30 C (Johnson and Davis, 1978).

A large female may produce as many as 10,000 eggs, which hatch in 4 to 7 days; however according to Meyer *et al.* (1973), many producers prefer to use 1 year old brood stock, since the ovaries of older fish are often largely destroyed by a protozoan parasite, *Pistophora ovariae*. Under natural conditions golden shiners spawn near areas with large amounts of vegetation or detritus, to which the adhesive fertilized eggs cling for incubation and hatching. Culturists often provide spawning mats for egg collection. The original spawning mats were made from Spanish moss spread between two layers of welded wire, but modern spawning mats are constructed of grasslike synthetic materials held by frames (Johnson and Davis, 1978). Each mat is approximately 1.0 to 1.3 m long and about 70 cm wide. Mats should be placed along the shoreline, with the deep end supported in a manner that will keep the mat horizontal (Meyer *et al.*, 1973).

Extensive culturists stock about 15 to 20 kg/ha of brood fish and may plant grass along the shorelines for natural spawning beds rather than utilizing spawning mats. Intensive culturists, on the other hand, utilize from 400 to 800 kg/ha of brood stock and usually rely on spawning mats (Meyer et al., 1973; Johnson and Davis, 1978).

Three types of production system are utilized for golden shiners (Johnson and Davis, 1978). In the first the eggs are transferred from the brood pond to grow-out ponds. Mats are checked 2 to 8 times daily and are removed from the brood ponds for incubation and hatching in grow-out ponds when they are laden with eggs. The second method is called the open or free spawning technique and involves the use of low stocking rates of brood stock (as mentioned above for extensively stocked ponds). Eggs are allowed to hatch and fry are grown to market size in the brood pond. The third method, fry transfer, consists of gathering fry from brood ponds in which open spawning occurred or from grow-out ponds that were intentionally stocked with excessive numbers of eggs by means of the egg transfer method. When fry reach a size that results in about 1800 fish per kilogram, they are collected with fine mesh seines, traps, or lift nets and moved into grow-out ponds.

Fry should be placed in grow-out ponds at densities of from about 375,000 to 625,000 per hectare. This is not accomplished by counting thousands of fish; rather, a sample of fry displacing a known volume of water (e.g., 100 ml) is counted and a conversion factor obtained for use on subsequent volumetric determinations. Growth of golden shiners is most rapid when they are stocked at about 190,000 per hectare, but that may not be an economical utilization of pond space. Slower, but acceptable growth is obtained within the range suggested above (Meyer et al., 1973).

Fathead Minnow Reproduction

Fathead minnows achieve a length of only about 10 cm, but they grow rapidly with a life span of 1 to 3 years (Meyer et al., 1973). Spawning occurs within the same temperature range as that for golden shiners (18 to 30 C). The female fathead minnow deposits eggs on the underside of nearly any type of solid material, including boards, plants, and rocks (Meyer et al., 1973). Johnson and Davis (1978) recommended the use of wooden soft drink cases because the partitions separate males and prevent fighting. We have used a large metal drum with holes cut in it as a fathead minnow spawning container with good success. Culturists often provide floating boards under which the fish can attach their eggs.

A female may lay from 200 to 500 eggs per spawn (Meyer et al., 1973) and may produce more than 4000 eggs during a single season. One fish was observed to spawn 12 times in a period of 11 weeks (Dobie et al., 1956). The eggs hatch in 4 to 7 days, depending on water temperature (Dobie et al., 1956; Meyer et al., 1973).

In the extensive method of culture, the eggs are allowed to remain in the brood ponds and hatch. In that case the adult fish are stocked at rates of 1250 to 5000 per hectare. If the adults are overstocked, the result may be overpopulation and stunting of the offspring. Since female fathead minnows are smaller than the males, brood fish stocking should not be done on the basis of size alone. The two sexes are easily distinguished during the breeding season, when the males become dark while the females remain silvery (Meyer et al., 1973).

Intensive culture involves the collection of young fish from the brood ponds, which had been stocked with approximately 50,000 adults per hectare in the ratio of five females for each male. A single hectare of water stocked in that fashion could produce up to 7.5 million young fathead minnows. Seines and lift nets are used for collection of the young. Those fish are subsequently stocked into fertilized ponds at densities of 250,000 to 750,000 per hectare (Meyer et al., 1973).

GOLDFISH

The goldfish (Carassius auratus), imported from Eurasia as an ornamental and aquarium fish (Dobie et al., 1956), is still extensively utilized for those purposes, in addition to its popularity in some areas as bait. General culture practices for goldfish are the same as those outlined for golden shiners and fathead minnows, but certain features of goldfish culture are somewhat unique.

Goldfish have been selectively bred for countless generations in the Orient. The history of selective breeding in this species probably surpasses that of nearly any other fish in culture. Goldfish vary in color from white to black, with many varieties featuring orange pigmentation. One United States producer supplies varieties with such names as comets, fantails, black moors, calico fantails, and shubunkins (Anonymous, 1972). Then there is the common variety, known to nearly everyone who has ever owned a fish bowl. Although goldfish seldom exceed about 10 cm in a home aquarium, in nature weights of more than 2 kg have been reported (Dobie et al., 1956).

Meyer et al. (1973) cautioned that the prospective goldfish producer should consider all the specialized aspects of the business before becoming involved—for example, the source of brood stock (the highly competitive business makes acquisition of good quality brood fish difficult and expensive), the need for many small ponds if several varieties are to be reared, and the need to evaluate special handling facilities, shipping costs, and proper market development, since all these factors differ to some extent from the requirements of other aquacultural ventures. In addition, Meyer et al. (1973) stressed that goldfish should be reared only in well water. Artesian spring water may also be successfully employed, provided it is of the proper temperature.

Goldfish normally begin spawning activity when the water temperature reaches about 15 C in the spring and continue to spawn through the summer if the fish do not become overcrowded (Dobie et al., 1956). Eggs, numbering 2000 to 4000 per female, are spawned on grass, roots, leaves, and similar objects, where they adhere and hatch in 6 to 7 days at 15 C (Dobie et al., 1956) and as soon as 48 hours at 30 C (Meyer et al., 1973). Most commercial producers utilize spawning mats of the type described earlier for bait minnows.

In preparation for spawning, brood fish that have been first treated for external parasites are stocked at the rate of 150 to 200 fish per newly filled brood pond. The brood ponds usually measure as small as about 8 x 8 m and as large as 15 x 15 m. These ponds should measure about 1.0 m at the deep end and should have a gently sloping bottom that provides approximately 70 cm of water at the shallow end. On the day the brood fish are stocked, 50 kg/ha of inorganic fertilizer (20-20-0) and 300 kg/ha of organic fertilizer should be added to stimulate a plankton bloom (Meyer et al., 1973). Goldfish consume zooplankton and aquatic insects as their natural foods. The use of organic fertilizer in conjunction with the inorganic treatment may be important in the stimulation of zooplankton production.

Goldfish spawn in cycles with intervals of 2 to 5 days. Spawning usually begins at dawn and continues until midmorning (Meyer et al., 1973). Most culturists remove egg-laden spawning mats after 9 A.M. on days when spawning occurs, replacing the mats with new ones. The mats can be carried from the brood ponds to the rearing ponds stacked on the bed of a truck; however on hot days they should be covered with wet burlap to prevent damage to the eggs. Egg-laden spawning mats are placed in about 5 to 25 cm of water near the shore of the rearing ponds, where they lie on the sloping bottom. This orientation should allow wave action to gently wash the eggs (Meyer et al., 1973), which apparently provides circulation around the developing embryos.

If the goldfish culturist wishes to produce bait-size fish, Meyer et al. (1973) recommended the placement of about 60 to 150 egg-laden spawning mats per hectare of rearing pond. If the producer is interested in rearing fish for the aquarium trade (thus requiring smaller individuals), between 675 and 800 egg-bearing spawning mats should be placed in each hectare of rearing pond.

Parasites and diseases affecting goldfish are similar to those discussed for minnows. Meyer *et al.* (1973) stressed the importance of treating brood stock for parasites and maintaining careful watch over all the fish for external parasites following the spawning season.

Information on feeding goldfish has been compiled by Meyer *et al.* (1973), who suggested that after the spawning season, brood stock should be fed a ration high in fish meal (twice the normal level) at 1 to 2% of body weight daily. Fry feeding is of a supplemental nature. A good fry rearing pond will have been fertilized with 20-20-0 inorganic fertilizer and with organic fertilizer at the rates presented for brood ponds. Commercial fish fry feeds (which must be very finely ground) are suitable as supplemental feed to the plankton in the ponds and should be fed at a rate of 1 kg/ha daily for 3 days, beginning immediately after hatching. The feed is distributed around the edges of the ponds. After the first 3 days of feeding, the rate is increased by 75 to 80 g daily. Feed should not be offered as a single meal but should be distributed several times daily. Small pellets may be used once the fish reach a length of 2.5 cm. Table 9.4

Table 9.4 Feed Types for Various Stages in the Life Cycle of Goldfish as Suggested by Meyer *et al.* (1973)

Life Stage	Feeding Recommendations
Brood stock	Commercial-type fish feed containing twice the normal amount of fish meal
Fry	Commercial-type fish feed with 38 to 40% protein; ground powder fine
Starter feed (small fingerlings)	Commercial-type fish feed containing 38 to 40% protein; pellets of small diameter, cut off short; 1 to 2 months
Grower feed (to market size)	Commercial-type fish feed with pellet size increased from that of starter feed; 30 to 32% protein is sufficient
Maintenance feed (holding ration)	Amount of animal protein is reduced to half of that of grower feed, with concomitant increase in carbohydrate level
Winter feed	Amount of alfalfa meal and animal protein increased an unspecified amount above those in the grower ration

outlines the feeding program suggested by Meyer *et al.* (1973) for goldfish of all ages.

CENTRARCHID CULTURE

Largemouth Bass

The largemouth bass (*Micropterus salmoides*) is not widely raised by commercial fish culturists in the United States, despite past and continuing interest by some private fish producers (Bardach *et al.*, 1972). The vast majority of largemouth bass now produced are hatched in public (federal or state) hatcheries, then distributed for stocking into private farm ponds and public waters of various descriptions.

United States hatcheries have been producing largemouth bass since the nineteenth century [reviewed by Snow (1975)], as documented in Table 9.2. Now as in the past, most hatcheries supply fingerlings of between 30 and 80 mm (Hutson, 1976b; Snow, 1975). Larger sizes (up to 20 cm) may be produced for corrective restocking of unbalanced fish ponds and for use as test animals (Snow, 1975).

Two basic production sytems for various species of centrarchid basses have evolved over the years. Both utilize earthen ponds, and their development has been described by Lydell (1904), Davis and Wiebe (1930), Langolis (1931), Howland (1932), Surber (1935), Wiebe (1935), Meehean (1936, 1939), Davis (1967), and others. The two methods, known as the extensive or spawning-rearing pond system and the intensive or fry transfer method, were described in detail most recently by Snow (1975), and most of the discussion below is based on that paper except when other authors are cited.

Extensive Culture Method. The extensive culture method for largemouth bass is seldom employed today unless the number of fingerlings required is very small. The method involves placement of brood fish into clean ponds at 25 to 100 per hectare in the spring. Fertilization is used if adequate natural zooplankton levels are not present. The adults spawn and the young fish are allowed to remain in the spawning pond with the brood fish until the fry grow to distribution size. Then they are collected by seining, trapping, pond drainage, or some combination of these techniques.

Problems associated with the extensive method of largemouth bass culture include the lack of control over the number of fry produced in the spawning ponds (except for the manipulation of numbers of brood fish stocked—and because of variability in fecundity that cannot be a very precise means of control) and the tendency of the brood fish to eat the fry and fingerlings.

Intensive Culture Methods. The intensive method provides the large-mouth bass culturist with much better control of all aspects of fry and fingerling production. Spawning ponds are specially prepared in the spring to receive brood fish, or the preparation may begin in the fall, when the empty ponds are disced several times over a period of time, then smoothed prior to the addition of water in the spring (Hutson, 1976b). Snow (1964) recommended a preflooding treatment of brood ponds with 11.2 kg/ha of 80% active ingredient simazine if the production of filamentous algae or rooted aquatic macrophytes has been a problem.

Prior to stocking in the brood ponds, adult bass are maintained in holding ponds. Bass begin to breed in Texas at 1 year of age, at which time they may weigh in excess of 200 g. The Texas Parks and Wildlife Department maintain brood stock in ponds at densities of 250 to 375 per hectare, feed them twice daily with chopped liver at a daily rate of 3% of body weight, and stock the holding ponds with golden shiners, fathead minnows, and goldfish as forage. Brood fish are generally utilized for four or five seasons, and the replacement rate is between 10 and 30% annually. Fish larger than about 2.5 kg become hard to handle and are usually replaced by younger fish (Hutson, 1976b).

The male largemouth bass prepares a shallow nest in the pond bottom in which the eggs are laid, fertilized, and incubated; when this procedure begins in the holding ponds, the spawning pond should be filled. Nest building activity usually begins when the water temperature reaches between 17 and 20 C.

When the water temperature is consistently above about 18 C, spawning will begin. As soon as eggs are found in the holding ponds, the adult bass should be transferred to brood ponds. Brood fish are stocked at about 100 to 250 per hectare, with a recommended ratio of two or three males for each female (Bishop, 1968). Sexing of the adults is relatively simple (Snow, 1963).

A female may spawn from one to five times during the 6 to 8 week spawning season and may produce between 5000 and 25,000 total fry. The male guards the eggs and fry for about 3 weeks. Fry hatch from 72 to 96 hours after spawning, depending on water temperature (Hutson, 1976b), and can first be seen on the nests 5 to 7 days following egg deposition. Once the fry have become free swimming, they can be trapped (Hutson, 1976b) or, since they remain in schools at that time, one or two persons capture them with nets. The treatment of ponds to eliminate unwanted vegetation, coupled with the lack of fertilization, often makes for clear water, and discovery and capture of fry are much less difficult than might otherwise be the case.

Rearing pond preparation should begin about the time eggs are first deposited in the spawning ponds. A pretreament with simazine (Snow, 1964) may be utilized as required. The rearing ponds should be filled 10 to 20 days before fry are stocked. These ponds are fertilized to produce zooplankton on which the fry will feed. Inorganic fertilizer has been used alone, or in combination with organic fertilizer. Liquid inorganic fertilization has also been effective (Hutson, 1976b). Snow (1975) described a technique whereby rye grass was planted in drained ponds during the fall and fertilized to support its growth. After flooding of the rearing ponds in the spring, the rye grass decomposed and provided a source of food for zooplankton.

If 40 to 80 mm fingerlings are desired, the fry may be stocked in rearing ponds at about 100,000 to 200,000 per hectare. Only a few weeks are required for fry to reach distribution size, at which time they can be captured by trapping, seining, pond draining, or a combination of those methods.

If larger fingerlings are desired, the density of fish in the rearing ponds should be reduced to about 25,000 to 30,000 per hectare. The time required before harvest and distribution will, of course, increase as compared with the production of smaller fish. Fertilization is practiced when large fingerlings are produced, but nutrient application in such cases is aimed at producing phytoplankton along with zooplankton (Chapter 3). Phytoplankton is utilized by the aquatic insects as food. Those insects in turn are preyed on by the bass fingerlings. The food habits of young largemouth bass in ponds have been studied by Rogers (1967). The most important items were found to be copepods, cladocerans, and midges. Copepods and cladocerans dominated the food of bass less than 15 mm long, whereas larger fish consumed mostly midges.

Cannibalism has been a problem for some culturists utilizing the intensive culture technique. Hutson (1976b) cited overstocking, stocking mixed sizes of fry, inadequate production of food organisms, and delayed harvest (leading to disparity in the size of fingerlings) as contributing to the problem.

Snow (1973, 1975) presented details of a system somewhat more intensive than that just described. This system utilizes either ponds or raceways as culture units, and the fish are supplied with prepared diets. Bass have not traditionally been reared on prepared feeds, but the Oregon Moist Pellet (OMP) has proved satisfactory (this diet is discussed in Chapter 5). Zooplankton are still produced to feed the early fry, but the OMP can be offered as soon as the fish are large enough to consume it. Training of bass to accept the OMP has not been too difficult.

The high intensity method of bass culture employs nylon spawning mats (Chastain and Snow, 1965), which are removed from the spawning ponds after use by the brood fish. The eggs are taken from the mats and incubated in hatching jars. The fry are maintained in rearing tanks and can be trained to accept prepared foods at the postlarval stage as indicated earlier. Alternatively, the fry are stocked in rearing ponds provided with natural food through fertilization and prepared diets are introduced later. *Disease, Parasite, and Predator Control.* Snow (1975) suggested routine treatment of largemouth bass brood stock for external parasites with 25 mg/l of formalin in ponds, plus medicated feed when prepared diets are utilized. Bass are subject to various diseases and parasites, and significant losses, particularly of fry and fingerlings, have occurred. Davis (1967) cited incidences of myxobacteria and such parasites as *Ichthyophthirius, Costia, Trichodina, Trichophyra,* and *Scyphidia.* Treatments may be similar to those of catfish and other species (Chapter 7).

Predaceous insects can be a significant problem for bass fry and young fingerlings. The backswimmer (*Notonecta*) is a particularly troublesome predator (Davis, 1967). Diesel fuel applied to the water produces a thin film that will kill air-breathing insects (including the backswimmer and several other important predators) very quickly, almost on contact. Fish are not damaged by this treatment, and the diesel fuel dissipates within a day or two. Meehean (1937) recommended the use of kerosene at about 100 to 120 l/ha, and a similar concentration of diesel fuel should be appropriate. The liquid should be added at the upwind end of the pond and will quickly spread to cover the pond surface with a thin film.

Sunfish Production

Davis (1967) listed the bluegill (*Lepomis macrochirus*), redear sunfish (*L. microlophus*), and green sunfish (*L. cyanellus*) as the most widely cultured sunfish species. The green sunfish is generally not stocked (except as a hybrid, as noted below), since it is not highly regarded by fishery managers in most parts of the United States. Green sunfish, in regions where they occur, seem to appear naturally in ponds and lakes, regardless of whether they were stocked into those waters. Texas is somewhat unique in that the Texas Parks and Wildlife Department provides hybrid sunfish for stocking, whereas most states utilize nonhybrid fishes.

Crosses between green sunfish males and redear sunfish females have been shown to produce about 80% males that grow rapidly and maintain the high ratio of males to females after their release and subsequent reproduction in the wild (Hutson, 1976a). Additional studies on this and other crosses are required to help solve the problems of sunfish overpopulation and stunting, which often lead to imbalance in bass-bluegill ponds. Although Texas is perhaps the only state utilizing hybrid sunfish extensively, the approach is not new. Ricker (1948) was the first to produce hybrid sunfish when he crossed bluegills with redear sunfish. That hybrid was produced for the same reason cited today; that is, to reduce overpopulation in ponds.

Bluegills can be used as a model for description of sunfish culture, which in many respects is the same as that of the culture of bass under the extensive system. Bluegill females initially spawn at the age of 1 year and may spawn more than once during a season (Davis, 1967). Sunfish nest in the manner of largemouth bass, and it has been reported that as many as 17,000 fry can be found per nest (Carbine, 1939).

Spawning ponds are often stocked with an equal number of males and females, although better results may be obtained with two males for every three females (Davis, 1967). Traditionally, the young are allowed to remain in the brood ponds with the adults for rearing. Blosz (1948) recommended a stocking rate of about 250 brood fish per hectare, which could be expected to produce up to 375,000 fry. Ponds should be fertilized to provide food for the fry. Bluegills and other sunfishes will accept prepared foods. When present in channel catfish rearing ponds, sunfish can often be observed competing for feed.

Diseases and parasites, and their treatment, are similar for sunfishes and largemouth bass. As in the case of all fish diseases and parasites, early detection and treatment are of primary importance.

STRIPED BASS

The striped bass (*Morone saxatilis*) represents a somewhat unique warmwater culture species in that it is an anadromous fish that has been introduced into landlocked freshwater lakes as a sport fish (reproducing populations have been uncovered in five lakes in four states, and more than 75 lakes in several states have been stocked). It is also highly prized as a marine sport fish. The native range of striped bass was along the Atlantic coast and in the Gulf of Mexico from Florida to Louisiana (Bonn et al., 1976), but it has been extended through the stocking programs of various states, and a few individuals have been taken in the marine waters of Texas (Hoese and Moore, 1977).

The inclusion of striped bass in this discussion could not have been very complete as recently as a decade ago. Impetus in striped bass culture increased greatly after the development of induced spawning techniques by Stevens (1966). At present, numerous state and federal hatcheries are producing large numbers of striped bass fingerlings for stocking. In 1967 the number produced was only 978,000. This increased to 13.1 million by 1975 (Bonn et al., 1976).

Production of striped bass is based on capturing wild brood stock, spawning them through hormone injection, hatching the eggs in captivity, and producing distribution-sized fingerlings in ponds. Some of the details of these techniques, as outlined below, are based on the work of Bonn *et al.* (1976); other suitable references are included as pertinent.

Brood Stock Capture and Handling

Brood stock must be captured as sexually mature adults during the limited period when the fish are nearly ready to spawn. Extreme care must be taken in handling the brood stock to ensure against damage that might prevent them from producing viable gametes.

When striped bass concentrate on the spawning grounds, they can be most efficiently collected by electrofishing. Other methods that have been employed include gill netting (nets should be checked and fish removed at 15 minute intervals), bow nets (resembling oversized dip nets), trap nets, pound nets, fyke nets, hoop nets, and hook and line fishing.

Once captured, the fish should be placed immediately in a large, aerated transportation tank and taken to the hatchery, where they can be held in live boxes or holding nets. Transportation tank water should contain 3 to 10°/oo salinity (obtained through the use of sodium chloride) and may be treated with 100 mg/l nitrofurazone to prevent infection of the adult animals during hauling. It has been recommended that no more than two or three 6.8 kg brood fish be carried in a 320 l capacity transportation tank.

Intramuscular injections of hormones are given to all brood fish before spawning, generally below the dorsal fin. These injections should be given at the time of capture. Only human chorionic gonadotropin (HCG) and follicle-stimulating hormone (FSH) have been used successfully. Females should receive doses of from 275 to 330 IU of HCG per kilogram of body weight in a single injection. Multiple injections (as are sometimes used with channel catfish) lead to premature egg expulsion (Bayless, 1972). Unlike channel catfish and other species (Chapter 6), male striped bass are also injected. Males should receive a dosage of 110 to 165 IU of HCG per kilogram of body weight (Bishop, 1974).

Spawning Procedure

If the brood fish are not injected with hormones at the time of capture, they should be injected as soon as possible after reaching the hatchery. Between 28 and 30 hours after injection, an egg sample should be obtained from each female by inserting a catheter of 3 mm outside diameter approximately 5 cm into the urogenital pore. The catheter should not enter the egg mass, or tissue destruction will result. The catheter will collect a few eggs, which can then be examined under a dissecting microscope. The stage of egg development will indicate to the aquaculturist how many hours must pass before ovulation occurs. Bonn *et al.* (1976) provided a series of photographs to assist hatchery personnel in making this determination. The females must be stripped at the proper time to obtain successful fertilization of the eggs. If the culturist waits too long (perhaps as little as 60 minutes following ovulation), a low hatch percentage will be obtained. Manual palpitation of the abdomen of the female will cause the expulsion of eggs from the vent when ovulation has begun.

Spawning is usually accomplished through the combined efforts of several people. Prior to stripping, the female should be anesthetized by spraying a 1 mg/l solution of quinaldine on the gills. Once ovulation has been verified, the female is held by two workers while a third applies pressure to the abdomen to force the eggs out and into a spawning pan. A male (usually smaller than the female and capable of being handled by a single person) is similarly stripped of milt, which is also allowed to flow into the pan. It is advisable to utilize two males for every female stripped to ensure good fertilization. Milt should be added within 2 minutes of egg expulsion. After the milt has been added, water is introduced into the pan and the mixture is stirred to ensure that the sperm are placed in close approximation with all the available eggs.

Within the normal water temperature range for spawning (about 15 to 20 C), microscopic examination of the eggs should reveal that the embryos are in the two- to four-cell stage of development about 2 hours after egg fertilization. The percentage of developing eggs can also be detemined at that time.

Hatching

Hatching may follow the hatching jar procedure first outlined by Worth (1884). Each 2 l jar is supplied with about 100,000 eggs, and water is added continuously at the rate of 40 to 120 ml/minute. The jars are designed to accommodate an upward current of water. Dead eggs become buoyant and are washed from the jars in the current. Well water is recommended for hatching. Under optimum conditions the water is between 16 to 18 C, has a pH between 7.5 and 8.5, is saturated in dissolved oxygen, contains no more than 10 mg/l of carbon dioxide, and has a total ammonia level of less than 0.5 mg/l.

Bayless (1972) reviewed the literature on incubation times and reported that between 16 and 18 C striped bass eggs will hatch in 48 to 56 hours (with the time required increasing as temperature decreases). Hatching success in the neighborhood of 50% is considered good.

A tank spawning method has been described by Bishop (1974). Brood fish are injected with HCG as described previously, and at least two males are put in a tank with a single female. The fish may spawn in the tank if left undisturbed. Eggs can be allowed to hatch in the tank, or they may be siphoned into aquaria for hatching.

Fry Holding and Feeding

As sac fry (up to 4 days after hatching), striped bass do not eat and can be held at high densities in aquaria. They should be kept suspended in the water column by continuously flowing a current of water through the tanks. After 4 days posthatching the density should be reduced from a maximum of 2500 fry per liter to about 250 fry per liter. At this time the fry may be placed in troughs and maintained for an additional 9 to 15 days before stocking in ponds.

Striped bass fry will begin to accept feed as early as 5 days after hatching—the time required for the mouth parts to become fully developed. *Artemia salina* nauplii are usually provided as the initial food for striped bass fry (details on the culture of *A. salina* are presented in Chapter 5). The fish should be offered brine shrimp nauplii at 4 to 8 hour intervals.

Pond Culture

Many culturists advise drying ponds through the winter and may disc them once or more times, to expose organic material on the bottom for oxidation. Prior to stocking, rye grass may be planted in the manner and for the purpose described for largemouth bass fry (Snow, 1975). Lime may also be utilized in dry, unplanted pond bottoms to sterilize the soil. Lime should be applied at the rate of 1000 kg/ha. When the possibility of the development of filamentous algae or rooted aquatic macrophytes exists, the dry pond bottoms can be treated with 10 kg/ha of simazine or 4 l/ha of 2,4-D herbicide.

Ponds should be filled just before stocking when possible. Well water is preferred, as in nearly all strategies of warmwater aquaculture. Ponds may be fertilized at the time of filling. Striped bass culturists do not feel that it is necessary to prepare the pond well in advance of introducing the fish. Fertilization is aimed at producing a zooplankton bloom, as was true in the culture of largemouth bass. Organic fertilization is often utilized, with or without inorganic fertilization.

Such organic materials as hay at 500 to 1000 kg/ha; alfalfa, bermuda, or milo pellets or meal at 200 to 500 kg/ha alone or at 100 to 400 kg/ha in combination with hay; soybean and cottonseed meals at 250 kg/ha; or animal manure at 500 kg/ha are among the treatments that have been successfully employed.

Inorganic fertilization alone should not be relied on for the production of plankton blooms in striped bass ponds, since inorganic fertilizers will stimulate phytoplankton to excess in the opinion of striper culturists. Superphosphate (0-20-0) or triple superphosphate (0-46-0) is sometimes used in conjunction with organic fertilization at rates of 10 to 50 kg/ha.

Before stocking the striped bass fry, ponds should be checked to ensure that there is an abundance of small zooplanktonic animals, that the water temperature exceeds 21 C during daytime, that pH is in the range 6.5 to 9.5, and that predatory insects are not abundant. When the culturist is satisfied that all these conditions have been met, fry may be stocked. This is usually done during early morning, and the fish should be tempered to the temperature of the receiving ponds.

Fry may be stocked at densities of from 125,000 to 500,000 per hectare. The optimum stocking density is considered to be about 200,000 fry per hectare. The fish remain in the rearing ponds until they are distributed to lakes.

Intensive Culture Techniques

High intensity culture of striped bass in tanks and raceways is still in the developmental stage, but a few things have been learned about the technique. As in the case of fry that are ultimately placed in rearing ponds, brine shrimp are the first food offered. When the fish are 14 to 21 days old, however, prepared diets in the form of particles about the size of brine shrimp nauplii are offered, in combination with *Artemia salina*. This is continued until about 28 days after hatching, when the live food can be discontinued.

Once the fish are readily accepting prepared food, the frequency of feeding should be increased from the every 3 hours or so that brine shrimp were fed. Some culturists utilize automatic feeders that drop in a known amount of feed at intervals as short as 10 minutes. When the fish reach 45 days of age, food should be offered only three times daily. High quality trout and salmon diets provide adequate nutrition for striped bass fry. The feed should contain no less than 38% protein.

Diseases and Parasites

Four genera of bacteria are commonly associated with striped bass. These are *Flexibacter, Aeromonas, Pseudomonas,* and *Vibrio.* When water hardness is greater than 50 mg/l, copper sulfate can be utilized at 0.5 to

1.0 mg/l to combat columnaris disease. Lower dosages are adequate in soft water. Diquat at 2 mg/l cation concentration can be used as a bacteriocide, and there is an indication that potassium permanganate at 3 mg/l is helpful in the control of bacterial infections.

The antibiotic nitrofurazone is effective against *Flexibacter, Aeromonas,* and *Pseudomonas* when used as a 1 to 6 hour bath at a concentration of 100 mg/l. Furanace may be utilized for 1 hour at 7.5 mg/l or as a several day bath at concentrations of from 0.05 to 0.1 mg/l. High concentrations of Furanace may be lethal to young striped bass. Oxytetracycline is effective on hemorrhagic septicemia when fed at the rate of 8.8 g of active ingredient per kilogram of feed.

Ectoparasites of the types seen on many of the other fishes discussed also attack striped bass. *Trichodina* can be treated effectively with 3 mg/l of potassium permanganate, although a second treatment may be required in ponds that have high levels of organic material (Hughes, 1975). Copper sulfate has been added to ponds at concentrations of 0.5 to 1.0 mg/l, to control ectoparasites. Dylox as 0.25 mg/l active ingredient seems to be effective against both *Ergasilus* and *Lernaea.* The acanthocephalan *Pomphorhynchus rocci* can be controlled by feeding di-*N*-butyl tin oxide at the rate of 220 mg/kg of fish for 5 days.

Formalin at 150 mg/l as a dip or 1 hour bath has been utilized to control protozoan parasites. Fish with bacterial infections should not be treated with formalin, since the additional stress may prove to be lethal. Malachite green has been administered at 66 mg/l for 10 seconds in dip treatments to control the fungus *Saprolegnia.*

Chemicals that have not been cleared for food fish are, in general, suitable for use on striped bass and the other species discussed in this chapter. Malachite green may be a notable exception, since in 1978 the U.S. government notified all its laboratories to discard existing supplies and discontinue the use of that chemical. In any event the array of chemicals that can be used on nonfood fishes is larger than that approved for fish that go directly for human consumption.

LITERATURE CITED

Adron, J. W., A. Blair, and C. B. Cowey. 1974. Rearing of plaice (*Pleuronectes platessa*) larvae to metamorphosis using an artificial diet. *Fish. Bull.* 72: 353–357.

Allan, P. F. 1952. *How to grow minnows.* Privately published by P. F. Allan, Fort Worth, Tex. 63 p.

Altman, R. W., and W. H. Irwin. 1957. *Minnow farming in the southwest.* Oklahoma Department of Wildlife Conservation, Oklahoma City, 35 p.

American Fisheries Society. 1970. *A list of common and scientific names of fishes.* American Fisheries Society, Special Publication 6, Washington, D.C. 150 p.

Anonymous, 1972. Ozark's goldfish breed in artificial nests. *Am. Fish Farmer,* February. 5 ff.

Bailey, W. M., M. D. Gibson, S. H. Newton, J. M. Martin, and D. L. Gray. 1976. Status of commercial aquaculture in Arkansas in 1975. *Proc. Southeast. Assoc. Fish Wildl. Agencies,* **30:** 246–250.

Bardach, J. E., J. H. Ryther, and W. O. McLarney. 1972. *Aquaculture.* Wiley-Interscience, New York. 868 p.

Bayless, J. D. 1972. *Artificial propagation and hybridization of striped bass, Morone saxatilis (Walbaum).* South Carolina Wildlife and Marine Resources Department, Special Report. 135 p.

Bishop, H. 1968. Largemouth bass culture in the southwest. In *Proceedings of the North Central fish culture workshop, Ames, Iowa,* pp. 24–27.

Bishop, R. D. 1974. The use of circular tanks for spawning striped bass, *Morone saxatilis. Proc. Southeast. Assoc. Game Fish Comm.* **28:** 35–44.

Blosz, J. 1948. Fish production program, 1947, in the southeast. *Prog. Fish-Cult.* **10:** 84–87.

Bonn, E. W., W. M. Bailey, J. D. Bayless, K. E. Erickson, and R. E. Stevens. 1976. *Guidelines for striped bass culture.* Southern Division, American Fisheries Society, Washington, D.C. 103 p.

Brice, J. J. 1898. *Report of the Commissioner.* U.S. Commission of Fish and Fisheries, Washington, D.C. 340 p.

Carbine, W. F. 1939. Observations on the spawning habits of centrarchid fishes in a deep lake, Oakland County, Michigan. *Trans. North Am. Wildl. Conf.* **4:** 275–287.

Chastain, G. A., and J. R. Snow. 1965. Nylon mats as spawning sites for largemouth bass, *Micropterus salmoides. Proc. Southeast. Assoc. Game Fish Comm.* **19:** 405–408.

Davis, H. S. 1967. *Culture and diseases of game fishes.* University of California Press, Berkeley. 332 p.

Davis, H. S., and A. H. Weibe. 1930. *Experiments in the culture of black bass and other pond fish.* U.S. Bureau of Fisheries, Bureau of Fisheries Document 1085, Washington, D.C.

Dobie, J. R., O. L. Meehean, S. F. Snieszko, and G. N. Washburn. 1956. *Raising bait fishes.* U.S. Fish and Wildlife Circular 35. 124 p.

Guidice, J. J. 1968. The culture of bait fishes. In *Proceedings of the commercial bait fish conference.* Texas A&M University, College Station, pp. 13–18.

Hoese, H. D., and R. H. Moore. 1977. *Fishes of the Gulf of Mexico.* Texas A&M University Press, College Station, 327 p.

Howland, J. W. 1932. Experiments in the propagation of spotted black bass. *Trans. Am. Fish. Soc.* **62:** 185–188.

Hughes, J. S. 1975. *Striped bass, Morone saxatilis (Walbaum), culture investigations in Louisiana with notes on sensitivity of fry and fingerlings to various chemicals.* Louisiana Wildlife and Fisheries Bulletin 13. 46 p.

Huner, J. V. 1975. The biological feasibility of raising bait-sized red swamp crawfish, *Procambarus clarkii* (Girard), in Louisiana. Ph.D. dissertation, Louisiana State University, Baton Rouge. 184 p.

Huner, J. V. 1976. Raising crawfish for food and fish bait: A new polyculture crop with fish. *Fisheries,* **1:** 7–9.

Huner, J. V., and J. W. Avault, Jr. 1976. Sequential pond flooding: A prospective management technique for extending production of bait-sized crawfish. *Trans. Am. Fish. Soc.* **105:** 637–642.

Hutson, P. L. 1976a. Hybrid sunfish. In *Proceedings of the fish farming conference and annual convention of the Catfish Farmers of Texas.* Texas A&M University, College Station, p. 32.

Hutson, P. L. 1976b. Largemouth bass culture. In *Proceedings of the fish farming conference and annual convention of the Catfish Farmers of Texas.* Texas A&M University, College Station, pp. 27–31.

Johnson, S. K. 1978. *Maintaining minnows—A guide for retailers.* Texas Agricultural Extension Service Publication MP-1320. Texas A&M University, College Station. 19 p.

Johnson, S. K., and J. T. Davis. 1978. *Raising minnows.* Texas Agricultural Extension Service Publication MP-783 (revised). Texas A&M University, College Station. 15 p.

Langolis, T. H. 1931. The problem of efficient management of hatcheries used in the production of pond fishes. *Trans. Am. Fish. Soc.* **61:** 106–115.

Lydell, D. 1904. *The habits and culture of black bass.* U.S. Fish Commission, Bulletin 22, pp. 39–44.

Meehean, O. L. 1936. Some factors controlling largemouth bass production. *Prog. Fish-Cult.* (old series), **16:** 1–7.

Meehean, O. L. 1937. Control of predaceous insects and larvae in ponds. *Prog. Fish-Cult.* (old series), **33:** 15–16.

Meehean, O. L. 1939. A method for the production of largemouth bass on natural food in fertilized ponds. *Prog. Fish-Cult.* (old series), **47:** 1–19.

Meyer, F. P., K. E. Sneed, and P. T. Eschmeyer. 1973. *Second report to the fish farmers.* U.S. Bureau of Sport Fish and Wildlife, Resources Publication 113. 123 p.

Prather, E. E., J. R. Fielding, M. C. Johnson, and H. S. Swingle. 1953. *Production of bait minnows in the southeast.* Alabama Agricultural Experiment Station, Auburn University, Auburn, Ala. 71 p.

Ricker, W. E. 1948. Hybrid sunfish for stocking small ponds. *Trans. Am. Fish. Soc.* **75:** 84–95.

Rogers, W. A. 1967. Food habits of young largemouth bass (*Micropterus salmoides*) in hatchery ponds. *Proc. Southeast. Assoc. Game Fish Comm.* **21:** 543–553.

Shelbourne, J. E. 1964. The artificial propagation of marine fish. In F. S. Russell (Ed.), *Advances in marine biology,* Vol. 2. Academic Press, New York, pp. 1–83.

Smith, E. R. 1968. Minnow pond construction and water quality. In *Proceedings of the commercial bait fish conference.* Texas A&M University, College Station, pp. 7–11.

Snow, J. R. 1963. A method of distinguishing male bass at spawning time. *Prog. Fish-Cult.* **25:** 49.

Snow, J. R. 1964. Simazine as a preflooding treatment for weed control in hatching ponds. *Proc. Southeast. Assoc. Game Fish Comm.* **18:** 441–447.

Snow, J. R. 1973. Controlled culture of largemouth bass fry. *Proc. Southeast. Assoc. Game Fish Comm.* **26:** 392–398.

Snow, J. R. 1975. Hatching propagation of the black bass. In H. E. Clapper (Ed.), *Black bass biology and management.* Sport Fishing Institute, Washington, D.C., pp. 344–356.

Stevens, R. E. 1966. Hormone-induced spawning of striped bass for reservoir stocking. *Prog. Fish-Cult.* **28:** 19–28.

Stickney, R. R., D. B. White, and D. Perlmutter. 1973. Growth of sea turtles on natural and artificial diets. *Bull. Ga. Acad. Sci.* **31:** 37–44.

Surber, E. W. 1935. The production of bass fry. *Prog. Fish-Cult.* (old series), **8:** 1–7.

Wiebe, A. H. 1935. *The pond culture of black bass.* Texas Game, Fish and Oyster Commission, Bulletin 8. 58 p.

Worth, S. G. 1884. Report on the propagation of striped bass at Weldon, N.C. in the spring of 1884. *Bull. U.S. Fish Comm.* **4:** 225–230.

APPENDIX 1

Conversion Factors for Units of Weight and Measurement in the English and Metric Systems (Capitalized Units are Those Most Often Utilized by Aquaculturists)

Unit of Measurement	Conversion Factor[a]
ACRE-FOOT	43,560 cubic feet
	1233.5 cubic meters
ACRE	43,450 square feet
	0.4 hectare
Angstrom (Å)	10^{-10} meter
CENTIMETER (cm)	0.39 inch
	0.01 meter
	10 millimeters
Cubic centimeter (cm^3)	0.06 cubic inch
	1 milliliter
CUBIC FOOT (ft^3)	7.48 gallons (U.S. liquid)
	0.03 cubic meter
	28.32 liters
Cubic inch (in^3)	16.39 cubic centimeters
CUBIC METER (m^3)	35.3 cubic feet
	264.2 gallons (U.S. liquid)
	1.3 cubic yards
	1000 liters
FOOT (ft)	0.3 meter
	12 inches
GALLON (U.S. liquid) (gal)	231 cubic inches
	0.13 cubic foot
	3.8 liters
Gram (g)	0.04 ounce (avoir.)
HECTARE (ha)	2.47 acres
	10,000 square meters
INCH (in)	2.54 centimeters
KILOGRAM (kg)	2.2 pounds (avoir.)
	1000 grams

Unit of Measurement	Conversion Factor[a]
KILOMETER (km)	0.6 mile
	1000 meters
LITER (l)	0.26 gallon (U.S. liquid)
	1.06 quarts (U.S. liquid)
METER (m)	39.37 inches
	1000 millimeters
	100 centimeters
Micron (μ)	10^{-6} meter
MILLIMETER (mm)	0.04 inch
	0.1 centimeter
MILE (statute) (mi)	5280 feet
	1.6 kilometers
Mile (nautical)	6080 feet (U.S. Navy)
OUNCE (U.S. fluid) (oz)	1.8 cubic inches
	29.57 cubic centimeters
Ounce (apoth.) (oz)	31.1 grams
Ounce (avoir.) (oz)	28.47 milliliters
Pint (U.S. liquid) (pt)	0.47 liter
	473.2 cubic centimeters
POUND (avoir.) (lb or p)	453.6 grams
Pound (apoth.) (lb or p)	373.2 grams
Quart (U.S. dry) (qt)	1.1 liters
QUART (liquid) (qt)	0.9 liter
Square centimeter (cm^2)	0.16 square inch
	100 square millimeters
Square foot (ft^2)	0.09 square meter
Square inch (in^2)	645.2 square millimeters
SQUARE METER (m^2)	10.8 square feet
Square yard (yd^2)	0.8 square meter
TON (metric) (t)	1000 kilograms
	2205 pounds
Ton (short) (t)	2000 pounds
	907.2 kilograms
Ton (long) (t)	2240 pounds
	1016 kilograms
Yard (yd)	0.91 meter
	3 feet

[a]All values have been rounded to the nearest 0.1 or 0.01 unit. This is generally adequate for aquacultural conversions; however more precise conversion factors may be required under certain circumstances.

APPENDIX 2

Equivalent Temperatures Between 0 and 40 C on the Celsius and Fahrenheit Scales[a]

Celsius	Fahrenheit	Celsius	Fahrenheit
0	32	20	68
1	33.8	21	69.8
2	35.6	22	71.6
3	37.4	23	73.4
4	39.2	24	75.2
5	41	25	77
6	42.8	26	78.8
7	44.6	27	80.6
8	46.4	28	82.5
9	48.2	29	84.2
10	50	30	86
11	51.8	31	87.8
12	53.6	32	89.6
13	55.4	33	91.4
14	57.2	34	93.2
15	59	35	95
16	60.8	36	96.8
17	62.6	37	98.6
18	64.4	38	100.4
19	66.2	39	102.2
		40	104

[a] Formulas for conversions between Celsius and Fahrenheit are as follows: $C = \frac{5}{9} (F - 32)$; $F = \frac{9}{5} (C + 32)$.

GLOSSARY

AD LIBITUM   With respect to the feeding of aquaculture animals, the presentation of feed until satiation is reached, after which feeding is discontinued.

ADENOSINE DIPHOSPHATE (ADP)   A high energy phosphate compound in living cells.

ADENOSINE TRIPHOSPHATE (ATP)   A high energy phosphate compound in living cells.

AEROBIC   A condition in which the environment contains free oxygen.

ALGAE   Members of the artificial plant grouping Thalophyta (plants lacking true stems, leaves, and roots) which possess chlorophyll. Include almost all sea weeds. See also filamentous algae.

ALKALINITY   The capacity of water for neutralizing strong acids; or, the total amount of carbonate and bicarbonate present in the water.

AMBIENT   The environment surrounding an organism. The ambient environment is not affected by the presence of the organism.

AMINO ACID   One of a group of nitrogenous organic acids that serves as one of the structural units of protein.

ANABOLISM   The assimilation and storage of food energy in protein, lipid, or carbohydrate form.

ANADROMOUS   Describing aquatic animals such as salmon and certain members of the family Clupeidae, which mature and live most of their lives in the sea, but return to fresh water to spawn.

ANAEROBIC   A condition in which the environment contains no free oxygen.

AQUACULTURE   The rearing of aquatic organisms under controlled or semicontrolled conditions.

AUFWUCHS   Aquatic organisms that are found attached to substrates but not penetrating them.

AUTOMATIC FEEDER   A device that dispenses feed particles at preselected times. Most automatic feeding devices are electrically operated.

AUTOTROPH   An organism that manufactures its own food from inorganic constituents, often through the use of energy obtained from light.

BENTHIC Describing the portion of the aquatic environment inhabited by organisms that live on or in the sediments.

BENTHOS Organisms that live on or in the sediments in aquatic environments.

BENTONITE An expandable clay often used to seal ponds.

BICARBONATE ALKALINITY The portion of total alkalinity contributed by bicarbonate ions.

BLOOM A phenomenon associated with phytoplankton in which a rapid increase in the population of one or more species occurs, allowing those species to dominate the community.

BIOCHEMICAL OXYGEN DEMAND (BOD). The decrease in the dissolved oxygen of a water sample incubated in the dark under certain conditions over a specified period. Often erroneously referred to as the biological oxygen demand.

BIOFILTER The component of closed recirculating water systems in which the removal or detoxification of certain dissolved compounds occurs as a result of microbiological activity. The most important reaction is nitrification of ammonia to nitrate.

BIOMASS The total weight of the organisms contained in a sample or expressed on the basis of the weight of organic material present per unit area or volume.

BRINE Water of a salinity higher than about 40 parts per thousand.

CAGE CULTURE Culture proceeding in chambers generally constructed of wire or netting around rigid frames, floated or suspended in large water bodies such as rivers, lakes, or bays.

CARBOHYDRATE One of the three energy-containing classes of food, carbohydrates contain carbon, hydrogen, and oxygen in the ratio $C:H:O = 1:2:1$. Examples are sugars and starch.

CARBONATE ALKALINITY The portion of total alkalinity contributed by carbonate ions.

CARNIVORE An animal that feeds exclusively on the tissues of other animals.

CATABOLISM The transformation of energy stored in protein, lipid, or carbohydrate form into free energy.

CATADROMOUS Describing aquatic animals such as the American eel, which mature and live most of their lives in fresh water but return to the sea to spawn.

CHLORINITY The weight of chlorine in grams contained in 1 kg of seawater after the bromides and iodides have been replaced by chloride, or 0.3285233 times the weight of silver equivalent to all the halides in a seawater sample. Chlorinity is determined by titration as one means of ascertaining the salinity of a water sample.

CHLOROPHYLL   Green plant pigment that can be separated into three types through spectral analysis. Chlorophyll is important in photosynthesis.

CHROMOSOME   The portion of a cell nucleus that carries the inheritance factors (genes).

CLOSED SYSTEM   The type of aquaculture water system wherein the water is conserved throughout most or all of the growing season. In most instances the water is recirculated through a culture chamber, a primary settling chamber, a biofilter, and a secondary settling chamber on each pass through the system.

COMMENSAL   A type of symbiotic relationship in which one species gains benefit while the other is neither harmed nor benefited.

COMMUNITY   The assemblage of organisms, both plant and animal, inhabiting a natural physiographic area.

COMPENSATION DEPTH   The depth of water at which photosynthetic oxygen production is balanced by respiratory uptake. The compensation depth is generally assumed to correlate with the depth at which light is 1% of its incident intensity.

COMPLETE DIET   An aquaculture diet that satisfies all the nutritional requirements of the target species.

CULTCH   Any substrate placed into the environment to attract the attachment of oyster larvae when they leave the plankton community and become benthic.

CULTURE CHAMBER   Any vessel utilized to hold and grow aquaculture organisms. Some examples are tanks, cages, silos, ponds, and raceways.

DEMAND FEEDER   A device that dispenses small amounts of prepared feed when activated by the aquaculture animals.

DEMERSAL   Organisms (generally fishes and/or their eggs) that exist on or near the bottom in aquatic environments.

DENITRIFICATION   The chemical reduction of nitrate to elemental nitrogen through a nitrite intermediate, by certain types of microorganisms.

DETRITUS   Finely divided settleable material of organic and inorganic origin, which is found suspended in the water column.

DIATOMS   Members of the algae class Bacillariophyceae. This group is characterized by single-celled plants with overlapping valves impregnated with silica.

DIGESTIBILITY   The degree to which a foodstuff is broken down into particles or forms suitable for absorption following ingestion.

DIPLOID   The somatic number of chromosomes, equal to twice the number of chromosomes found in the gamete of a species.

DIOECIOUS   Organisms in which the sexes are distinct.

DISSOLVED OXYGEN (DO)   The amount of elemental oxygen present in a solution. In the aquatic environment the DO level is affected by temperature, salinity, and altitude and is largely controlled by photosynthesis and respiration.

DIURNAL   Daily; especially in reference to actions that are completed within 24 hours or recur every 24 hours.

DYSTROPHIC   Describing a lake that has high levels of organic acids, low calcium carbonate, and low nutrients. Dystrophy is one of the final stages in lake succession.

ECDYSIS   Molting; the shedding of the exoskeleton to allow for growth in crustaceans.

ENVIRONMENT   The total of all internal and external conditions that may affect an organism or community of organisms.

EPILIMNION   The upper, mixed layer of a stratified body of water.

EPIZOOTIC   An outbreak or epidemic of a disease or parasite in a population of organisms.

ESTUARY   A semienclosed coastal water body with free connection with the open sea, in which seawater is diluted to some degree by fresh water.

EURYHALINE   Organisms that can adapt to wide variations in salinity.

EUTROPHIC   Describing lakes that contain abundant levels of nutrients, resulting in high levels of organic production. Eutrophication is an intermediate stage in lake succession and can be accelerated by the activities of humans.

EXOPHTHALMIA   A condition in which the eyeballs begin to protrude from the head (usually seen in fish) as a result of the accumulation of fluid or gases at the back of the eye socket. Exophthalmia ("popeye") is a symptom of gas-bubble disease and may be caused by certain other afflictions.

EXTENSIVE CULTURE   Low intensity aquaculture such as is practiced in ponds by subsistence culturists. Extensive culture is characterized by large water areas in which low densities of culture animals are maintained, controlled to a limited extent by the culturist.

EXTRUSION   The process by which diets are prepared by passing the ingredients through a die under high pressure and temperature. When the feed leaves the die it expands, and a pellet that will float is formed.

FATTY ACID   An organic acid containing carbon, hydrogen, and oxygen, which is a member of the class of foods known as lipids.

FECUNDITY   The number of eggs produced annually by a female animal or per unit body weight of a female.

FEE FISHING   An enterprise in which catchable organisms are stocked into a pond or lake and customers pay for the privilege of fishing. Payment may be by the animals caught, by the hour, or by the weight of the catch.

FILAMENTOUS ALGAE   Species of algae in which individual cells are connected in long strings. Such algae often lead to problems when they occur in culture chambers.

FILTER MEDIUM   A substrate in a mechanical or biological filter that helps in water treatment, either by trapping suspended particles or by providing a surface for the attachment of beneficial microorganisms.

FINGERLING   A fish larger than a fry but not of marketable size. Though not rigidly defined, fingerling fishes are generally between about 2 and 25 cm long.

FISH PROTEIN CONCENTRATE (FPC)   A high protein flour prepared from dried ground whole fish. FPC was originally manufactured as a means of providing an inexpensive, complete protein source for humans.

FLASHING   The action often exhibited by parasitized fishes when they swim erratically and expose the sides or abdomen to the surface, causing a flash of reflected light.

FLOATING FEED   Prepared feed pellets produced by the extrusion process. Such pellets are made under conditions of high heat and temperature and will float at the water surface for extended periods.

FOOD CHAIN   A sequence of organisms, each of which provides food for the next, from primary producers to ultimate consumers, or top carnivores.

FOOD CONVERSION EFFICIENCY   The reciprocal of food conversion ratio times 100, expressed as a percentage.

FOOD CONVERSION RATIO   In aquaculture, the amount of food fed divided by weight gain. The lower the FCR, the more efficient the animal is at converting feed into new tissue.

FOOD WEB   The complex feeding relationships that exist in any community of organisms.

FOULING ORGANISMS   Organisms that attach to such man-made surfaces in the aquatic environment as nets, water pipes, cages, and ship hulls and result in either nuisance or structural damage to the object to which they are attached. Examples are oysters (in certain instances), barnacles, sponges, bryozoans, and corals.

FRESH WATER   Water having a salinity of less than 0.5 part per thousand.

FRY   Newly hatched fish that have the external characteristics of the adult but are smaller than fingerlings.

GAMETE   A mature reproductive cell, either male (sperm) or female (ovum).

GAS-BUBBLE DISEASE   A malady that results from the supersaturation of atmospheric gases (especially nitrogen) in water. Gas-bubble disease often affects fish in discharge canals below power plants in the winter when cold water is rapidly heated by passage through the power plant condensers. Fish exposed to supersaturated water may exhibit exophthalmia and other symptoms similar to the bends experienced by divers who ascend from depth too rapidly.

GENE   The portion of a chromosome that carries a particular inheritance factor.

GENOTYPE   The sum total of the genes contained in the chromosomes of an individual organism, both those that are expressed and those that are latent.

GYNOGENESIS   The development of an ovum following sperm penetration, but without fusion of the gametes.

HANGING CULTURE   See string culture.

HAPLOID   The total number of chromosomes present in germ cells (sperm or ova).

HARDNESS   The concentration of divalent cations (especially calcium and magnesium) present in a water sample.

HATCHING JAR   A bottlelike container used to incubate fish eggs. Usually a current of water is passed through the jar to maintain the dissolved oxygen level and carry away metabolites.

HATCHING TROUGH   A raceway utilized to hatch the eggs of fishes. In catfish culture hatching troughs are fitted with paddle wheels. The eggs are placed in baskets and the action of the paddle wheels causes a gentle motion of the egg masses, allowing for the passage of highly oxygenated water over each egg.

HERBIVORE   Any animal that feeds exclusively on plant material.

HERMAPHRODITE   An animal that has reproductive tissues capable of forming both male and female gametes.

HETEROSIS   Hybrid vigor; the result of breeding distantly related individuals may sometimes be an increase in growth rate, food conversion efficiency, dress-out percentage, or some other desirable characteristic.

HETEROTROPH   An organism that obtains its nourishment only from preformed organic matter.

HETEROZYGOUS Describing the condition in which one of the pair of genes responsible for a particular trait is dominant and the other is recessive.

HOMOIOTHERMIC Warm-blooded animals; those that maintain a nearly constant body temperature regardless of the ambient temperature.

HOMOZYGOUS Describing the condition in which both genes of a pair responsible for a particular trait are either dominant or recessive.

HYPERTONIC Refers to a solution having an osmotic pressure greater than that of a solution with which it is being compared.

HYPERVITAMINOSIS Nutritional disease symptoms related to an excess level of one or more vitamins. Hypervitaminosis has been reported only with respect to the fat-soluble vitamins.

HYPOLIMNION The deep water in a stratified lake. The hypolimnion underlies the thermocline and may become anaerobic in the summer.

HYPOTONIC Refers to a solution having an osmotic pressure less than that of a solution with which it is being compared.

HYPOVITAMINOSIS Nutritional disease symptoms produced by a deficiency of one or more vitamins in the diet. Hypovitaminosis can occur as a result of deficiencies of either fat-soluble or water-soluble vitamins.

"ICH" The common name for the parasitic disease caused by the protozoan *Ichthyophthirius multifiliis*.

INTENSIVE CULTURE The rearing of aquaculture organisms in extremely high densities with a great measure of control in the hands of the culturist. Tanks, raceways, silos, and cages are examples of culture chambers utilized in conjunction with intensive culture systems.

ISOCALORIC Describing diets that contain the same number of calories but not the same types and/or proportions of food.

ISOTONIC Refers to a solution that exhibits the same osmotic pressure as a solution with which it is being compared.

KELP A member of the plant group Laminariales, which is composed of usually large, blade-shaped, or vinelike brown algae.

LARVA An embryo that becomes independent before assuming the characteristics of the parent animal.

LENTIC Designates standing water (e.g., lakes or ponds).

LIPID One of the three classes of energy-containing foodstuffs; lipids are portions of plant or animal tissues that are soluble in nonpolar solvents (e.g., ether, benzene, and chloroform).

LITTORAL   The zone on the seashore that lies between the high and low tide extremes.

LOTIC   Describing a flowing water environment (e.g., a stream or river).

MACROPHYTES   Large, primarily rooted higher aquatic plants (not including the kelps).

MARICULTURE   Marine aquaculture.

MARINE   Describing water with a salinity of between 17.0 and 40.0 parts per thousand.

MECHANICAL FILTER   A device that removes suspended materials from water. The medium in such filters is often sand or gravel, although screens, various types of fiber, and other materials may also be utilized.

MESOHALINE   Water having a salinity in the range of 3.0 to 10.0 parts per thousand.

METABOLISM   A group of processes that occur in living organisms by which nutrition, synthesis of new tissue, and respiration are achieved. See also anabolism, catabolism.

METABOLIZABLE ENERGY (ME)   The amount of energy that can be extracted from a foodstuff and utilized for metabolism. Metabolizable energy may be considered to be equivalent to the physiological fuel value of a food.

METAMORPHOSIS   The change in configuration of an organism from the larval stage to the appearance of the adult animal.

MONOCULTURE   The rearing of a single species in an aquaculture chamber.

MONOSEX CULTURE   The rearing of one sex in an aquaculture chamber, to the exclusion of the other.

NAUPLIUS   A limb-bearing early larval stage common in crustaceans; plural, *nauplii*.

NICOTINAMIDE ADENINE DINUCLEOTIDE PHOSPHATE (NADP)   A compound important in biological oxidation reactions.

NITRIFICATION   The oxidation of nitrogen from ammonia through nitrite to nitrate.

NITROGEN FIXATION   The process by which certain bacteria are able to transform elemental nitrogen into ammonia.

OLIGOHALINE   Water having salinity in the range of 0.5 to 3.0 parts per thousand. Also, organisms which can only tolerate a narrow salinity range.

OLIGOTROPHIC   Describing lakes that contain low levels of nutrients and have low organic productivity.

OMNIVORE An animal that consumes both plant and animal material in its normal diet.

OOGENESIS The process by which a mature ovum is produced.

OPEN SYSTEM An aquaculture water system in which water continuously flows through the culture area and is discarded after a single pass.

OSMOREGULATION The process by which organisms are able to maintain an internal salt balance different from that which occurs in the external medium.

OSMOSIS The migration of fluids through semipermeable membranes from solutions of lower concentration to those of higher concentration of dissolved particles.

OVERTURN The complete mixing of a lake, which occurs when thermal stratification is broken (generally in the spring and fall).

OVIPAROUS The type of reproduction in which eggs are released to develop and hatch outside the body of the female.

OVOVIVIPAROUS The type of reproduction in which eggs develop within the body of the female, but the developing embryo receives no nourishment from the mother.

OZONATION The sterilization of culture system water through the addition of ozone.

PARASITISM A symbiotic relationship in which one organism lives on or within the other and obtains its nourishment at the expense of the host.

PELAGIC Organisms that inhabit the water column.

PERIPHYTON The autotrophic component of the *Aufwuchs* community.

"PER MILLE" Equivalent to "parts per thousand" in salinity measurement.

pH The negative logarithm of the hydrogen ion concentration expressed in gram equivalents.

PHENOTYPE The characteristics inherited by an individual which are visible.

PHOTOAUTOTROPH An autotrophic organism that requires light energy in the synthesis of new organic matter.

PHOTOSYNTHESIS The elaboration of organic matter from carbon dioxide and water in the presence of chlorophyll, light, and certain enzymes.

PHOTOTAXIS The response of an animal to light; phototaxis may be positive, negative, or neutral).

PHYSIOLOGICAL FUEL VALUE See metabolizable energy.

PHYTOPLANKTON The plant constituents of the plankton community.

PLANKTON   Aquatic organisms that are suspended in the water column and are at the mercy of currents.

PLEOPOD   One of the swimming appendages of crustaceans.

POIKILOTHERMIC   Cold-blooded animals; those whose internal body temperature approximates that of the external environment (includes virtually all aquaculture animals).

POLYCULTURE   The rearing of two or more noncompetitive species in the same culture chamber.

POLYHALINE   Refers to water having salinity in the range of 10.0 to 17.0 parts per thousand.

POPEYE   See exophthalmia.

POPULATION   The total collection of organisms of a single species in a natural physiographic region.

ppm   Parts per million.

ppt   Parts per thousand (also "per mille" and $^o/_{oo}$).

PRACTICAL DIET   A prepared diet made from readily available, natural feed ingredients such as grains, meats, and meat by-products.

PREPARED DIET   A diet manufactured from grains, fish meal, meats, and meat by-products, or purified ingredients, as opposed to a natural diet of living organisms or detritus found in the aquatic environment. Also called artificial diet.

PRIMARY PRODUCTION   The elaboration of organic matter from inorganic materials by photoautotrophic organisms.

PRIMARY PRODUCTIVITY   The rate at which tissue is elaborated by primary producers.

PRIMARY SETTLING CHAMBER   A unit in a closed recirculating water system in which waste feed, fecal material, and other suspended particles are allowed to settle before the water flows into the biofilter.

PRODUCTION   The elaboration of organic matter by the organisms in a specific area or volume over a given period.

PROTEIN   A member of a class of chemical compounds that is one of the three energy-bearing food groups; contains carbon, nitrogen, hydrogen, oxygen, and sulfur; and is composed of a series of amino acids.

PURIFIED DIET   A prepared diet composed of individual amino acids, fatty acids, simple sugars, vitamins, and minerals.

PVC   Polyvinyl chloride; a plastic from which pipes and other plumbing materials utilized by aquaculturists are manufactured.

RACEWAY   A culture chamber that is generally long and narrow. Water enters one end and exits the other in most cases.

REDD   The nest of salmonid fishes.

REFRACTOMETER   An instrument that measures the refractive index of liquids. A refractometer can be utilized to measure the salinity of water because refractive index and density are directly related, and salinity can be determined from density.

RESPIRATION   The oxidation-reduction process by which the chemical energy in food is transformed into other kinds of energy. Oxygen is consumed and carbon dioxide released in respiration.

SAC FRY   A newly hatched fish whose protruding abdomen is filled with yolk material from which the animal obtains nourishment for the first few days of life.

SALINITY   The measure of the total amount of dissolved salts in a sample of water, in parts per thousand, by weight when all the carbonates have been converted to oxide, bromide and iodide have been replaced by chloride, and all organic matter has been oxidized. Salinity also equals $0.03 + 1.805 \times$ chlorinity.

SATUROMETER   A device that is used to measure the degree of gas saturation in water.

SECONDARY SETTLING CHAMBER   A chamber similar to the primary settling chamber of closed recirculating water systems but located between the biofilter and the culture chamber.

SEMICLOSED SYSTEM   A water system in which some of the water is replaced intermittently or continuously, while part of the fluid is recirculated through settling basins and a biofilter.

SEMIDIURNAL   Occurring twice within 24 hours (e.g., semidiurnal tides are those that both rise and fall twice daily).

SEMIPURIFIED DIET   A prepared diet composed partly of highly purified ingredients (e.g., amino acids or fatty acids) and partly of more practical ingredients (e.g., corn starch).

SILO CULTURE   Culture systems employing culture chambers that utilize more vertical than horizontal space.

SINKING FEED   Prepared feeds made with a pellet mill that forces the ingredients through a die under fairly low temperature and pressure and produces pellets that sink when placed in the water.

SOMATIC   Cells other then reproductive cells in the body of an organism. Somatic cells contain the diploid number of chromosomes.

SPAT   Newly settled oysters.

SPERMATOGENESIS   The maturation of sperm cells within the male of a species.

STANDING CROP   The biomass present in a body of water at a particular time.

STENOHALINE   Organisms that can exist only within a narrow range of salinity.

STRING CULTURE   The rearing of attached forms (such as mussels and oysters) from ropes or lines suspended from frames or floating platforms in ponds, embayments, or in open water. Also called hanging culture.

SUBSISTENCE CULTURE   Extensive pond culture, which is simplified to the extent that individuals or small groups of people can rear aquatic organisms for their own consumption at a rate that provides a relatively continuous supply.

SUPPLEMENTAL DIET   A prepared diet formulated to provide protein and other nutrients in excess of those obtained from natural food organisms in the environment.

SUSPENDED SOLIDS   Particles larger than 0.45 micron which are found in the water column.

SYMBIOSIS   A relationship between two species.

SYNERGISM   A relationship between two or more parameters.

TANK   A relatively small culture chamber. A culture tank may be round, square, rectangular, or another shape.

THERMOCLINE   A steep temperature gradient in a lake or ocean.

TOTAL ALKALINITY   The alkalinity due to the presence of both carbonate and bicarbonate ions.

TRIGLYCERIDE   A molecule composed of glycerol and three fatty acids.

TROPHIC LEVEL   The position an organism occupies in the food web (e.g., herbivore, omnivore, or carnivore).

TURBIDITY   A degree to which the penetration of light in water is limited by the presence in the water of suspended or dissolved substances.

UV STERILIZATION   The sterilization of water by passing it near sources of ultraviolet radiation.

VITAMIN   An organic compound that is required in small amounts for normal growth and health.

VIVIPAROUS   The type of reproduction in which the egg develops within the body of the female and the developing embryo obtains nourishment from the mother.

WINKLER TITRATION   The determination of dissolved oxygen in water by chemical titration.

WINTERKILL   The loss of animals in a lake, pond, or other water body as a result of heavy ice cover, which restricts photosynthesis and leads to an oxygen depletion. Also, a condition in which oysters are frozen and die when exposed to cold temperatures at low tide in intertidal beds.

ZOOPLANKTON   The animal component of the plankton community.

ZYGOTE   The cell formed by the uniting of male and female gametes.

Index

Abalone, 131
Ablation, eyestalk, 223, 240
Acanthocephala, 285, 340
Achromobacter sp., 94
Acid, acetic, 267-268, 284
　amino, 93, 167-168, 183-188
　ascorbic, 176, 180
　aspartic, 185-186
　capric, 190
　caproic, 190
　caprylic, 190
　fatty, 189-193
　folic, 176, 180
　glutamic, 185-186
　lauric, 190
　linoleic, 190-193
　linolenic, 190-193
　myristic, 190
　nicotinic, 175, 180
　oleic, 190-193
　palmitic, 190
　pantothenic, 175, 179
　stearic, 190
　uric, 93
Acriflavin, 308
ADP, *see* Diphosphate, adenosine
Aequipectin irradians, 2, 131
Aeration, in algae culture, 163
　in biofiltration, 43, 47
　in brine shrimp culture, 165
　in disease treatment, 266
　in feed suspension, 195
　by flow regulators, 39
　in gas elimination, 30
　in live-hauling, 306
　methods of, 53-55, 73
　in oxygen depletions, 123-124
　to precipitate iron, 30

　supplemental, 40, 42, 153
Aeromonas hydrophila, 277, 279
Aeromonas liquefaciens, see Aeromonas
　hydrophila
Aeromonas salmonicida, 277
Aeromonas sp., 287, 308, 339-
　340
Aflatoxin, 196-197
Agasicles sp., 109
Agnatha, 223
Airlifts, 39-40
Alanine, 184, 186
Algae, benthic, 86, 90-91
　in biofiltration, 43
　blue-green, 43, 197
　control of, 155-156
　culture of, 163-164
　dinoflagellates, 197
　as feed, 88-89, 161, 163-165
　and fertilizer, 102-103
　filamentous, 1, 78, 86
　and light, 143-144
　nuisance, 88-89
　toxic, 197
　in waste recycling, 74
Algin, 196
Alkalinity, 129, 147-148, 156
Alloglossidium corti, 286
Alosa sapidissima, 323
Ambloplites rupestris, 323
Ammonia, accumulation, 36, 48
　adsorption, 49
　in biofiltration, 46-47
　chemistry, 93-98
　ionization, 96
　measurement, 98
　and metabolism, 172
　nitrification, 43, 94

361